Atlas of the Food System

Teresa Marat-Mendes •
Sara Silva Lopes • João Cunha Borges •
Patrícia Bento d'Almeida

Atlas of the Food System

Challenges for a Sustainable Transition of the Lisbon Region

Springer

Teresa Marat-Mendes
Iscte-Instituto Universitário de Lisboa
Lisbon, Portugal

Sara Silva Lopes
Iscte-Instituto Universitário de Lisboa
Lisbon, Portugal

João Cunha Borges
Iscte-Instituto Universitário de Lisboa
Lisbon, Portugal

Patrícia Bento d'Almeida
Iscte-Instituto Universitário de Lisboa
Lisbon, Portugal

ISBN 978-3-030-94832-0 ISBN 978-3-030-94833-7 (eBook)
https://doi.org/10.1007/978-3-030-94833-7

This Springer imprint is published by the registered company Springer Nature Switzerland AG
The registered company address is: Gewerbestrasse 11, 6330 Cham, Switzerland

Foreword by Peter J. Larkham

Mapping, Cities and Food

It seems relatively unusual for academic works to have Foreword, and some seem to function merely as introductions to the structure and content of the volume. This volume does not need that: it is, simply, an atlas relating to food in one city-region. Instead, I wish to emphasize the importance of the ideas of food, cities and mapping. How we think about them, and the use we make of that knowledge, will be vital for our future.

Food is one of the most critical issues faced by people today. Many readers of this volume will live comfortable urban lives in westernized industrial societies and might question this assertion. Yet there are many in the world not in such fortunate positions. Many who do live in those favoured societies are themselves not in fortunate positions: poverty, homelessness and poor environmental quality are more common than we might wish or even recognize. So, worldwide, there are many who do not get enough to eat; and many others whose diets are inappropriate—obesogenic or otherwise unhealthy, or deficient in various components. These concerns lead to the issue of 'food security'. As defined by the United Nations Committee on World Food Security, this is the expectation that all people, at all times, have physical, social and economic access to sufficient, safe and nutritious food that meets their food preferences and dietary needs for an active and healthy life.

This ventures into the territory of human rights. Whose responsibility is it to ensure that this right is delivered; that everyone benefits from food security? Roles and responsibilities are mixed, from the individual to communities; municipal, regional and local governance; corporations; and even international organizations such as the UN. There are tools to guide activity, including the Sustainable Development Goals, and organizations such as the UN World Food Programme and the International Food Policy Research Institute. But activities are inevitably affected by political values and concerns at all scales, consideration of economic costs and benefits, the inevitability of some form of climate change, and a global population that continues to grow—according to the Population Division of the UN Department of Economic and Social Affairs forecasts, from a current population of 7.8 billion to some 10.9 billion at the end of this century: that is a growth of approximately 81 million people per year. How can we provide the right food at the right time and the right place to this population? We can't do it for everyone at the present, so clearly things will need to change.

If food is a problem, how about cities? Many of them are growing too, and again the Population Division forecasts make sobering reading. Some 55% of the current world population lives in urban areas, and this is likely to increase to 68% by 2050. That could add a further 2.5 billion urban dwellers, and nearly 90% of that increase is taking place in Asia and Africa. Yet, in some areas including Japan, the Republic of Korea and Eastern Europe, we are seeing deurbanization: cities shrinking as a result of economic contraction, natural disaster, low fertility and emigration. In some 'comfortable' countries, there is population movement out of crowded cities to even more comfortable rural areas: yet such migrants usually remain socially and economically dependent on urban areas, and very rarely produce significant amounts of food. The urban size/population problem is clearly diverse, sizeable and growing.

Urban agriculture has been suggested, and in some locations actively promoted, as part of the solution to food insecurity in cities. It can exist at many scales and intensities, from individuals using their gardens, allotments or waste/under-used spaces, to intensive industrial-scale production underground, on rooftops, in purpose-built multistorey farms using smart technology, controlled climates, hydroponics and other technologies. It exists in city centres, suburban sprawls and extends out into peri-urban areas. However, its productivity is currently relatively low, especially in the more individual endeavours, which seem to have produced most publicity. The more industrial enterprises are more productive but usually focus on specific crops such as salads. Overall, it is suggested that urban agriculture could meet 15–20% of global food demand. While this is undoubtedly a useful contribution to food security, it is only part of the solution, and its products often do not reach disadvantaged urban communities.

Even at its best, urban agriculture is hardly a solution to some of the 'food deserts' that are evident in many cities even in prosperous countries. The problem may not be an absolute lack of food, but lack of access to a healthy diet and/or preference for unhealthy foods, leading for example to land-use planning policies restricting hot-food takeaways in some UK cities, especially close to schools.

Such food deserts can be readily demonstrated by mapping, and this brings us to my third issue. Maps provide an extremely valuable, and quickly perceived, presentation of potentially large and complex data sets with a spatial dimension. They can be a key tool in promoting understanding and can inform evidence-driven policy development and implementation. Recent examples using the benefits of IT for the production and dissemination of maps can be extremely effective, such as with Amsterdam's Energy Atlas produced through the EU-funded TRANSFORM programme. Yet there are still benefits to paper maps and atlases. Although static and not continuously updatable, they form a permanent record, not susceptible to IT hardware problems or software updates. Many readers still appreciate the tangible artefact, which can convey a perception of physicality, certainty and reliability. And a well-designed map, accurately and effectively portraying appropriate data, often has many of the qualities of a work of art.

This *Atlas of the Food System* provides a constructive and informative response to these issues. It presents clear mapping of the city-region's various food systems in a range of ways —highlighting, for example, the diversity and spread, but very small extent, or urban agriculture spaces in some of the municipalities. Adding photography to the cartography allows consideration of the physical contribution of growing spaces to urban form and design, and hence our perception and use of spaces. This is timely as our current global crisis reminds us of the health and well-being contributions of such spaces, and proximity to nature, for urban dwellers. But Part IV shows that this is more than a mere representational Atlas. It goes far beyond this, in seeking to make sense of the cartographic data and present proposals for better integrating urban agriculture into a sustainable urban planning system.

While this Atlas focuses on just one city-region, albeit with a wealth of detail and constructive proposals, we need more such initiatives to even begin to address the large-scale problems mentioned here. The scale of problems, and range of potential solutions, needs to be higher in the consciousness of everyone, from those renting or owning small urban plots, to global corporations active in food systems, and to governance at all scales. The physical artefact of an Atlas can help here: it forms a permanent reminder. But we also need real-time data collection, analysis and display, to better understand how, and indeed how fast, the situation can change; and we need policy-makers to design *and implement* suitable policies. The future will not be the same as today, and today is not the same as the past. This Atlas pushes us to learn, think and plan. "Learn from yesterday, live for today, hope for tomorrow. The important thing is not to stop questioning" (Einstein).

Peter J. Larkham
Birmingham City University
Birmingham, UK

Foreword by June Komisar and Joe Nasr

Food Systems at the Intersection of Place-Based Design and Spatial Planning

The spatial history of an urban landscape can help to envision the future and understand the past from a visual and spatial perspective. This atlas of Lisbon's regional food system can help students, scholars, and practitioners understand the agricultural history of locations that now feature dense urban developments and conurbations of smaller towns at the city's periphery, the connectivity of food spaces and structures to the city and its hinterland, and the way in which such spaces and structures can be strengthened as assets or infrastructure for the resilience of human settlements.

This atlas illustrates the materiality/physicality of food and its systemic nature—in the sense of being a fundamental urban system, on par with transportation, gas supply or sewerage. These dual dimensions are receiving increasing attention as aspects of an understanding of food-related problems and solutions. The materiality of food—its presence in specific spaces and its shaping of places—and its systemic nature (as seen in the emergence of the 'city-region food system' concept)—mean that the production and improvement of urban areas must be founded on as deep an understanding of food systems as that of the relation between transportation systems and urban form, for example.

This book is a welcome addition to recent literature about urban agriculture and food systems as it shows in a compelling, visual way how deeply the food system is entrenched into our landscapes, and indeed how much food systems profoundly shape urban form. The powerful visual picture of the variety of forms agricultural land takes in different regions illuminates what relationships between food systems and design exist, as well as how planning for food growing, processing, and distribution is vital to considerations about the future of any city or region. The mapping and historical analysis of food systems are crucial components of understanding how agriculture, trade routes, markets and dense urban developments have worked together in the past, and how they have shaped the environments we live in now. Creating a sustainable and resilient future depends on a robust understanding of the relationships between the population, the built form that shelters them, and the food system that sustains them.

The design and planning components of our teaching and research initiatives are aided by the new literature that is beginning to enrich the field (this atlas being a good example)—a literature that intersects the analysis of space and place with that of urban and regional food systems. For some time, we have been undertaking research, teaching, advocacy and service work about the role of design and planning in the shaping of localized food systems. Notably through our Carrot City initiative, we have documented how the functioning of such systems goes hand in hand with sustainable aspects of architectural and planning proposals and projects that range from building design that integrates community food hubs to productive urban landscapes that act as green infrastructure. We will briefly illustrate these connections through some activities in which each of us has been involved.

Beginning with a strong interest from students in the role design plays in many aspects of the food system, June has been teaching architecture with a focus on projects that incorporate aspects of urban agriculture, food preparation, food literacy and/or distribution. Students have demonstrated remarkable ingenuity in finding opportunities to incorporate urban agriculture and other food system components within the confines of existing or proposed built form. They have proposed urban agricultural education hubs, farms above arterial roads, markets under elevated highways, and growing spaces on existing and new rooftops, combining the mitigation of urban heat island effects with the provision of growing spaces. They have designed markets, food halls, and food production facilities, enhancing traditional food-centred districts. A crucial component of such proposals is an understanding of the big picture, from the presence—and at times the disappearance—of traditional agricultural lands to locations that lack access to healthy food.

In Joe's case, two experiences with participatory localized food mapping in Toronto connect with the effort underlying this atlas. He participated in an effort to map and inventory 'Food by Ward', seeking to document the 'food assets' located in each ward of the City. This started as a collectively constructed tool to inform candidates running for municipal office about the presence of food infrastructure across the city, while highlighting gaps and needs that these candidates can be pressed to address. After the election, the existence of such documentation remained useful to show the richness of and the deficiencies in the food-system landscape. In parallel, Joe helped Toronto Urban Growers, the city's urban agriculture, to map and describe urban agriculture across the city. Through this effort, individual actors in this landscape started to see themselves as part of a broad landscape and movement. These two different efforts show that mapping urban and regional food systems is multifunctional—beyond painting a picture and telling a story of food systems, such mapping can serve to educate, politicize and give shape to a collective, systemic identity, as well as to highlight gaps and deficiencies in such systems.

This atlas can help us understand how food systems, from agricultural lands to spaces for trading, shape the world we live in. It provides a powerful message about how food is a part of the most basic aspect of our culture and environment. Although centred on one region, Lisbon, this book can spark a trend of spatial and visual studies of the food systems in major cities across the planet, a vital resource for designers and planners of our infrastructure at many scales.

June Komisar
Joe Nasr
Ryerson University
Toronto, CA, Canada

Acknowledgements

Producing a book with this magnitude depends on the support of many people and institutions to whom we are extremely grateful. Thus, the authors would like to thank the following persons and entities (organized alphabetically) who have kindly supported this work, while granting:

i. the permission to use and publish original archived sources, included in this Atlas;
ii. the provision of cartographic elements and information on urban agriculture, included in this Atlas;
iii. the permission to use and publish photographs taken by the authors, included in this Atlas;
iv. the offer of personal photographs and respective permissions, included in this Atlas;
v. the supportive discussions and observations to specific topics included in this Atlas;
vi. the financial and work conditions that supported the research work conducted for this Atlas.

Águas do Tejo Atlântico (Grupo Águas de Portugal); ALDI Portugal; Ana Mélice Dias; Arquivo Municipal de Almada; Arquivo Municipal de Lisboa; Arquivo Municipal de Loures (Técnicas Catarina Bernardo and Dora Marques); Arquivo Municipal de Palmela; Arquivo Municipal de Vila Franca de Xira (Técnico Superior José Rocha); Arquivo Municipal de Cascais (Técnica Superior Sandra Santos); Barbara Koole; Biblioteca Francisco Keil do Amaral (Ordem dos Arquitetos Secção Regional Sul); Câmara Municipal de Alcochete (Geógrafo Carlos Aniceto); Câmara Municipal de Almada (Divisão de Inovação, Clima e Energia—Dr.ª Sara Dionísio e Dr. Nuno Lopes; Departamento de Cultura—Dr. Paulo Reis e Dr.ª Otília Rosado); Câmara Municipal da Amadora (Divisão de Informação Geográfica—Técnico Superior Fernando Ferreira); Câmara Municipal de Cascais; Câmara Municipal de Lisboa (Divisão de Gestão de Informação Georreferenciada—Geografa Margarida Coelho da Silva); Câmara Municipal de Loures (Gabinete de Planeamento—Coordenadora Paula Vidal Pereira); Câmara Municipal de Mafra (Departamento de Urbanismo, Obras Municipais e Ambiente—Dirigente Carla Romana); Câmara Municipal da Moita (Gabinete de Desenvolvimento Urbanístico—Engenheira Rita Costa Barros); Câmara Municipal do Montijo (Gabinete de Informação Geográfica—Coordenador Francisco Silva Cardoso); Câmara Municipal de Odivelas (Divisão de planeamento urbanístico—Chefe de Divisão Florinda Lixa); Câmara Municipal de Oeiras (Gabinete de Inteligência Territorial—Geografo Miguel Jeremias); Câmara Municipal de Palmela (Departamento de Administração e Desenvolvimento Organizacional—Técnica superior Patrícia Franco); Câmara Municipal de Seixal (Divisão de Desenvolvimento Estratégico—Eng.º Geógrafo Francisco Figueiredo); Câmara Municipal de Sesimbra (Unidade Técnica de Sistemas de Informação Geográfica—Técnica superior Rute Ferreira); Câmara Municipal de Setúbal (Departamento de Urbanismo—Chefe da Divisão de Planeamento Urbanístico Vasco Raminhas da Silva); Câmara Municipal de Vila Franca de Xira; Casa dos Frangos de Moscavide; DINÂMIA'CET; Direcção-Geral do Território (Divisão de Informação e Gestão Territorial (DIGT) e Direção de Serviços de Ordenamento do Território (DSOT)—

Eng.º Geógrafo Francisco M. Sequeira); DOCAPESCA Lota de Setúbal; Fabiano Lemes; Filipa de Castro Guerreiro; F.I.T.- Fomento da Indústria do Tomate SA; Fundação para a Ciência e a Tecnologia; Iscte—Instituto Universitário de Lisboa; Ivor Samuels; Joana Benedito; Joana Ramanauskaitė; José Manuel Henriques; Laboratório Nacional de Engenharia Civil (LNEC); Małgorzata Hanzl; Maltibérica—Sociedade Produtora de Malte S.A.; Mercado Abastecedor da Região de Lisboa (MARL); Michael Hebbert; Ordem dos Arquitectos (OA), Portugal; RAPORAL STEC; Raul Lopes; RIBERALVES; SCC Sociedade Central de Cervejas e Bebidas, S.A.; Robert Fincham; Sistema de Informação para o Património Arquitectónico (SIPA)—Forte de Sacavém (Técnica Superior Cátia Taveira Martins); SIMARSUL (Grupo Águas de Portugal); Stephen Stead; Time Out; Thomas Weith; Tropical Vanity.

The authors wish to acknowledge the support of Iscte-Instituto Universitário de Lisboa and DINÂMIA-CET, where the research for this Atlas was developed, as well as the Institutions and the research teams involved in Project SPLACH, in particular his coordinator, Paulo Pinho, for the stimulating research environment. Our acknowledgement extends to the Portuguese Science Foundation (FCT) for funding the research produced over Grant POCI-01–0145-FEDER-01643 (together with Fundo Europeu de Desenvolvimento Regional, FEDER, and PORTUGAL PAC2020), Grant DFA/BPD/117167/2016 (together with European Social Fund, ESF), Grant SFRH/BD/148556/2019, Grant DFA/BD/5568/2021 and Grant PTDC/EMS-ENE/2197/2012. This research has fed into this book.

The authors are also grateful to Peter J. Larkham, June Komisar and Joe Nasr who have enthusiastically contributed with their forewords to this Atlas. The authors also wish to acknowledge all the support and best guidance provided by Springer.

To our families, with our endless gratitude!

Teresa Marat-Mendes
Sara Silva Lopes
João Cunha Borges
Patrícia Bento d'Almeida

Introduction

Food Systems (FS) encompass the entire range of actors and their interlinked value-adding activities involved in the production, aggregation, processing, distribution, consumption and disposal of food products that originate from agriculture, forestry or fisheries, and parts of the broader economic, societal and natural environments in which they are embedded. The food system is composed of sub-systems (e.g. farming system, waste management system, input supply system, etc.) and interacts with other key systems (e.g. energy system, trade system, health system, etc.).

Therefore, a structural change in the food system might originate from a change in another system; for example, a policy promoting more biofuel in the energy system will have a significant impact on the food system.

A Sustainable Food System (SFS) is a food system that delivers food security and nutrition for all in such a way that the economic, social and environmental bases to generate food security and nutrition for future generations are not compromised. This means that:

—It is profitable throughout (economic sustainability);
—It has broad-based benefits for society (social sustainability); and
—It has a positive or neutral impact on the natural environment (environmental sustainability).

Food and Agriculture Organization of the United Nations (FAO)
in *Sustainable Food Systems—Concept and Framework* Brief

Cities and the Problem of Resources

The work presented in this Atlas is rooted on the ambition to disclose the relationships between territory, urban form and basic resources taking an evolutionary perspective. It stems from a research journey aimed to examine both the morphological and the socio-metabolic aspects of the territory, towards a more sustainable urbanism (Marat-Mendes 2002, 2020). A first moment implied establishing a methodological approach to assess processes of physical change, interlinking them with those occurred in the historic-political, economic and environmental realms, and connecting their different outcomes (Marat-Mendes 2002, 2004). Afterwards, socio-metabolic changes were inferred through an observation of the key transitions verified in the built and the natural environments, understood as consequences of social needs, aspirations and activities. As key resources, food and water allow a particularly clear vision of how these realities intertwine.

Relationships of this complexity are highly localized and can hardly be understood generally—a degree of specificity must be sought (Marat-Mendes et al. 2015, 2016, 2021). Two research projects, financed by the Portuguese Foundation for Science and Technology, proved fundamental to apply our morphological and socio-metabolic concerns to the study of a particular territory—the Lisbon Metropolitan Region. These were Project MEMO—Evolution of the Lisbon Metropolitan Area Metabolism[1] and Project SPLACH—Spatial Planning for Change.[2] These have allowed our engagement with the study of the historical processes of territorial transformation, through the spatial analysis of its operating systems. Here, we highlight the articulation of habitat forms of the Lisbon Region and its food system. While the metropolitan food system has been observed before (Oliveira et al. 2014), the spatial component may be analyzed further, while its historical formation has often remained scarcely explored. Moreover, the recent publication of the Portuguese Agricultural Census of 2019 (INE 2021) adds new information of the dynamics of food production in the country. This Atlas is motivated by the conviction that surveying the current reality of food-related activities in their territorial aspect is paramount to enable architects, planners, decision-makers and ultimately civic society to fully acknowledge how food can contribute to transform our current habitats into more sustainable ones.

The spatial implications of the food system are, to an extent, obvious. Fields, crops, farms and livestock occupy large areas, usually in the countryside, while markets and other food-access facilities are a familiar part of most neighbourhoods. No house is built—at least today—without a kitchen. And yet, we seem oddly oblivious to the origin of our food, or to all the space, labour and organization necessary for it to get to us (Steel 2008; Parham 2015). The food system – the pattern across which food is produced, transformed, distributed, traded, consumed, and disposed of (Pothukuchi and Kaufman 2000; Steel 2008)—is a basic and vital need for human societies. Thus, if a society cannot feed itself in a sustainable manner, it can hardly aspire to be sustainable at all. Despite this obvious importance, the food system is hardly a key concern for those dealing with the spaces we inhabit, from planners to urban designers to architects.

The links between food systems and urban morphology are gaining new scholarly attention and if, despite all their differences, the leading schools of urban morphology (English, Italian and French) have focused unanimously on open spaces, street systems and buildings, these can be identified with the phases of the food system: production in open spaces, distribution in roads and streets and buildings for retail activities (Salvador 2019)—to which we may also add industrial transformation and consumption spaces.

Cities shelter immense populations that need to be fed. Currently, this is ensured by a global food system, which, while profitable for the companies operating in it, are highly harmful to the environment. Food supply chains have extended physically over such huge time-space distances, that new (and incontrollable) power relations necessarily emerged between local and global, as well as between producers, retailers and consumers, with additional problems arising from the food miles, i.e. the distance food travels from farm to fork, and resulting CO_2 emissions (Spaargaren et al. 2014; Hardman and Larkham 2016). In its spatial dimension, beyond its specific phases, the food system also includes a set of activities, actors and institutions (Pothukuchi and Kaufman 2000) and power relations (Moorcroft 1972; Spaargaren et al. 2014; Wekerle and Classens 2015). Indicators such as employment, sales, values, wages, food expenditure and consumption in activities of industries, retail, wholesale and agriculture are important components of the food system as well (Pothukuchi and Kaufman 2000). This explains why studies of the food system often take a holistic and multi-scale approach, extending for issues of political economy, public health, environmental impacts and cultural capital (Pothukuchi and Kaufman 2000; Steel 2008; Brinkley 2013; Parham 2015). Among the labyrinthine complexity of scales, institutions, practices and values implied in the food system, food access has become conspicuously taken for granted, to the point that we have unlearned how to assess it properly (Steel 2008). Yet, it is obvious that between an urban allotment garden and an intensive monocultural exploration, or between a small grocery shop and a franchise supermarket, or between cooking a meal at home or

[1]Grant PTDC/EMS-ENE/2197/2012.
[2]Grant POCI/-01-0145-FEDER-016431.

ordering it from an online delivery platform, differences are not only in economic scales, practices, social status or values. There are important spatial consequences and contexts whose understanding is fundamental. Indeed, the Food and Agriculture Organization of the United Nations (FAO 2019) has called for a transformation of metropolitan and urban food systems, but without this spatial reach, it is impossible to conceive how or towards which this system can change.

In face of acute problems arising in our exploration of natural systems and the organization of social activities and needs, according to globalized circuits, our cavalier indifference towards the way our basic needs are ensured must end. Social relations and links between social activities and nature cannot be amended, improved or changed without acknowledging just how deeply food is implied in all aspects of these relations and links (Steel 2020). During 1960s, when concerns for the environmental impact of urbanization started to be problematized, the engineer Abel Wolman (1892–1989) proposed a methodology for quantifying stocks and flows of materials and energy necessary for the functioning of specific environments (Wolman 1965). To this heuristic understanding of the city as a functioning organism Wolman called urban metabolism. This concept was later retrieved by the Vienna School of Social Ecology, which articulated metabolic functioning with the social activities which co-determine them, constructing ideal types of socio-ecological systems (Fischer-Kowalski 1998; Fischer-Kowalski and Weisz 2016). According to this school, human and natural systems interact and coevolve over time, having substantial impacts upon one another, with causality working in both directions. Socio-metabolism, thus, proposes a historical approach to society–nature relations (Fischer-Kowalski and Weisz 2016).

To understand how our current food system has come to be, one must look back into how it formed. The city has long been distinct from the countryside insofar as its key activity is trade, not agriculture (Weber 1966). Nevertheless, the relationship between city and countryside was for long one of interdependence. Lack of transportation systems able to withstand long distances and of sophisticated techniques for food preservation meant that production and consumption had to be relatively close by (Costa et al. 2016). All of this started to change with modernity. The Industrial Revolution implied dramatic changes in modes of production, nature exploration, social organization and cultural values, all of which were to impact food systems forever. If the discovery of agriculture slowly had taken place everywhere in the world, granting a gradual transition from hunter-gatherer societies to agrarian societies, the transition or discontinuity from these to industrial societies happened abruptly and took only a few decades to generalize, at least in the Western world, "[s]upported by a unique combination of a land-use system with a high surplus rate, specific patterns of natural resource endowment, technological breakthroughs in coal extraction and metallurgy, institutional change and population growth" (Krausmann et al. 2016).

The 'founding fathers' of modern sociology have studied modernity from different perspectives, namely, the importance of capitalism by Karl Marx (1818–1893), industrialism by Émile Durkheim (1858–1917) and rationalization by Max Weber (1864–1920) (Giddens 1990). To these, one may add another decisive transformation, namely, the emergence of a new metabolism, dependent on fossil fuels and the associated high-density energy demand (Fischer-Kowalski and Haberl 1997; Krausmann et al. 2016). Such increase allows for production systems—and also of distribution and commerce—to grow larger and larger, contributing (however disparately) to the economic growth of enterprises and nation states. One such example is the shift from water base transportation (river and seas) to train transportation, which triggered a great expansion of provision hinterlands of cities, allowing for resources to be brought from greater distances (Niza et al. 2016). If the twentieth century saw the fruition of many changes envisioned from the eighteenth century onwards, the transformation of food systems is certainly among the most drastic.

The (Silent) Role of Food in Architecture and Urban Design

A recent upsurge of research exploring the relations between space, architecture and food has been verified in the past two decades (for example Steel 2008, 2021; Parham 2015; Hardman and Larkham 2016; Marot 2019) and has rightly been hailed for shedding light upon a subject that seldom finds its way to architectural theory. Yet, such a detachment between food and architectural or urban design is a historically contingent phenomenon. Food growing has always been an important land-use at the borderlines of cities around the world, with boundaries often being marked by ploughed land and crops as well as grazing areas and clusters of houses, farms, cattle stalls, fields and gardens (Kostof 1991; Parham 2015). These spaces formed a foodshed, a surrounding territory from which the city brought a key part of its food supply (Getz 1989; Parham 2015; Salvador 2019), but such organization was to change paradigmatically throughout modernity. Within the indictment of North American regional planning scholar Clyde Weaver (1945–2015) that urban planning ended up translating the capitalist organization to space (Weaver 1984), architectural culture has often played an abiding role, which is not to say that architects, urban planners and urban designers have not shown a concern with these problems. In reality, such concerns have been verified throughout modernity, and one of the most immediate merits of recently published research is precisely that it retrieves such proposals from historiographical obscurity.

While the Industrial Revolution spawns rapid transformations from 1760 onwards, it also boosted urban growth to an unprecedented rate. With cities and industries physically expanding over territories previously used for food production, a certain 'tradition' of seeking solutions to minimize the growing contrast between city and countryside also emerged. François Cointeraux (1740–1830), a French bricklayer and Professor of Rural Architecture, promoted adobe and mud-brick construction, seeking innovative approaches to new rural houses (which he hypothesized could apply to cities too), alongside techniques for improving agriculture produce. To this hand-in-hand consideration of architecture and agriculture, Cointeraux called 'Agritecture', suggesting a planned incorporation of agriculture in metropolitan space (Marot 2019).

The utopian Socialist writer Charles Fourier (1772–1837) envisioned a new habitat model, the Phalanstére—a large collective housing estate—surrounded by an agroforestry area, which supplied inhabitants with food (Beecher and Bienvenu 1972). Fourier's utopian plan also detailed dietary prescriptions, made to increase pleasure and intellectual stimulus (Barthes 1971).

Ildefons Cerdà's (1815–1876) Eixample plan for Barcelona (1860) envisioned a new possible society, as an outcome of new industrialized modes of production. This new city was predicated upon a new General Theory for Urbanization, adopting new infrastructures to accommodate both people and the mobility of goods, as well as the flows of water and sewers, and an urban design that conceived the city as whole with several units articulated with the existing territory (Marat-Mendes 2002). The blocks were conceived to accommodate buildings and vegetable gardens, aiming towards an urban–rural city (Marat-Mendes 2002).

Ebenezer Howard's (1850–1920) Garden City model consisted of a network of urban cores, linked by trainways, and separated by rural buffers, whose agriculture exploration fed the regional community (Howard 1902). At more or less the same time, when industrial transformation and transportation by train were changing the landscape and forms of settlement, Scottish social geographer Patrick Geddes (1854–1932) surveyed regional territories and proposed alternatives to continuous urbanizations or conurbations. In his view, territories were to be organized according to the production and maintenance of basic needs, while social and labour groups should manage the territory accordingly, in specific territorial formations (Geddes 1915).

A somewhat underappreciated figure of German Modernism, landscape architect Leberecht Migge (1881–1935) sought to condense elements from the Garden City with the

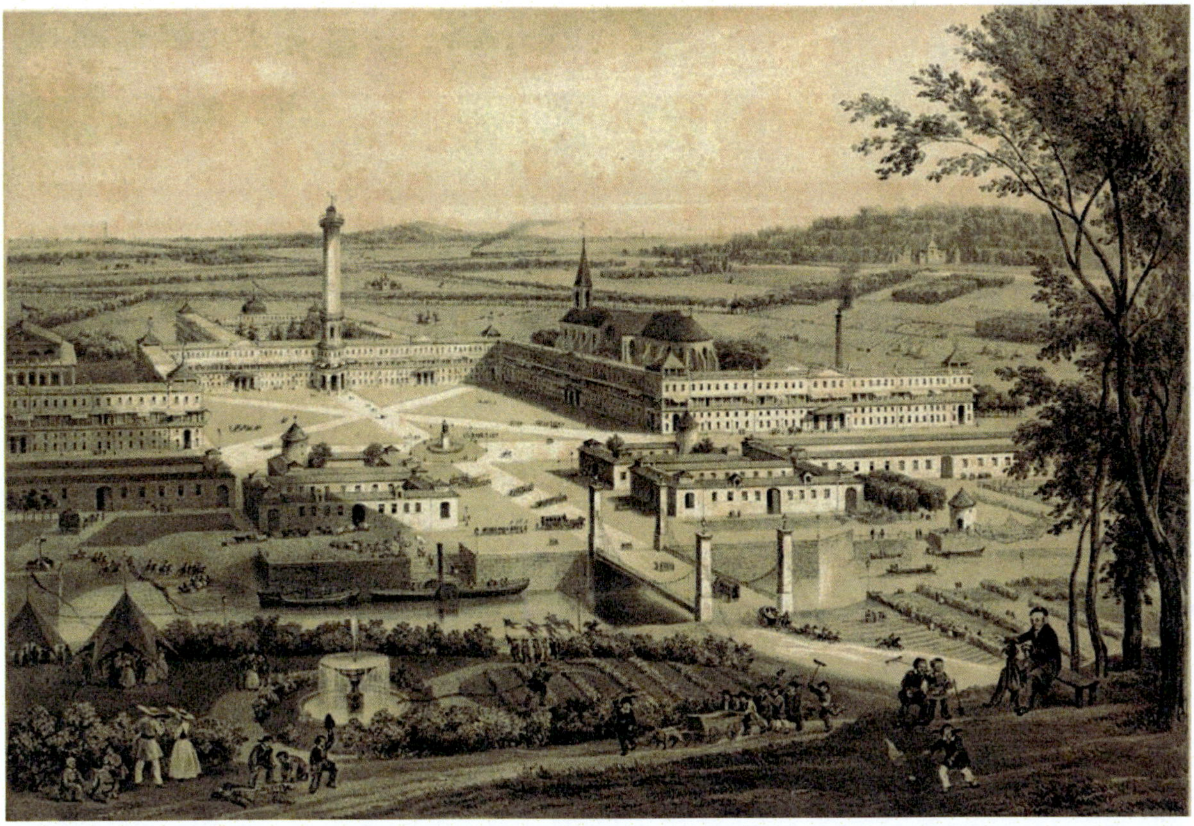

Fig. 1 View of a Phalanstére by H. Fugére, inspired by Charles Fourier—no date, nineteenth century. Public domain

internationalist design of *Siedlungen* (affordable housing), also being a pioneer in the acknowledgement of the links between urban planning and the metabolic cycle (Marot 2019). He proposed the productive garden as a set of open-air rooms integrally belonging to the housing unit, and using domestic waste as compost (Marot 2019). The Horseshoe Estate in Berlin (1925–33) he planned with Bruno Taut (1880–1938), Matin Wagner (1885–1957) and Ottokar Wagler (1881–1954) remains one of his most emblematic achievements, which predicted a fusion of urban and rural that many cities, we shall see, only sought to promote at the dawn of the twenty-first century.

Prior to World War Two (WW2), Howard's influence can hardly be overestimated in several European countries, but his vision was seldom fully materialized, and are instead was assimilated through the interpretations of architects such as Barry Parker (1867–1947), Raymond Unwin (1863–1940) or Louis de Soissons (1890–1962), who designed Garden Suburbs, lacking the agricultural belts and the circular food system.

In most Western countries, planning structures merely handle these productive transformations by organizing space according to the schemes most convenient for capitalist organization, creating a spatial segregation between production and consumption. Soon, such segregation produced a hierarchy where the urban, as consumer, controls the rural, as producer (Weaver 1984). Meanwhile, proposals for modern cities were discussed and formulated in the *Congrés Internacionaux d'Architecture Moderne* (CIAM), especially those prior to WW2, whose urban principles are epitomized by the 'Chartre d'Athénes', produced in CIAM 4 (1934) and published 10 years later by Le Corbusier (1887–1965) as his own work. The need to produce or access food was never directly mentioned in these, tacitly assumed to be a problem for industrial organization. Interestingly, in his manifesto book 'Vers une architecture' (Le Corbusier 1923), Le Corbusier collected several images of grain silos,[3] which he discussed in purely formal terms, ironically omitting the important *function* they had in modern food systems.

In the paradigmatic visions of CIAM modernism, the city became independent from its productive hinterland. The previous models—from Fourier to Howard—which incorporated food provision did not become dominant, as the influence of the urban paradigms of Le Corbusier and the 'Chartre d'Athénes' started to generalize. However, in the USA, the situation is somewhat different. The leading figure of architectural modernity, Frank Lloyd Wright (1867–1959), proposed an ideal vision of settlement, the 'Broadacre City' (1934–35). This was developed throughout the depression time, advancing an organic urbanism as a critique of the industrial city. It consisted of an indigenous organic settlement across a seemingly boundless cultivated plain with a transportation and communication network infrastructure, organized through a grid, with a county government administering a population of landowning citizen-farmers (Waldheim 2010). In order to be functioning and productive, this city, which Wright later rebaptized as 'Usonia' proposed that each inhabitant was to be granted a child-birth right to one acre of land, and modern houses were placed among kitchen gardens and small farms, interspersed with light industry, commercial centres, markets, civic buildings and infrastructure (Waldheim 2010).

With the physical destruction that followed WW2 in many of the belligerent countries, the physical structure of human habitat had to be extensively reconstructed or even built entirely anew, prompting architects to advance critical approaches to the relationship between habitats and communities, at several scales. The rationalist principles of pre-WW2 CIAM meetings come to be questioned and often repealed (Smithson 1974). This does not translate into a greater sensibility towards food-production, although an intensified consciousness of the importance of the local scale and of local conditions—environmental, physical, communitarian—is generally used as a tool for critiquing the excessive rationalism of the 'Chartre d'Athénes'.

Ludwig Hilberseimer (1885–1967) is an important example of an architect whose pre-WW2 work reflected the key aesthetic and functional principles inherent to the 'Chartre d'Athénes', as testified by his iconic drawings for the *Hochhausstadt* (High-Rise City) from 1924. Yet after WW2, his attention shifted from such abstract urban proposals towards the studies he had been developing since the late 1920s on European precedents to the Garden City. This led him to give greater emphasis to lower housing densities and to study the relations between transportation systems and different settlement units within a region (Waldheim 2010). In 1949, he published a book where he posits his key ideas for a 'New Regional Pattern', a horizontal network of suburban settlements with housing, farms, light industry, commerce and civic buildings, whose physical structure was dependent upon topography, hydrology, vegetation, wind patterns and other natural phenomena (Waldheim 2010). While the influence of the Broadacre City is explicit, Hilberseimer refrains from using the abstract grid and instead accepted territorial conditions as a predetermined landscape feature and a physical structuring element.

From 1959 to 1964, Dutch visual artist Constant Nieuwenhuis (1920–2005) proposed several images for a new form of human habitat, which he called New Babylon—a mosaic collage of urban settlements on elevated corridors spreading over agricultural land. This utopian vision was embedded in Constant's political ideas, namely his proximity to the International Situationist, a leftist group of intellectuals who sought to combine Marxist thought with the artistic avant-garde. Accordingly, Constant envisioned a city where labour was fully automated and the people's lives were nomadic and leisure-based—however, even in such context, the agrarian aspect of New Babylon was still somewhat of a backdrop for the urban element (Marot 2019).

[3]In the chapter titled 'TROIS RAPPELS A MM. LES ARCHITECTES' [sic].

A similar problem can be observed in the Agricultural City project (1960) by Japanese architect Kisho Kurukawa's (1934–2007) with its urban mat raised above stilts over agricultural land (Marot 2019). Kurokawa designed this proposal in response to a typhoon in Ise Bay, and it demonstrated the architect's involvement with Japanese Metabolism. The Metabolist Manifesto of 1960 advanced future designs for a coming world, envisioning solutions for problems different and extreme territorial situations, where design and technology—applied to architecture, urbanism and beyond—were understood as "a denotation of human vitality", and therefore, a metabolism (Noboru Kawazoe, quoted in Koolhaas and Obrist 2011).

The obsession with commercial culture inherent in 1960s Pop Art placed great emphasis on the ways food was publicized and commercialized. Suffice to think of the emblematic images by Andy Warhol (1928–1987) of Campbell's soup cans, Coca-Cola bottles or the banana on the cover of the Velvet Underground and Nico album (1967), all of which had a lasting effect on art. But these are not all. The oversized sculptures of Claes Oldenburg (b.1929) often represented popular food items of North American society. The aesthetics of food advertisement were often a source of materials for the early collages of Eduardo Paolozzi (1924–2005), like 'Refreshing and delicious' (1972) with its Coca-Cola ads. When realist painting gained a new breadth in North American art, some of its most important examples were images of food, including the gumball machines of Charles Bell (1935–1995) or the diners and restaurants in the urban landscapes of Richard Estes (b.1932). Such imagery testifies to the taste of the new consumer society in the 1960s, somewhat echoing the still life paintings with food and flowers, which popularized in art of late Mannerism, the Baroque and the Rococo, from the marked scenes of Pieter Aertsen (1508–1575) to the bourgeois meals of Jean Simeon Chardin (1699–1779) and Frans Snyders (1579–1657). In Portugal, Baroque painter Josefa de Óbidos (1630–1684) often painted this type of scene, constructing a sense of abundance through the lush details of fruits, bread and sweet pastries.

By the time, Pop Art and this new consumer society were emerging in the West, Portugal was still living under a dictatorship, which among other serious restrictions, limited the access of the Portuguese to the cultural developments in foreign countries. However, a kind of 'post-pop' emerged later in the country, where the visual strategies of pop aesthetics were matched with a self-conscious political stance (Vasconcelos and Rosas 2018). Food and food advertising also had a role in this national 'post-pop', from the drawings made from chocolate wrappers by Lourdes Castro (b.1930), to the colourful paintings of Teresa Magalhães (b.1944), Eduardo Batarda (b.1943) or Carlos Carreiro (b.1946) where food items were often the central motifs.

In Portugal, by the 1960s and 1970s, balconies in city homes were still often used for cultivations, mostly among inhabitants whose origins were on the countryside but had migrated to cities. Home cultivation—now carried in the context of urban housing—was a way to face the expensiveness of urban life by spearing money through food cultivation for self-sustenance. Such practices are testified by surveys on housing habits, conducted at the National Laboratory for Civil Engineering (*Laboratório Nacional de Engenharia Civil*—LNEC) in a research team led by architect Nuno Portas (b.1934) (Portas and Gomes 1963; Pereira and Portas 1967; Pereira and Gago 1974; d'Almeida et al. 2020; Marat-Mendes et al. 2021a). This practice seems to have slowly withered as these populations increasingly aspired to the common practices of the modern bourgeois home.

When British architect Colin Moorcroft (b.1947) organized the small anthology 'Designing for survival' (Moorcroft 1972), invited by the *Architectural Design* journal, he highlighted water, agriculture and food access as key concerns for territorial organization, architecture and urban design. Moorcroft collected studies, projects and articles from architects, engineers and architecture students, some of which unknown to the public, but all sharing an interest in the contribution that architectural and urban design can give towards countering the evident environmental dangers of the epoch. Concerns over agrarian territories are directly understood for their productive aspect, and their relationship with cities is in all senses more than a mere landscape problem. It is interesting to notice the materials Moorcroft collects to present this idea, some of which stand out to the contemporary reader for anticipating many problems (and solutions) of current literature on these subjects.

From Nick Roberts, an architecture student at the London Architectural Association, Moorcroft presented a set of proposals for spatial, productive and community organization resulting from a study on the Welsh village of Peterchurch. Such proposals were meant to reverse urban decay by improving the self-sufficiency of small communities through the collective development of a network of farm communes and village cooperatives. The article, titled 'Alternative rural development' focused on creating and growing small-scale production of clothing, furniture, food and biotechnic equipment, part for local consumption and part for trade, to ensure links with similar communities. Food was the priority production, with distribution centres and cooperative markets integrated in the village fabrics, since, as the author states, "For neighbourhood groups to have true political autonomy it is essential that they should have a measure of control over their basic resources, food and energy" (Nick Roberts, quoted in Moorcroft 1972: 425). Communal store provisions and sharing-based economy were the touchstones of this organization. The territory was reconceived for its potential for food production, articulating horticulture, semi-intensive agriculture, agroforestry and grazing areas, as well as ponds for fish, ducks, algae, reeds and grass on riverbanks.

A study by Dave Harrison, titled 'Food and cities: levels of manipulation', also presented by Moorcroft (1972), submitted that both city and countryside are fundamentally transformed with the growth of the agrifood industry, with two interrelated possibilities arising: individual or communal food production and independent systems for buying and distributing, both resulting in changes of diet. Another article, by Malcom B. Wells (1926–2009), was provocatively titled 'An ecologically sound architecture is possible'. The author enlists 15 things that wild land does

> (1) creates pure air; (2) creates pure water; (3) stores rain water; (4) produces its own food; (5) creates rich soil; (6) uses solar energy; (7) stores solar energy; (8) creates silence; (9) consumes its own wastes; (10) maintains itself; (11) matches nature's pace; (12) provides wildlife habitat; (13) provides human habitat; (14) moderates climate and weather and (15) is beautiful (Malcom B. Wells, quoted in Moorcroft 1972: 434).

Following a sensibility inspired by US Libertarian writer Henry David Thoreau (1817–1862), Wells submits that architecture can only match number 13 (providing human habitat) and does so at any cost, failing to see "the miracle in everything around us" and thus constituting an "ecological failure" (Malcom B. Wells quoted in Moorcroft 1972: 434).

'Agronica', the 1993 utopian project of Italian architect and former Archizoom member Andrea Branzi (b.1938), also proposed a settlement model that made use of food supply as a key activity, with the architect reanimating "the long tradition of using the urban project as a social and cultural critique" (Waldheim 2010). The name of the city derived from the junction between 'agriculture' and 'electronics', and the urban proposal sought to hybridize urban buildings, making them partly fluid, nomad and changeable according to seasonality and to the porosity of agricultural fields and activities, with agricultural posts as structuring landscape elements (Marot 2019). The frailty of buildings and their linear development emphasized the permanence of agrarian land, the key to the whole urban model, emphasizing the importance of the environmental aspect of urbanism (Waldheim 2010).

Throughout most of modernity, and in spite of the original ideas prompted by several architects, the conception of the city as a non-productive territory became unquestionably dominant in the West and consistently ensured an estrangement of urbanites from the origin of their food (Steel 2008). This is concomitant with the estrangement of the food system from most spatial planning policies, identified nearly two decades ago by Pothukuchi and Kaufman (2000) in the US context, but which might as well apply to other realities in the Global North (Marat-Mendes et al. 2021).

Some positive signs started to emerge from the growing research and policies on sustainability, prompted by the publication of the Brundtland Report (UN 1987) and, less conspicuously, by the report of the Club of Rome (Meadows et al. 1972), which advocated for the displacement of economic growth as the chief societal goal.

In the late 1990s, when Richard Rogers was commissioned to conduct studies on sustainable urban form, the concept of metabolism acquired great importance. Rogers (1998) proposed the compact city theory, explaining its potential to approach a desirable circular metabolism, i.e. where outputs such as waste and materials could be recycled and reused within the production system. Yet while Rogers' use of metabolic notions inspired a new breadth of interest on this subject, one fundamental aspect was left out of both the compact city model and many of its critiques: the food supply.

The growing detachment between these concerns and urbanism is surprising, considering that the importance of food in relations between societies and nature has been acknowledged in other areas consistently. This includes the natural sciences (for instance, Darwin 1871; Odum 1975; Alier 1977; Fischer-Kowalski 1998) but also social anthropology, which has noted the role of food in symbolic cultural values and even in the organization of non-Western societies into clans and phratries (Frazer 1910, 1911; Lévi-Strauss 1962, 1964; Douglas 1966; Descola 2001). A key work regarding these issues is Colin Turnbull's (1924–1994) powerful ethnography with the Iks of North Uganda, a people devastated by hunger (Turnbull 1972).

Indeed, behaviours and values have a fundamental impact on food systems (Timmer 2010), which extends to concerns on sustainability and sustainable use of cities (Tibbs 2011; Lo 2016; Marat-Mendes and Borges 2019). It is not by chance that in the several articles collected by Moorcroft (1972), habits and lifestyle were often mentioned and observed as the keys to achieving more sustainable habitats and construction technologies. Sustainability transitions are often dependent upon domestic and community behaviours (Lo 2016), and the generalized conditions of material wellbeing in the West seem to have given rise to a growing concern for non-material values, favourable to ecological preservation (Tibbs 2011).

Two examples that are particularly relevant to understand contemporary food systems are the works of German photographer Andreas Gursky (b.1955) and Canadian photographer Edward Burtynsky (b.1955). Among Gursky's uncanny photographs are included images of the food system economies of scale, like the depressing livestock facilities of the American prairie, or the colourful but crushing commercial stands of mass-produced food items, like his iconic photograph '99 Cents' (1999). Also focusing on industrial production, Burtynsky often depicts the Chinese context, the great competitor of the USA in the capitalist West. In his photographs, the limit between aesthetics and documentation is blurred, as in 'Manufacturing #17: Deda Chicken Processing Plant, Dehui City, China' (2005), where we see an endless sea of factory workers packing chickens. These people seem reduced to mere objects, no humanity about them, smashed by the gigantic assembly line they integrate.

In contemporary culture, food has a very significative place (Marat-Mendes and Borges 2019). Suffice to think of highly popular cooking contests like 'Chopped' and 'Masterchef', to entire TV networks like 'Food Network' and countless YouTube channels, from recipe repositories like 'Tasty' to food tourism to successful cooking shows like 'Babish Culinary Universe' and 'How to cook that', some with incredible audience reach, and with its protagonists publishing successful cookbooks.

The book and film series about psychiatrist/cannibal Hannibal Lecter, created by writer Thomas Harris (b.1940), spawned the popular and highly-acclaimed films 'Manhunter' (1986) by Michael Mann (b.1943), 'The silence of the lambs' (1991) by Jonathan Demme (1944–2017), 'Hannibal' (2001) by Ridley Scott (b.1937) and 'Red Dragon' (2002) by Brett Ratner (b.1969). This was revamped in 2013–2015 in a NBC series developed by Bryan Fuller (b.1969), which displays detailed (and highly meaningful) accounts of food preparation. Julia Ducournau's (b.1983) film 'Raw' (2016) focuses on the life of a vegetarian girl who accidentally eats meat and starts feeling uncontrollable urges, which radically shift her eating habits, turning her into a cannibal. 'Butcher's Block' (2017), the third season of Scy-Fy horror anthology series 'Channel Zero', created by Nick Antosca (b.1983) tells the story of a once-prosperous meat factory, which functioned as a cover-up for a cannibal family. 'Swallow' (2019) by Carlo Mirabella-Davis (b.1981) focuses on the anxiety-eating habits of its protagonist, who often eats pins, thumbtacks and batteries, eventually diagnosed with pica (a disease that leads to the eating of inedible objects). The series 'Servant' (2019-ongoing) created by Tony Basgallop (b.1968) for AppleTV has a chef as its co-protagonist, and the scenes of sophisticated food preparation lead to a lugubrious climax at the end of season 1, where a special declicacy is prepared with an umbilical chord. Yet the most striking example of food being assimilated into the themes of recent film or television is perhaps the film 'Upstream color' (2013) by Shane Carruth (b.1972) an enigmatic and powerful example of art reflecting upon the 'invisible' role of food in our lives, as the film tells a wild love story derived from an intricate poisoning scheme to which the nearly-complete food system is instrumental.

Hunger also has great symbolic charge in contemporary culture. We highlight recent novels by Portuguese writers, namely 'Estuário' (Estuary) by Lídia Jorge (b.1946), and 'Um bailarino na batalha' (A Male Dancer in the Battle) by Hélia Correia (b.1949), both of which take the dystopian aspects of the present and (more directly or less) anticipate a future marked by depletion and deprivation, in Correia's (2018) case announced already by the very contemporary tragedy of refugees crossing the seas to reach Europe, in Jorge's (2018) by the deprivation that remains in a significative part of the African continent.

Another aspect of great relevance is the increased attention placed upon dietary regimes, with a growth of vegetarianism and veganism, calling for a reconsideration of animal welfare and rights, as well as the environmental pitfalls of industrial livestock. A particularly compelling observation of this problem can be (prophetically) found in an essay by American fiction writer and essayist David Foster Wallace (1962–2008) titled 'Consider the Lobster. In this now iconic literary essay, Wallace chronicles his own attendance to the Maine Lobster Festival, followed by a meditation on how lobster was once a disgusting food served to prison inmates but evolved to be "posh, a delicacy, only a step or two down from caviar" (Wallace 2004: 238) as soon as it was discovered that if the animal was boiled alive, its meat would taste good. By describing the cooking process in painstaking detail, Wallace challenges the reader to consider this meal ethically.

At a time when such concerns have moved to the forefront of food debates, alongside a recognition of the importance of eating vegetables, the way food is produced and distributed deserves reconsideration, especially in cities, where food tends to arrive from the outside.

Being entangled with the city and its organization, the problems that are prone to public debate must return to the architectural agenda, if we consider architecture from a sociologically rooted perspective (Almeida 1964). Responding to particular social challenges, creating an ethically charged urban space that is formally relevant from the perspective of aesthetics was once key for architectural creation, at a time when architects actively participated in planning and design (Almeida 1964; Marat-Mendes et al. 2021b).

Indeed, the imbrication of ethic and aesthetic phenomena—central to architecture and urban design—is not a frivolous problem. Quite the contrary: it is a historically contingent interplay involving deliberation, infrastructuring and forms of governance (Gandy 2004). It is in this relationship that, we submit, creative solutions will arise to make urban food systems more sustainable. It is the design-led approach to everyday life that produces the most innovative answers (Ballantyne-Brodie and Telalbasic 2017). In such framework, design responds to the constant balance between supply and demand, using already existing decision-making channels. Collaborative approaches to design can galvanize service systems, with designers helping to empower citizens and institutions in creating new food services effectively define new cultural meanings with potential to influence policies and new practices for local decision-making (Ballantyne-Brodie and Telalbasic 2017).

One fundamental example of how the food system, considered from a local perspective, can inspire architectural and urban design can be found in the model of Continuous Productive

Urban Landscapes (CPUL), which proposes corridors of different types of agriculture, making urban ecological structures more productive, and linking them with the countryside (Viljoen et al. 2005; Viljoen and Bohn 2014). This paradigmatic vision, brought forward by architects, is one of the most interesting examples of a disruption with the modern partition between the city and the hinterland. The ability of urban agriculture to tap into resource and waste streams, reducing the costs and requirements with infrastructure is also highlighted in this territorial strategy (Viljoen and Bohn 2012). The proponents of the CPUL concept also acknowledge the barriers posed by planning instruments, as thus suggest its development may depend upon the creation of an 'action plan' to articulate general planning solutions with particular design solutions, allowing an adequate transition from the larger scale of a collective vision to the particular places and contexts where particular proposals are developed by smaller groups (Viljoen and Bohn 2012).

An international project titled CitAgra, gathering researchers from several universities, and coordinated by Tomasz Jelenski from the Krakow University of Technology, which started in 2017, focused on studying the potential of urban cultivation as a catalyst for socio-economic sustainability, harnessing urban forestry and agriculture towards the organization of Nature-Based Solutions for regenerating neighbourhoods and enhancing urban environmental services (Jelenski 2019). Among its chief, concerns were increasing intensity of underused urbanized land, regeneration, adaptive reuse of industrial estates, enhancing green areas and creation of biologically active surfaces of different types. Importantly, CitAgra was meant to counteract the existing trends towards sprawl, fragmentation and gentrification, thus its proposed agro-urban solutions were meant to be economically profitable from a social perspective as well as self-sustained, based on local public–private partnerships, with public participation, seeking sustainable regeneration through small-scale initiatives (Jelenski 2019). Blue–green networks were conceived to balance urban cores and green belts were extruded from river valleys, industrial buffers and agrarian wastelands (Jelenski 2019).

These more recent proposals may have a relevant follow-up in future territorial strategies, since urban trends such as guerrilla gardening, small-scale food producing and processing (Reynolds 2008), rooftop food production (Rodriguez 2016; Marat-Mendes 2017; Orsini et al. 2017), family urban farms (Olufemi 2016) and edible districts (Ghosh 2016) have emerged in scholarly research—and beyond it—as design solutions and as forms of urban activism towards sustainability.

A Transition of Metropolitan Food Systems?

Current international agenda, including the '2030 Agenda for Sustainable Development' (UN 2015), the 'New Urban Agenda—Habitat III' (UN 2016) and the 'Urban Food Agenda' (FAO 2019) all urge planners and policymakers to tackle unsustainable patterns of spatial organization, reconsidering the real potential of green spaces and seeking to link the city with the countryside, while accepting food-production as a valid element within urban areas. Thus, the estrangement of urbanites from their food must no longer stand. This represents a formal recognition of the dangers inherent to the estrangement of urbanites towards the food system.

Food-related problems are rising all around the world, over land conflicts, food riots, price surges and human rights (Sonnino 2013), owing to the globalized reliance on cheap crops from third world countries, which erase biodiversity and traditional agricultural methods, while inflicting tremendous misery and poverty upon workers—all of which are problems already clear for decades (Castro 1946; Moorcroft 1972). Some cities are thus accepting a role as actors in their food systems (Sonnino 2013) and international agenda are increasingly emphasizing the need for local provision supply chains (FAO 2019). The Milan Urban Food Policy Pact is the most notable example of this situation, with impact at the municipal scale since its implementation in 2015.

Planning for a sustainable environment requires balance between social, economic and physical systems, which we must learn how to consider at several scales. Currently, 55% of the world population lives in urban areas, and this is expected to increase to 68% by 2050 (UN 2016). This growth rate and the resulting unbalance between urban and rural populations challenge us to revise and critique how cities are managed, planned, and how spatial planning —from land-uses to architecture—may contribute to a desirable sustainable transition.

Yet, as has been abundantly clear at least since Rogers' (1998) book 'Cities for a small planet', in order for spatial planning to positively contribute to this sustainable transition, it will not be enough to consider the socio-metabolic aspect of cities. The morphological dimension will also need to be considered (Baccini and Oswald 2003).

Marat-Mendes' (2002) comparative study of neoclassical urbanism proposes four 'ground-rules' of sustainable urban form: adaptability, continuity, flexibility and resilience. Each of these entails different types of urban transformation, including in design, form and materiality, but also in use or function, each having specific sustainability impacts. As cities and metropolitan regions continue to change in the twenty-first century, these 'ground-rules', as well as the historical and disciplinary analysis they encapsulate, may find application in the wider context of land-use and unbuilt elements of human habitats.

However, a sustainable transition through circular socio-metabolisms has gotten irregular acceptance within key instruments and policies of spatial planning in many countries. Most often, sectoral policies regarding energy consumption or transportation systems have been promoted, downplaying the fact that sustainability pitfalls exceed the problems of any specific sector. In some cases, this also exposes the limitations of planning structures based on 'management' rather than actual 'planning', ensuring relative liberty for urban development by the private sector, while delaying necessary integral transformations towards sustainability and resilience that require coordination and planning. Beyond the agenda established by the UN, other initiatives keep prompting this change, as is the case with the Green New Deal for Europe,[4] which proposes policies for transition, overturning austerity capitalism and the intensification of socio-economic inequalities.

One way we may kickstart this transition would be to go 'back to basics', rethinking cities and their systems from the very roots: housing, water, electricity, transportation, internet and food. While some of these are widely acknowledged in planning instruments, others, particularly water and food, are strangely absent (Marat-Mendes 2016). Yet all are susceptible of being changed towards new models, which promote balance and quality for environments and for people.

When Pothukuchi and Kaufman (2000) surveyed the relevance of the food system in American planning agencies, they concluded that food-related activities were usually understood as a responsibility of the private sector, from which public institutions must refrain. Such perspective lingers today, in spite of the fact we live in a world where one in eight people go hungry daily (Olufemi 2016), with States rushing to bail-out private companies and banks at the expense of public sector services, prompting new forms of poverty, including at the nutritional level (Morgan 2013).

In the transition from an industrial to a post-industrial context, the city, once a material processor, becomes an information processor (Ravetz 2000). This post-industrial transition, which many cities are still undergoing, does not however change its need for food—nor for industrialized goods. Regardless of the dominant economic structure, urbanites always need to eat. Thus, all aspects relating to the origin and production of food, nutritional behaviours, economic exploration systems, labour, domestic activities and even leisure have to be considered when envisioning new habitats (Reisch et al. 2013). Moreover, one still needs to question how, in a world that produces such immense wealth, is it possible that food is not yet

[4]https://ec.europa.eu/info/strategy/priorities-2019-2024/european-green-deal_en.

acknowledged as a basic and fundamental human right—which it is not, considering that hunger has still not been mitigated.

Food geographer Roberta Sonnino (b.1968) has proposed that relationality is the most important feature of the recently emerging foodscape, since cities are starting to articulate their food system with wider sets of public goods and accepting a role as actors in food access. This author also points out the critical importance of the food system in fuelling local resilience, including through sustained food security and short supply chains (Sonnino 2013, 2016). This vision is in continuance with long-standing debates about strengthening strategies for local development through municipal empowerment (Lopes 1991) or otherwise, and Sonnino (2016) clearly frames in cultural and political terms the inclusion of the food system in urban planning, particularly at the municipal scale.

However, in most countries at least, official public policies on the food system are scant, fragmentary and seldom move beyond the local scale. The prevalence of the private sector remains unchallenged, and scholarly research seems to have moved quicker than political change. Since the agri-food industry is highly fragmented and presents great resistance to change (Brinkley 2013), it passes its fragmentation to other realms, including the spatial, and specifically the urban.

Achieving a more sustainable system raises a wide array of problems (spatial, social, political, economic, institutional and environmental), confirming how fundamental to our lives it really is. Thus, it is entangled in the paradigmatic problems of our time, including the rise of neoliberal politics, social inequality and the power of increasingly unregulated markets (Lipovetsky 2004), all of which have very significative impacts upon cities and their surroundings (Drago 2017). Any desirable change in spatial planning and food policies must, therefore, encompass a larger rethinking of institutional structures and power distribution underlying urban systems (Cohen and Ilieva 2015; Purcell and Tyman 2015).

Our time seems marked by the sensation that while we have outgrown the institutions that regulate our lives, we are still struggling to figure out what new institutions can be envisioned, making creativity (and even radicalism) more important than ever (Bartlett 2017). Urban systems—deeply implicated in the division of society in space—and food systems—ensuring a human basic need—provide important points of view to envision new conceptual frameworks and corresponding institutions, responding more closely to the demands of populations across several scales of space and of governance (Marat-Mendes et al. 2015; Oswald and Baccini 2003).

Planning for Change

The planning situation in many European countries has changed from a rationalist perspective, in which the institutions of the State define planning and its implementation, towards more open-ended processes where government is replaced by governance and the State acts as a facilitator and mediator between actors (Ferrão 2011). Yet, if the perspective of governance is in principle meant to allow for everyone to participate or partake in planning decisions and processes, in practice governance has mostly been a decision-making process, through which the private sector more easily imposes its needs and interests upon space (Ives 2015). A tension has long been felt in the practices of urban design and architecture, namely between bottom-up flexible policies allowing for diffuse local change, and a centralized planning system uncomfortable with informal change, a tension which the spatial fragmentation of food systems illustrates perfectly (Brinkley 2013).

Recent studies conducted on South Africa demonstrate how the imposition of these interests is particularly harmful to traditional and emerging economies and informal food systems, as small shops and productions are increasingly replaced by shopping malls and economies of scale, with important impacts upon employment and livelihood for locals

(Frayne et al. 2009; Battersby 2017). As such, informal and emergent practices within the food system may be threatened by a planning system where the lead role is, in practice, in the hands of the private market.

Weaver (1984) argued that the very basis of most planning structures amount to a spatial translation of the capitalist system, thus it is little surprising that such informal or emerging practices are difficult to integrate in regulatory planning. For instance, a study on urban agriculture in the Lisbon hinterland suggests that an economic upscaling should be a part of its future development (Delgado 2018). However, this could also translate into a normalization or transformation of urban agriculture into a form of agri-food business uncapable of competing with contemporary economies of scale or, in a more depressing alternative, it could lose its popular character in order to be able to compete.

The problem of economic scale is often a central theme for food system studies. In general, there have been three major lines of scholarly approach: (i) the continuation of dominant agri-food business models; (ii) the call for alternative food systems and (iii) a 'third way' that moderately accommodates elements from the two previous models (Viljoen and Bohn 2012). Most planners tend to think in terms of large scale, which is typically not very welcoming to more 'localized' initiatives for the urban food system (Viljoen and Bohn 2012). With regards to agriculture, although its practice could still be verified in worldwide cities throughout the twentieth century and particularly at times of crisis (Iaquita and Drescher 2010), only recently has it been acknowledged by international institutions as a legitimate part of urban food systems or as a key practice to be promoted and protected in the context sustainable transitions in both the territory and the layout of food access (FAO 2019). If the modern dynamic of spatial organization divided production and consumption between countryside and city (Weaver 1984), we currently witness the unsustainable consequences of such severance. The ever-growing food industry has been for decades pushing production facilities further and further away from consuming cities—thus creating a need for long-distance transportation, air pollution, poverty and land conflict in third world countries where food production physically takes place (Castro 1946; Moorcroft 1972; Spaargaren et al. 2014).

Other than consumption and production spaces, an important part of urban planning for better food systems is waste disposal. British geographer Matthew Gandy (b.1965) has pointed out how modern technology and modern society brought about a new emphasis on infrastructure, constructing a 'hidden city' of sewages and hydraulic systems running underground and allowing for sanitary conditions to be attained (Gandy 2004, 2014). Other forms of infrastructure—roads, highways, reservoirs, garners, waste management stations, etc—remain visible and fundamental for the functioning of cities. In the food system, visible and 'hidden' structures and infrastructures interplay. Access to water is a key aspect of agriculture, whose soils are regulated by municipal plans, often without considering such important aspects for sustainability as land-use intensification (Erb et al. 2016). Factories and logistics centres mediate between produce and its public availability in all different kinds of selling points. But such factories are treated, in most spatial planning instruments—and often on popular perception—as generic elements of the industrial fabric. Planners stipulate industrial areas without tackling the eventual importance of food transformation within a regional scope, with the destination of produce depending upon the will and the interests of owners.

A similar situation applies to food commerce, although this is arguably the most visible aspect of the food system. Planning regulations do not usually establish any fundamental difference between the commerce of food and of other items. Restaurants, neighbourhood shops, supermarkets, hypermarkets, taverns, wholesalers, delis, bakeries, quick-service restaurants and coffeeshops (Pothukuchi and Kaufman 2000; Moudon et al. 2013) are essential parts of any city's economy, but their high-potential role for resilience policies, by forming a network of food access and thus preventing the formation of food deserts, is unacknowledged in most planning structures.

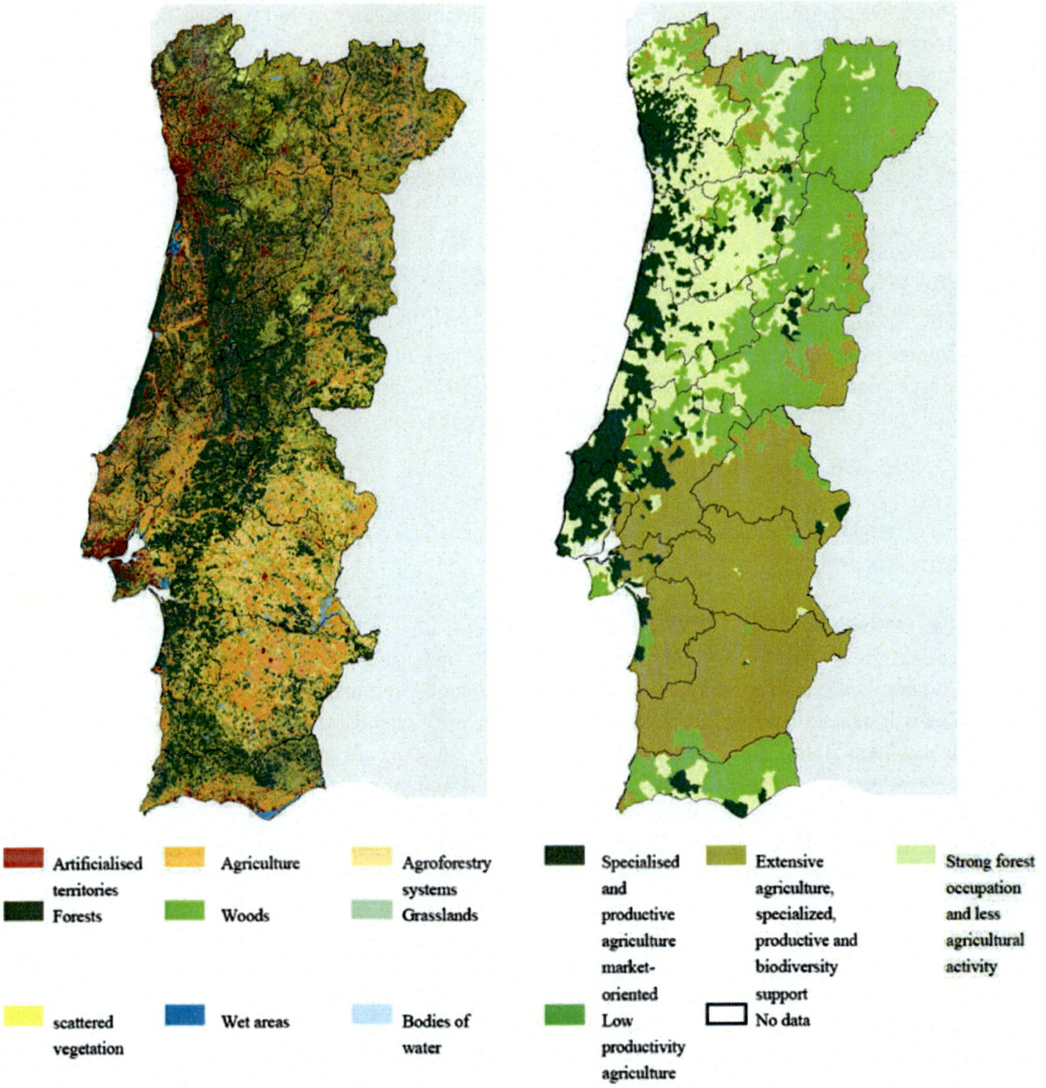

Fig. 2 Land Use and Occupancy Chart Mainland Portugal—Megaclasses (left) and Territorial profiles for agriculture (right). Source: PNPOT—Programa Nacional de Ordenamento do Território, 2019

Urban planning, urban design and architecture, as practices that interpret and define (at several scales) forms of spatial organization may reinvent themselves by emphasizing the importance of civic engagement and social movements, some of which already advanced (unprofessional) design solutions for existing problems (Ballantyne-Brodie and Telalbasic 2017). In some cases, the pathway to sustainability implies taking these actions seriously, learning from 'niches' and popular activities, understanding how they may be scaled-up and, even more importantly, how they may be articulated, so that different phases of the food system are integrated within urban territories, instead of segregated (Crowley et al. 2021).

As such, designing solutions for a more sustainable food system must start from an understanding of specific territories and a study of the activities already taking place in them. Creativity, willingness and adequate use of available resources—including space itself—are paramount for sustainable transitions. But so is knowledge on the conditions of the territory, past and present, from which new pathways can be derived.

In this Atlas, we propose that the food system offers a sustainable alternative for urbanism, one that promotes self-reliance and increased resilience, while reclaiming many of the positive principles inherent to architectural solutions across modern history which have included a concern for food. Following the recommendation of the FAO in their 'Urban Food Agenda' (FAO 2019), we must re-learn how to observe (and design) cities, not contained within their administrative limits, but as elements within a city-region. An assessment at the city-region scale is key to imagine a re-localization of food supplies, with a particular impact on the possibility of decreasing food miles, i.e. the distance run by foodstuffs from production to consumption (Brinkley 2013) and to decrease the detachment of urbanites from the origin of their food. This demands not only a reconsideration of metropolitan rural spaces but also more creative approaches to green spaces within urban settlements themselves, which may prove important to cover part of familiar consumption (Viljoen et al. 2005). In order for a sustainable transition of urban food systems, what is necessary is innovative policies, which acknowledge and correct systemic problems in food-consumption patterns (Reisch et al. 2013).

The Lisbon Region

In this Atlas, we will focus on the food-related activities of one specific area, the Lisbon Region. Its limits were established for the first time in 1959,[5] precisely when the dictatorship wished to order a Regional Spatial Plan—which would eventually be finished in 1964 but not approved, as will be seen in Chap. 2. In 1991, the same territory would form the basis of the Lisbon Metropolitan Area (*Área Metropolitana de Lisboa*—AML), an intermunicipal association meant to promote integrated policies regarding the development of the region. In the 1950s, this territory comprised 16 municipalities, distributed around the Tagus River: Lisboa, Loures, Vila Franca de Xira, Oeiras, Cascais, Sintra and Mafra on the Northern Bank, and Almada, Seixal, Barreiro, Moita, Montijo, Alcochete, Sesimbra, Palmela and Setúbal on the Southern Bank. Since then, however, two processes of administrative autonomation gave rise to two additional municipalities: Amadora became independent from Oeiras in 1979 and Odivelas from Loures in 1998. However, either the role of regions and of intermunicipal associations is unclear in the Portuguese political and administrative context.

In February 2019, the Portuguese Parliament welcomed the 'Forum for Public Policies', an event that gathered politicians (mostly from municipalities) and scholars to discuss, among other political problems, the need for regionalization in Portugal. One of the key moments of the Forum was the speech from Prime Minister António Costa (b.1961),[6] who saluted the discussion on regional governance, but noticed the need to safeguard it from political manoeuvres in the campaign for the upcoming election (to be held in October 2019). Costa was re-elected as Prime Minister, but 4 months later already hinted that regionalization is a subject after the next election (Almeida 2021).

To be sure, this debate is everything but new. And the same is true for the postponement of the debate itself. As architect Nuno Teotónio Pereira (1922–2016) noticed when regionalization failed in the first attempt, there is a strong tradition of regional identity, that is cultural, economic, and social (Pereira 1994). Teotónio Pereira dedicated several chronicles to the problem of the Portuguese regions back in 1994–1995, when this discussion was happening under the government led by then Prime Minister Aníbal Cavaco Silva (b.1939),[7] who opposed regionalization. Although this process was met with a change in political power, with

[5]Law 2099, published in Diário do Govêrno, 186/1959, Série I, 14-8-1959.

[6]Became Prime minister from 2015, re-elected in 2019.

[7]Elected Prime minister from 1985 to 1995.

António Guterres (b.1949)[8] being elected Prime Minister in 1995, regionalization never came to pass in Portugal. The 1998 referendum had the participation of less than half the population of a country that was not used to voting in referendums, and there was a clear lack of political will to change a system in which power was always strongly centralized in Lisbon (Pereira 1994). In the end, over 62% of voters refused regionalization, with some attributing the responsibility to the proposed regional organization. The debate was dead for more than a decade. The refusal crosses partisan lines, and regions seem to have been avoided by the two centrist parties who have traditionally governed the country since the 1976 Constitution. Arguments against regionalization often are based on the fact that Portugal is a relatively small country, or in the need to not enlarge the political class with more institutions and more titles being created. However, these arguments leave out the striking fact that, however small, Portugal is a territory with historically and geographically self-evident regional differences, but also with very marked contrasts between city and countryside, as well as between inland and coast.

Moreover, the argument that territorial extension is too small for regions ignores that, while these do not exist, power is highly centralized in Lisbon, and there is no intermediary scale between national and local political power. Some have defended that strengthening the powers of municipalities—a process which has begun with the 1974 Revolution itself—is enough to decentralize power, but this idea ignores the fact that, particularly in small municipalities, local power is extremely exposed to corruption: for instance in 2004–2008, the majority of legal processes for corruption was moved against local power (Lima 2011). In 2018, 48% of corruption processes were also brought against local power, the highest percentage ever (Marcelino 2019). So, it would seem that an intermediary scale of power, between the central State and the municipalities, would be a way to balance inequalities, increase political accountability, and create policies that make sense within the specificities of each region. Moreover, as Teotónio Pereira already had noticed, many problems extend beyond municipal borderlines, and therefore cannot be solved by municipalities themselves, as is specially the case with the territory (Pereira 1994). This indictment is abundantly confirmed by the study of the food system, especially in metropolitan regions.

The problem of the territory is the touchstone of regionalization, for while there are indeed social, cultural and economic differences between Portuguese territories, these differences are also expressed in terms of geography and of the corresponding settlement types.

Portuguese geographer Orlando Ribeiro (1911–1997) has noticed that the division of the Portuguese territory has been a constant aspect of its political formation and administration, finding traces of it back to the thirteenth century. He, furthermore, identifies in the fifteenth century historical chronicles of Fernão Lopes (1385–1460), the first Portuguese historian, a list of the then called Portuguese comarcas (counties): "antre Doiro [sic] e Minho, Trá-s-os-Montes, Beira, Estremadura, antre Tejo e Odiana [sic], leaving out the Kingdom of Algarve" (Ribeiro 1987: 102). The names of these counties were deeply descriptive of territorial elements like rivers and mountains. Across history, the particularities of these regions became ingrained in several cultural aspects, but the Portuguese democracy has not granted them political autonomy, instead distributing power at the municipal scale alone. In terms of spatial planning, and despite the fact that regional planning has existed in the Portuguese legislation since 1988, few plans have actually been created and thoroughly implemented. However, regional planning and regional communities have long been hailed as the keys to political autonomy and self-sufficiency. Looking back to the roots of regional spatial planning, Weaver (1984) highlights five key ideas that can be found from early thinkers like Charles Fourier to Paul Vidal de la Blanche (1845–1918), Frederic Le Play (1806–1882) and Patrick Geddes, namely: a) a revulsion with the industrial city; b) negative reaction to economic and political centralization; c) a belief in restoration of regional life; d) defence of mixed territories

[8]Elected Prime minister from 1995 to 2002.

Fig. 3 Lisbon Metropolitan Area (LMA) and its location in Portugal. Source: Authors

to avoid hierarchies between city and countryside and e) combination of manual and intellectual labour.

Also noteworthy in this context is the Survey on Regional Architecture in Portugal (which will be discussed in Chap. 4 of this Atlas), published in 1961 with the title 'Arquitectura Popular em Portugal' (Popular Architecture in Portugal). This survey, which focused on popular, traditional and vernacular architecture mostly on the Portuguese countryside, depicted a considerable variety of construction techniques and forms (AAVV 1961).

As recalled by architect Francisco Silva Dias (b.1930), who participated in the survey, collected materials demonstrated the variation and diversity of vernacular construction within the country, varying from region to region, and disproving the belief in a national style of architecture, held by the Government and, in a different way, by architects of a more 'aristocratic' taste such as Raul Lino (1879–1974) and Fernando Távora (1923–2005) (Esteves and Mestre 1987).

While regions are highly recognizable in the territory and in its traditions, there seems to be a lack of political will to address the decentralization of Portuguese power. In November 2019, the current government decided to relocate its 'State Secretariat for Inland Development' to the municipality of Bragança in Trás-os-Montes, to take it closer to the rural population. However, this effort is hardly capable of handling the problems of highly centralized power and the concomitant hierarchy between city and countryside. Without creating conditions for local communities to have a deciding voice over the territory they inhabit, this decentralization falls short of its political purpose.

To a certain extent, the Lisbon Region integrates the region called Estremadura, although the official formation of a Metropolitan Region around the capital city has somewhat effaced this integration. However, Orlando Ribeiro has noticed the specificity of this region and its internal differences in the following way:

> Traditional provinces, as is known, do not correspond today to administrative divisions, but they are clearly felt among educated people, as well as among their natives. Only Estremadura is an exception in this, perhaps because it is the most heterogenic and because, more than any other province, it had its limits redrawn, being cut northwards and eastwards and expanded to the south of the Tagus river, which was considered as its natural limit for a long time. Thus, it is usual to call the natives of the provinces by their ethnic designation (and they do it themselves too), but no one will call themselves estremenho, just as no one says they *go to Estremadura* or that they will *travel through Estremadura* (Ribeiro 1987: 102—author's italics).

With the 1991 creation of the AML, the relation of the Lisbon Region with the remainder of Estremadura is seldom noticed, a situation that was altered by the COVID-19 pandemic, with the administration of sanitary cordons and of the easement of lockdown measures. Currently, the Lisbon Region includes a vast diversity of activities and types of territories, ranging from the entirely urban—like the Lisbon city—to the entirely rural—like the Great Wetland of Tagus and the Montijo exclave.

In the above mentioned survey on regional (or popular) architecture, the country was divided into six zones—Minho, Trás-os-Montes, Beiras, Estremadura, Alentejo and Algarve—to which six team of architects were assigned. Estremadura corresponded to Zone 4, and its team was led by architects Nuno Teotónio Pereira, António Pinto de Freitas (1925–2014) and Francisco Silva Dias. It was delimited in a very liberal way, between Setúbal, Abrantes, Coimbra and Praia da Mira (AAVV 1961). It is interesting to recall the notes written by the architects about this area, particularly considering that this amounts to descriptions of this territory at such a recent time as the late 1950s:

> Vineyards and potatos occupy great extensions and, nearby Lisbon, the 'região saloia' [rural region] appears to be one of the neatest areas of polyculture. In basalt-based land, cereals are cultivated on open fields. Moorlands and pine forests cover the poorer soils (AAVV 1961: 10—free translation).

> In front of Lisbon [...] the Tagus enters with numerous arms through the lowlands. It is a zone of intimate communion between land and water. The tidal mills and the old boats are reflected in the same landscape. On the background, the industrial zones of Barreiro and Montijo (AAVV 1961: 13—free translation).

> [...] most land north of Tagus is polycultivated, which means labour and care all year long, demanding a communion between the land and the people, which favours dispersion: on the Centre, between hamlets and settlements, and on the North in small cores or on plain fields resulting from subdivision alongside circulation routes, through a special form, an ordered dispersion, profoundly reflected on the settlement structures – linear, almost without a core and often reduced to a single street marked by houses. [...] The

Fig. 4 Rural houses in Cascais (left) and threshing floor in Arneiro dos Marinheiros, Sintra (right). Source: Inquérito sobre a a Arquitectura Regional Portuguesa (1955–1960). Zona 4 (Arquitectos Nuno Teotónio Pereira, António Pinto de Freitas e Francisco Silva Dias) © Ordem dos Arquitectos

> South on the other hand is covered by uniform vegetation or large extensions of the same cultivation, be it rice or wheat. The majority of the population owns no land, and labour is required only a few times a year, for reaping or paring and little else, and this separation between labour and the field favours concentration (AAVV 1961: 17—free translation).

> When the core is located on a high place, streets have movement and include ramps and stairwells, alongside ingenious ditches for the fast drainage of floods. Houses are huddled as in nativity scenes. There is a constant use of curved lines, gables and terraces [...]. Streets curl or break in small segments for adaptation. Settlements have a nearly organic malleability (AAVV 1961: 26—free translation).

> To the structured and compact character of the hilltop or old inland settlements, the agrarian areas of disperse population oppose settlements of disperse elements. Each house is surrounded by a 'courela' [a strip-shaped vegetable garden] and often includes a haystack, a winery and a barn. The alleys and squares infiltrate or are defined between walls and hacks of each house. The street as a fundamental structuring element disappears. Now it results from the shape of properties [...] In the ill-defined squares, trails are open by footsteps and chickens are pecking and sheep graze. On a corner, or between houses, are the threshing floor which is individual or collective, as well as the church and a common well (AAVV 1961: 28—free translation).

These descriptions are found alongside a wide array of photographs of settlements, fields and landscapes, and chiefly of buildings. Among these, there are several that illustrate the social and economic aspects of life in this region. It is noteworthy that cities and towns are not heavily featured in the survey, which tended to focus more on the rural territories of Portugal. However, it is still visible that, even on municipalities just a few miles away from the capital city, there is still a predominance of a rural life and of a primary economy, where the landscape often includes windmills, haystacks, granaries, barns and other agricultural implements; where settlements are usually small and old, typical of a society where families stick to the same premises, inheriting and inhabiting the properties of their elders; where fairs and festivals are still influenced by the agricultural cycles and the seasons.

There are still many villages in the Lisbon Region, and even more if we consider Estremadura altogether. But the life they harbour is unlikely to match that which is portrayed in the survey. If the 1974 democratic Revolution brought about, among other things, a liberalization of the market and effective pathways for social mobility, it is also true that the key orientation was towards urbanization and an abandonment of rural life. This transformation of the society and the territory was already on the way in cities—in Lisbon and Porto, high-density construction was already being accepted by the conservative dictatorship at the time of the survey, as it was acknowledged that population growth in cities demanded an acceptance of properly urban-scale construction. Conversely, land used for food production decreased, and the food system—as well as dietary habits—were liberated from spatial constraints.

The idea of local or even regional cuisine dates back only to the late nineteenth century—until then, most recipe books were constructed over perceptions of quality, being rather international or national, although some products were promoted with an origin indication as a guarantee of certain features (Guerreiro 2018). In Portugal, the effect of the dictatorship, with its values of tradition, family and nationalism, placed much cultural emphasis on cooking as a feminine art and a cornerstone of domestic economy (Guerreiro 2018; Peniche 2020).

The National Secretariat for Information and Tourism (*Secretaria de Estado da Informação e Turismo*) promoted such vision through cooking shows and contests via television, as well as through the publication of books. A key example of this promotion can be encountered in 1961—coincidentally, the same year the Survey on Regional Architecture was published. Promoted by the SNI and by the National Broadcaster (*Radiotelevisão Portuguesa*—RTP), the Contest for Portuguese Regional Cooking and Confectionary (*Concurso de Cozinha e Doçaria Regional Portuguesa*) tested hundreds of recipes submitted by the general public and appreciated by a jury which included cook Maria de Lourdes Modesto (b.1930), culinary critic Daniel Constant (1907–1984), João Ribeiro (1905–1988), the chef of the Avis Hotel restaurant, and a culinary professor. The laureate recipes were published in 1971 in a book titled 'Recipes of Portuguese Cooking and Confectionary' (*Receitas de Cozinha e Doçaria Portuguesa*) a title that omitted the regional aspect (like the architecture survey had) despite its presence in the internal organization of the book.

Yet this discrete publication is far from the most representative outcome of the process initiated in 1961. Maria de Lourdes Modesto, whom the New York Times once called 'Portugal's Julia Child' and who hosted a live cooking show, started in 1961 to systematically collect, compare and perfect an extensive inventory of recipes, often sent by anonymous people from all over the country. The results of this research were finally published in book form in 1982, with the title 'Traditional Portuguese Cooking' (*Cozinha Tradicional Portuguesa*) which, with over ten editions, is not only one of the most successful cookbooks in Portugal but also remains a key reference for contemporary cooks. While the title again omits the regional character of the inventory, the book is organized by dividing the country into 11 regions—Entre Douro e Minho, Trás-os-Montes e Alto Douro, Beira Alta, Beira Baixa, Beira Litoral, Ribatejo, Estremadura, Alentejo, Algarve, Madeira and Azores—presenting for each a selection of representative recipes. Modesto (1982: 7) understood this work as a "survey on the Portuguese gastronomic heritage", highlighting dishes that best represent local traditions as expressed in family cooking.

In Modesto's survey, the Lisbon Region was included in Estremadura—as it had been in the 1961 architecture survey. The opening presentation of each region was written by António Manuel Couto Viana (1933–2010), a conservative and monarchic Portuguese poet. In the case of Estremadura at least, much of the presentation text focuses on the pivotal role of Lisbon in the formation and transformation of the Portuguese Kingdom—and oddly, the city's history stops during the dictatorship. However, the writer does find space to mention local food, highlighting the 'bohemian vegetable gardens' of the capital city, as well as the fishing activities in the Sado estuary, the cultivation of its bank with orange trees and vineyards, the cheese from Azeitão, and the wines from the 'Bacchic belt' of northern Lisbon, through Bucelas, Colares and Carcavelos (António Manuel Couto Viana, quoted in Modesto 1982: 193).

Afterwards, Modesto collected a total of 82 dishes typical of Estremadura, divided between starters (appetizers), soups, shellfish and fish, rice and 'açorda' (uncooked bread soup), poultry, meats, sweets and puddings, cookies and pastries, cakes and fruit jams. From these recipes, 49 are generically assigned to Estremadura as a whole, while 32 have specific origins, 8 in locations from the Lisbon Region, and 25 from outside it. Years later, Modesto donated her inventory to the Portuguese Association of Professional Cooks (Associação Portuguesa de Cozinheiros Profissionais), who made additional recipes available to the public (Modesto 2015). In Estremadura, there were 17 new recipes, 3 being general, 9 from the Lisbon Region and 5 from outside it. Between the book and the website, 100 recipes are available from Estremadura.

The largest set of recipes is from the fish and shellfish section, with 17 dishes, followed by the appetizers with ten. Meat counts only nine dishes, although to these one may add the four dishes of poultry (rabbit and chicken). However, out of four dishes of rice or 'açorda', three include shellfish. Eggs and olive oil are very frequent, the latter being used in 36 out of the 99 recipes, and the latter in 24. Wine, despite its omnipresence in the cultivations of the Lisbon Region, is used as an ingredient in only nine of these recipes.

Many of the dishes express the varied backgrounds they emerge from, for it is clear that while some of them employ ingredients of high price or in large quantities, others are clearly marked by the restraints felt by the popular classes. There is also an urban feel associated with dishes from the Lisbon city's cafés, most of which are meat dishes, in contrast with popular dishes made from ingredients sold by women street sellers. This expresses a coexistence of a traditional and popular world alongside a growing city that seeks to become more cosmopolite.

At the time of the Revolution, a considerable part of the Lisbon city was already urbanized, and several of its suburban cores were already clearly formed as well. Although the democratic regime opted for decentralizing spatial planning, attributing to municipalities the responsibility for creating a plan and coordinate land-use conversions, for several decades, the private sector urbanized a considerable amount of rural land, sometimes in direct contradiction with the dispositions of planning guidelines (Mourão and Marat-Mendes 2016). This resulted in the tremendous increase of urbanized land that characterizes the Lisbon Region today and that seems far from over. Rural landscapes and activities were pushed further away from urban areas, and segregation between urban and rural became undeniable.

From the several expressions that have recently come to characterize the complex and diversified forms of human habitat—megacity, megapolis, metropolis, conurbation—the concept that seems to apply more clearly to the Lisbon Region is that of conurbation.

This term was first coined by Geddes (1915) to designate identifiable territorial agglomerations that had formed because urban cores, under the pressures of industrialization and modernization, had grown to the point of merger. While the boundaries of these cores melded, city-regions or conurbations emerged. Such agglomerates included cities, towns and villages, as well as their interstitial space, filled with more neighbourhoods, housing estates, industrial belts and other forms of construction.

A train trip across Costa do Sol (Lisbon—Oeiras—Cascais), the northern line (Lisbon – Loures—Vila Franca de Xira) or the Sintra line (Lisbon—Amadora—Sintra) would leave no doubts as to the continuous character of urbanization in the Lisbon Region, broken only from time to time by green buffers, which, in some cases, are soils where urbanization is either impossible or expected in the future. In the Southern Bank of Tagus (Almada—Seixal—Barreiro—Moita—Montijo), this urban continuity, sometimes interrupted by aquifers, is also quite evident, with only the town of Alcochete clearly detached from the remaining riverside settlements. Thus, speaking of the Lisbon Region means speaking of a specific conurbation.

This situation poses important questions for the future. Will urbanization keep growing, thus continuing the dynamic verified across the twentieth century? At the beginning of the twenty-first century, the city of Lisbon was losing population (CMLisboa 2012), while its outskirts were gaining population: this testifies to a population movement across the metropolitan area, sometimes in search of more affordable housing. While it is important to notice that currently, the Lisbon Region is in a situation of housing surplus (Mourão and Marat-Mendes 2016), it is also true that access to housing has been increasingly difficult for a great part of the population, because of both speculation from the real-estate sector and the sequestration of a considerable part of the housing stock for touristic activities in the city of Lisbon. Moreover, it is known that housing construction is currently led by the priorities of the private market, which treats housing as a commodity, instead of a need of local populations. In that sense, it would be possible for the region to continue urbanizing, even if such growth is neither necessary nor desirable. Beyond the 'bubble' implied in such continued growth, this would further implicate the loss of more rural soils.

Another option would be to deflect growth as a key goal for territorial management, and to acknowledge the value and potential of the rural aspects of the Lisbon Region. Indeed, a considerable part of its territory is productive, and while the primary sector amounts to only a negligible part of the regional labour structure, the pursuit of sustainable territorial strategies would suggest that the role of rural activities must be revised in systematic relation with urban ones.

The history of the twentieth century in the Lisbon Region testifies to a tremendous cycle of change: changes in territorial organization, in society, in morphology and in metabolism—and this illustrates the potential of this territory for a wide array of activities and uses. This complex history is only valuable if we are able to draw lessons from it, and if it inspires us to conceive of new transformations, that can face the challenges of today, when the advantages of further urbanization and growth are everything but self-evident. Here, we submit that the food system and its transition history can be a great tool towards a new understanding of the territory of the Lisbon Region and envisioning a more sustainable future for it. And while this implies leaving certain ideas behind (namely the necessity of further growth), it also implies accepting new ideas, which can ensure valuable steps towards overall sustainability and inclusiveness.

Framework and Themes

The Atlas of the Food System is conceived as a survey, inspired by the works of architects like those who worked in similar works in Portugal (AAVV 1961; OA et al. 2006), and also following Geddes' (1915) proposal that urban design and planning must be based on a survey on the current conditions of the territory. In the late 1950s, architects went on the road and sought to understand the territory in its complexity and diversity, bringing together a wide range of concerns, from the architectural to the urban, from the economic to the political, from the social to the environmental. Indeed, ever since its first edition, it has been an outstanding source of inspiration for many architects (and occasionally for urban planners too) and critics, and in the early twenty-first century, it inspired the Portuguese Architects' Association (*Ordem dos Arquitectos*, OA) to create a new survey, this time focusing on architecture of the twentieth century, but following a methodology very similar to that of 1961. From neorealist to postmodern, many architects in Portugal took the photographs from a rural Portugal that seemed more and more distant, as much as images of both joyful and melancholic city life in Lisbon (Palla and Martins 1959) as inspirations to build a new Portugal, a new way of seeing the land and the people (Tostões 1997; Figueira 2016; Marat-Mendes and Cabrita 2016).

What we propose here is a survey focused on activities and land-uses from the food system in the Lisbon Region. We do not seek to exhaust our theme, but rather to present examples that seem both representatives of what exists and the territory, but also suggestive of new ways of understanding it. We provide a portrait, meant to account for a current reality and to imagine how this reality may evolve and improve in the future. To do so, we use varied means, from interpretation of planning instruments to maps, photographs and schemes. Our key aim is to inform future planning policies and practices towards an improved food system and a more sustainable territory, but this is pointless unless we accept that such policies, if they are to be effective, will need to be complex, multidisciplinary and will sometimes imply paradigmatic changes. Such is the challenge of 'our common future', as sustainability was once emblematically defined (UN 1987).

The first part of our Atlas explores the history of spatial planning in the territory of the Lisbon Region, highlighting the key morphological and metabolic implications of each plan and observing if and how, directly or indirectly, they incorporate the food system and its related activities and land-uses. Throughout the twentieth century, the regional territory was subjected to tremendous demographic, political and economic changes, and this prompted the emergence of spatial planning at different scales. Although the Portuguese legal framework and scheme for spatial planning underwent several stages of development, for over 100 years, it was subjected to several plans both integral and specific, ranging from detailed neighbourhood schemes to municipal, sub-regional and regional plans and foreplans.

These plans were conceived over a territory that was previously inhabited and which previously hosted a set of economic activities, all of which amounted to a specific landscape. All of these dimensions were interpreted by plans. When implemented, planning changes life, labour and landscape, although quite often, these change without the aid of plans. Despite frequently being undermined, spatial planning effectively contributed to change, and indeed it was a key element of socio-ecological transitions throughout the twentieth century. Socio-ecological transitions refer to paradigmatic changes in relationships between social and ecological systems, i.e. the way in which societies and nature intertwine and interlock (Fischer-Kowalski and Weisz 2016). Here, we seek to understand the specific relations between spatial planning and these socio-ecological transitions within the Lisbon Region territory.

In the second part of the Atlas, we aim to follow the example of Portuguese architects who in the past have dedicated to surveying the existing territory. Such exercise is meant to demonstrate the kind of structures and spaces to which architects and urban designers can contribute if they want to contribute to a more sustainable metropolitan food system. In the first moment, we describe the important role that photographic surveys have had in twentieth century Portuguese architecture, highlighting the metabolic changes that such surveys (unconsciously) testify to. In a second moment, we survey the current reality of the Lisbon Region, aiming for a visual characterization of food-related activities and land-use and they can currently be encountered. While all the phases of the food system can be found in the metropolitan territory, this does not suffice to call it a food system, since this would demand a degree of functional articulation that is in reality rather incipient. However, a survey based on territorial conditions can effectively expose the activities that take place within its confines as well as the spatial and architectural solutions they entail or call for.

The vision of the urban landscape as one solely focused on consumption is limiting and it strengthens the current hierarchies between city and countryside (or urbanites and rural dwellers). A problem, in Portugal, has a very considerable dimension. Such is the theme of the third part of this Atlas. Throughout the years, many studies have shown the real importance of urban agriculture in the Lisbon Region (Castel-Branco et al. 1985; Cabannes and Raposo 2013; Costa et al. 2016; Harper and Afonso 2016; Delgado 2018; Messina and Mourato 2018; Marat-Mendes et al. 2021a). The already existing practices in the regional urban centres contribute not only to the food system but also for landscape maintenance and a more sustainable use of urban space.

Urban agriculture has been studied for its importance in soil quality, social activities, landscapes and culture (Reisch et al. 2013; Cohen and Ilieva 2015), its role in 'Right to the City' and similar movements (Purcell and Tyman 2015; Wekerle and Classens 2015; Reynolds 2008; Kopnina 2018) and its fundamental contribution for 'designing for survival', i.e. for settlements with ecologically sound relations to their surrounding ecosystems (Moorcroft 1972). Its role in the urban patterns of Global South cities has been acknowledged for some time (Egziabher et al. 1994; Magdishima et al. 2013; Lynch et al. 2013; Chipungu et al. 2014), and it is one important instance in which Global North cities may learn from those of the Global South, where productive activities have always been a significant part of cities. The

Continuous Productive Urban Landscape proposal (Viljoen et al. 2005; Viljoen and Bohn 2012, 2014) advances an actual urban model to create cities with effectively productive spaces, while also continuing a well-established scholarly tradition on urban agriculture (Smit and Nasr 1992). This proposition constitutes a fundamental example of agricultural activities being integrated in the planning of cities, as long suggested by scholars (Philips 2013; Hardman and Larkham 2014; Napawan 2016; Orisini et al. 2017; Mees 2017; Proksch 2017).

Studies of urban agriculture in the Lisbon Region find it detached from planning policies (Oliveira and Morgado 2016; Marat-Mendes et al. 2021a), remaining a marginal practice (Cabannes and Raposo 2013; Delgado 2018). When officialized by municipalities, it depends on a limited range of urban agricultural typologies in the landscape of LMA municipalities (Dias 2018; Marat-Mendes et al. 2021). We draw from information of the planning and socio-ecological transition history research to understand how agriculture has emerged in these spaces, but also the role they may have in future socio-ecological transformations towards a more sustainable regional food system.

The fourth part presents a brief summary of the current situation of the Lisbon Region, taking notice of previous works on its food system (Oliveira et al. 2014; Niza 2017) but also the recent data made available to the public with the Agricultural Census of 2019 (INE 2021). The key conclusions of our own research presented in the first three chapters will also be presented, noticing the importance and the challenges of the food system as a research and teaching agenda for architecture and urban design. Some key recommendations are retrieved and applied to a specific zone within the Lisbon Region.

References

Alier JM (1977) Haciendas, plantations and collective farms. Frank Cass, New York

Almeida PV (1964) Da Utilidade Social da Arquitectura. Análise Soc 2(6):237–248

Almeida SJ (2021) Moção de António Costa aponta para regionalização em 2024. Público, May 27th. https://www.publico.pt/2021/05/27/politica/noticia/mocao-antonio-costa-aponta-regionalizacao-2024-1964335

Ballantyne-Brodie E, Telalbasic I (2017) Designing local food systems in everyday life through service design strategies. Des J 20(sup1):S3079–S3095. https://doi.org/10.1080/14606925.2017.1352816

Bartlett J (2017) Radicals—outsiders changing the world. Windmill, London

Beecher J, Bienvenu R (eds) (1972) The Utopian vision of Charles Fourier. Houghton Mifflin, Boston

Brinkley C (2013) Avenues into Food Planning: A Review of Scholarly Food System Research. Int Plan Stud 18(2):243–266. https://doi.org/10.1080/13563475.2013.774150

Cabannes Y, Raposo I (2013) Peri-urban agriculture, social inclusion of migrant population and Right to the City - Practices in Lisbon and London. City 17(29):235–250. https://doi.org/10.1080/13604813.2013.765652

Castel-Branco I, Saraiva M, Neto M (1985) As "hortas urbanas" em Lisboa. Sociedade e Território 3:100–107

Chipungu L, Magidimisha HH, Hardman M, Beesley L (2014) 'The importance of soil quality in the safe practice of urban agriculture in Zimbabwe, Kenya and South Africa', In: Land-Use Change Impacts on Soil Processes in Tropical and Savannah Ecosystems. CABI. 978–1–78064–210–9

CMLisboa - Camara Municipal de Lisboa (2012) Plano Director Municipal De Lisboa. Published in Diário da República 168/2012, Série II, 30–8–2012

Cohen N, Ilieva RT (2015) Transitioning the Food System: A Strategic Practice Management Approach for Cities. Environ Innov Soc Trans 17:199–217. https://doi.org/10.1016/j.eist.2015.01.003

Correia H (2018) Um bailarino na batalha. Relógio d'Água, Lisboa

Costa S, Fox-Kämper R, GoodR SI, Treija S, Atanasovska JR, Bonnavaud H (2016) The position of urban allotment gardens within the urban fabric. In: Bell S, Fox-Kämper R, Keshavarz N, Benson M, Caputo S, Noori S, Voigt A (eds) Urban allotment gardens in Europe. Routledge, London, pp 201–228

Crowley D, Marat-Mendes T, Falanga R, Henfrey T, Penha-Lopes G (2021) Towards a necessary regenerative urban planning. Insights from community-led initiatives for ecocity transformation. Cidades, Comunidades & Territórios [Online], Sp21 | 2021, https://doi.org/10.15847/cct.20505UIDB/03127/2020.83

Darwin C (1871) The descent of man: selection in relation to sex. Penguin, London

Delgado C (2018) Contrasting practices and perceptions of urban agriculture in Portugal. International Journal of Urban Sustainability and Development 10:170–185

Descola P (2001) Nature and society: anthropological perspectives. Routledge, London

Dias AM (2018) The Shape of Food: An analysis of urban agricultural shapes in Lisbon's Greater Area. MSc dissertation, Iscte—Instituto Universitário de Lisboa, Lisbon. http://hdl.handle.net/10071/19300
Douglas M (1966) ([2002]) Purity and danger. Routledge, London
Egziabher AG, Lee-Smith D, Maxwell D, Ali Memon P, Mougeot L, J. Sawio C (1994) Cities feeding people: an examination of urban agriculture in East Africa. International Development Research Centre, Ottawa, CA
Erb K-H, Fetzel T, Haberl H, Kastner T, Kroisleitner C, Lauk C, Niederstscheider M, Plutzar C (2016) Beyond Inputs and Outputs: Opening the Black-box of Land-use Intensity Chapter 4. In: Haberl H, Fischer-Kowalski M, Krausmann F, Winiwarter V (eds) Social Ecology. Springer, Switzerland, pp 93–124
Esteves J, Mestre V (1987[2004]) A partir de uma conversa com o arquitecto Silva Dias a propósito do inquérito à arquitectura regional portuguesa. In: Figueira J, Nunes J, Milheiro AV, Graça Dias M (eds) J-A Jornal Arquitectos, Antologia 1981–2004. Ordem dos Arquitectos, Lisboa, pp 96–99
FAO (Food and agriculture organization), (2019) FAO framework for the Urban Food Agenda. FAO, Rome. https://doi.org/10.4060/ca3151en
Fischer-Kowalski M (1998) Society's metabolism. The intellectual history of material flow analysis, Part I, 1860–1970. Indus Ecol 1:61–72
Fischer-Kowalski M, Weisz H (2016) The Archipelago of Social Ecology and the Island of the Vienna School Chapter 1. In: Haberl H, Fischer-Kowalski M, Krausmann F, Winiwarter V (eds) Social Ecology. Springer, Switzerland, pp 3–28
Frazer JG (1910) Totemism and exogamy. MacMillan, London
Frazer JG (1911) The Golden Bough, vol 3—Taboo and the perils of the soul. MacMillan, London
Gandy M (2004) Rethinking Urban Metabolism: Water, Space and the Modern City. City 8:363–379
Gandy M (2014) The Fabric of Space Water, Modernity, and the Urban Imagination. The MIT Press, Cambridge, Massachusetts, London, England
Geddes P (1915) Cities in evolution. An Introduction to the town planning movement and to the study of civics. Williams & Norgate, London
Ghosh JB (2016) Edible eastside. Urban Des (Food City) 140:31–33
Giddens A (1990) The consequences of modernity. Polity Press, London
Guerreiro FJB (2018) Uma Cozinha Portuguesa, com certeza: a 'Culinária Portuguesa' de António Maria de Oliveira Bello. Trilhas Da História 8(15):221–236
Hardman M, Larkham PJ (2014) Informal Urban Agriculture. The Secret Lives of Guerrilla Gardeners. Springer, London
Harper K, Afonso A (2016) Cultivating civic ecology: a photovoice study with urban gardeners in Lisbon. Portugal. Anthropology in Action 23(1):6–13
INE - Instituto Nacional de Estatística (2021) Recenseamento Agrícola 2019. Análise dos principais resultados. INE, Lisboa. https://www.ine.pt/xurl/pub/437178558. ISBN 978–989–25–0562–6
Ives A (2015) Neoliberalism and the concept of governance: renewing with an older liberal tradition to legitimate the power of capital. Mémoire(s), identité(s), marginalité(s) dans le monde occidental contemporain 14. https://doi.org/10.4000/mimmoc.2263
Jorge L (2018) Estuário. Dom Quixote, Lisboa
Koolhaas R, Obrist HU (2011) Project Japan: metabolism talks. Taschen, Koln
Kopnina H (2018) Plastic Flowers and Mowed Lawns – The Exploration of Everyday Unsustainability. J Urban. https://doi.org/10.1080/17549175.2018.1527780
Krausmann F, Weisz H, Eisenmenger N (2016) Transitions in sociometabolic regimes throughout human history, Chapter 3. In: Haberl H, Fischer-Kowalski M, Krausmann F, Winiwarter V (eds) Social ecology. Springer, Switzerland, pp 63–92
Le Corbusier (1923) Vers une architecture. Éditions Crès, Collection de "L'Esprit Nouveau", Paris
Lévi-Strauss C (1962[2012]) La Pensée Sauvage. Plon, Paris Lévi-Strauss C (1964[2009]). Mythologiques—Le cru et le cuit. Plon, Paris
Lima, IGS (2011) A Corrupção Participada na Administração Local em Portugal (2004–2008). MSc dissertation. Instituto de Ciências Sociais UL, Lisboa
Lipovetsky G (2004) Les temps hypermodernes. Poche, Paris
Lo AY (2016) Small is green? Urban form and sustainable consumption in selected OECD metropolitan areas. Land Use Policy 54:212– 220. https://doi.org/10.1016/j.landusepol.2016.02.014
Lynch K, Maconachie R, Binns T, Tengbe P, Bangura KS (2013) Meeting the urban challenge? Urban agriculture and food security in post-conflict Freetown, Sierra Leone. Appl Geogr 36:31–39
Magidimisha H, Chipungu L, Awuorh-Hayangah R (2013) Challenges and strategies among the poor: focus on urban agriculture in KwaMashu, Durban, South Africa. J Agric Food Syst Commun Dev 3(2):109–126
Magidimisha HH, Hardman M, Beesley L (2014) 'The importance of soil quality in the safe practice of urban agriculture in Zimbabwe, Kenya and South Africa', In: Land-Use Change Impacts on Soil Processes in Tropical and Savannah Ecosystems. CABI. 978–1–78064–210–9
Marat-Mendes T (2002) The Sustainable Urban Form. A Comparative study in Lisbon, Edinburgh and Barcelona. PhD Thesis, University of Nottingham, Nottingham UK
Marat-Mendes T (2004) Dimensioning the Sustainable Urban Fabric. In: Babalis D (ed) Ecological Design for an effective Urban Regeneration. Firenze University Press, Firenze, pp 27–40

Marat-Mendes T (2016) Physical, social and cultural dimensions of Urban Morphology: redressing the balance? Urban Morphol 20 (2):167–168
Marat-Mendes T (2020) Designing for sustainability: retrieving a systemic role for urban form. Urban Morphology 24(2):235–238
Marat-Mendes T (coord), Mourão J, d'Almeida PB, Niza S (2015) Water and Agriculture Atlas: Lisbon Region in 1900–1940. Atlas da Água e da Agricultura. Região de Lisboa em 1900–1940. Iscte - Instituto Universitário de Lisboa, DINÂMIA'CET-IUL, Lisboa
Marat-Mendes T, Borges JC (2019a) The urban planning transition history of the Lisbon Metropolitan Area. In: Proceedings of ASEOP 2019 Congress, Venice, 9–13 July 2019
Marat-Mendes T, Borges JC (2019b) The role of food in re-imagining the city—from the neighbourhood to the region. In: Juvara M, Ledwon S (eds) Proceedings of ISOCARP 2019 Congress, Jakarta, 9–13 September 2019
Marat-Mendes T, Borges J, Dias A, Lopes R (2021) Planning for a Sustainable Food System. The potential role of urban agriculture in Lisbon Metropolitan Area. Journal of Urbanism: International Research on Placemaking and Urban Sustainability. ID: 1880960 https://doi.org/10.1080/17549175.2021.1880960
Marat-Mendes T, d'Almeida PB, Borges JC (2021a) Concepts and definitions for a sustainable planning transition: lessons from moments of change. European Planning Studies. https://doi.org/10.1080/09654313.2021.1894095
Marat-Mendes T, d'Almeida PB, Borges JC (2021b) Concepts and definitions for a sustainable planning transition: lessons from moments of change. European Planning Studies. https://doi.org/10.1080/09654313.2021.1894095
Marcelino V (2019) Metade dos casos de corrupção tem origem em autarquias. Diário de Notícias, June 8th
Mees C (2017) Participatory Design and Self-building in Shared Urban Open Spaces: Community Gardens and Casitas in New York City. Springer, Switzerland
Messina T, Mourato JM (2018) Comunity Urban Food Gardens policy design: a study case with Almada (Portugal) and Paris (France). In Proceedings of the international scientific event, 26–27 April 2018, Lisbon, Portugal - Connections and missing links within urban agriculture, food and food systems, pp 57–61
Modesto ML (1982 [1990]) Cozinha tradicional Portuguesa. Verbo, Lisbon
Modesto ML (2015) Acervo Maria de Lourdes Modesto. https://acpp.pt/maria-de-lourdes-modesto
Moorcroft C (ed.) (1972) Designing for survival. Architectural Design, July: 413–433
Moudon AV, Drewnowski A, Duncan GE, Hurvitz PM, Saelens BE, Scharnhorst E (2013) Characterizing the food environment: pitfalls and future directions. Public Health Nutr 16(7):1238–1243. https://doi.org/10.1017/S1368980013000773
Napawan NC (2016) Complexity in urban agriculture: the role of landscape typologies in promoting urban agriculture's growth. Journal of Urbanism: International Research on Placemaking and Urban Sustainabilit 9(1):19–38. https://doi.org/10.1080/17549175.2014.950317
Niza S (2017) O sistema alimentar no contexto do metabolismo urbano da Área Metropolitana de Lisboa. In: Oliveira R, Amâncio S, Fadigas L (eds) Alfaces na avenida. Estratégias para (bem) alimentar a cidade. Universidade de Lisboa, Colégio Food, Farming and Forest, Lisboa, pp 42–47
Odum E (1975) Ecology. Holt Rinehart and Winston, New York
Oliveira R, Morgado MJ (2016) Planning the Urban Food System of the Metropolitan Área of Lisbon. A conceptual framework. In: Roggema R (ed) Agriculture in an Urbanizing Society: Proceedings of the Sixth AESOP Conference on Sustainable Food Planning, "Finding Spaces for Productive Cities" November 5–7, 2014. Leeuwarden, the Netherlands. Cambridge Scholars Publishing, Newcastle
Oliveira R, Morgado MJ, Martinho R, Maurício I (2014) O sistema alimentar urbano da Área Metropolitana de Lisboa – análise e diagnóstico. FCSH, Lisboa
Olufemi O (2016) Urban farming and family farms. Urban Des (Food City) 140(18):30
Orsini F, Dubbeling M, de Zeeuw H, Gianquinto G (eds) (2017) Rooftop Urban Agriculture. Springer, Switzerland
Palla V, Martins C (1959) Lisboa cidade triste e alegre, Circulo do Livro, Lisboa
Parham S (2015) Food and urbanism – the convivial city and a sustainable future. Bloomsbury, London
Peniche A (2020) A aventura das mulheres. Fernando Rosas, Francisco Louçã, João Teixeira Lopes, Andrea Peniche, Luís Trindade and Miguel Cardina, O século XX Portugues. Tinta da China, Lisbon, pp 245–292
Pereira M, Gago M (1974) O uso do Espaço na Habitação. LNEC, Lisboa
Pereira M, Portas N (1967) Inquérito à Habitação Urbana: Objetivos Gerais do Inquérito. LNEC, Lisboa
Pereira NT (1994 [1996]) O verdadeiro inconveniente da regionalização. Tempos, lugares, pessoas. Lisbon: Público, p 129
Philips A (2013) Designing Urban Agriculture: A Complete Guide to the Planning, Design, Construction, Maintenance and Management of Edible Landscapes.Wyley, London
Portas N, Gomes N (1963) Inquérito-Piloto sobre as necessidades familiares em matéria de habitação. Relatório: Programa, amostra e resultado da fase de inquirição. Lisboa, LNEC
Pothukuchi K, Kaufman JL (2000) The Food System. J Am Plann Assoc 66(2):113–124. https://doi.org/10.1080/01944360008976093
Proksch G (2017) Creating Urban Agricultural Systems: An Integrated Approach to Design. Routledge, London

Purcell M, Tyman S (2015) Cultivating Food as a Right to the City. Local Environ 20(10):1132–1147. https://doi.org/10.1080/13549839.2014.903236

Ravetz J (2000) Urban form and the sustainability of urban systems: theory and practice in a Northern Conurbation. In: Williams K, Burton E, Jenks M (eds) Achieving sustainable urban form. E & FN SPON, London, pp 215–228

Reisch L, Eberle U, Lorek S (2013) Sustainable food consumption: an overview of contemporary issues and policies. Sustainability: Science, Practice and Policy 9(2):7–25

Reynolds R (2008) On Guerrilla Gardening. A Handbook for Gardening Without Boundaries. Bloomsbury, London

Ribeiro O (1939) Inquérito ao habitat rural. IAC, CoimbraRibeiro O (1987) A formação de Portugal. Ministério da Educação, Lisboa

Rodriguez O (2016) Towards an indigenous, sunlit, rooftop food production. Urban Des (Food City) 140: 24–27

Salvador MS (2019) Shaping the city through food: the historic foodscape of Lisbon as case study. Urban Design International 24:80–93

Smit J, Nasr J (1992) Urban Agriculture for Sustainable Cities: Using Waste and Idle Land and Water Bodies as Resources. Environ Urban 4(2):141–152. https://doi.org/10.1177/095624789200400214

Smithson A (ed) (1974) Team 10 primer. MIT Press, Cambridge, Massachusetts

Spaargaren G, Oosterveer P, Loeber A (eds) (2012) Food practices in transition. Routledge, London

Steel C (2008[2013]) Hungry city – how food shapes our lives. Vintage, London

Steel C (2020) Sitopia. Chatto & Windus, London

Timmer CP (2010) Reflections on food crises past. Food Policy 35(1):1–11

Turnbull C (1972) The mountain people. Touchstone, New York

UN (United Nations) (1987) Our Common Future. http://www.un-documents.net/our-common-future.pdf

Underwood DK (1991) Alfred Agache, French sociology, and modern urbanism in France and Brazil. J Soc Arch Hist 50(2):130–166

Vasconcelos A, Rosas P (2018) Post-pop: beyond the commonplace. Fundação Calouste Gulbenkian, Lisboa

Viljoen A, Bohn K (2012) Planning and designing food systems, moving to the physical. In: Viljoen A, Wiskerke JCS (ed) Sustainable food planning: evolving theory and practice. Wageningen Academics, Gelderland

Viljoen A, Bohn K (2014) Second Nature Urban Agriculture: Designing Productive Cities. Routledge, London

Viljoen A, Bohn K, Howe J (2005) CPUL – Continuous Productive Urban Landscapes. Routledge, London

Waldheim C (2010) Notes toward a history of Agrarian urbanism. Places J. https://placesjournal.org/article/history-of-agrarian-urbanism/?cn-reloaded=1

Wallace DF (2004[2005]) Consider the lobster. In: Consider the lobster and other essays. Abacus, London, pp 235–254

Wekerle GR, Classens M (2015) Food Production in the City: (Re)negotiating Land. Food and Property. Local Environment 20(10):1175–1193

Wiskerke JCS (ed) Sustainable food planning: evolving theory and practice. Wageningen Academics, Gelderland

Wolman A (1965) The Metabolism of Cities. Sci Am 213(3):179–190. https://doi.org/10.1038/scientificamerican0965-178

Contents

About the Authors

Teresa Marat-Mendes is an Architect and Associated Professor at the Department of Architecture and Urbanism of Iscte- Instituto Universitário de Lisboa. She is a senior researcher at DINÂMIA'CET Research Center, where she coordinated the research teams for Project 'MEMO—Evolution of the Lisbon Metropolitan Area Metabolism. Lessons towards a Sustainable Urban Future' and Project 'SPLACH-Spatial Planning for Change'. She was awarded Women in Science in 2019, a recognition attributed by Ciência Viva for those women scientists in Portugal, whose outstanding work has been central to the national progress in science and in national technology recorded in the last decades. She is a member of the International Seminar on Urban Form (ISUF) Council and the former president and co-founder of the Portuguese Language Network on Urban Morphology (PNUM).

Sara Silva Lopes is an Architect and Research Assistant at DINÂMIA'CET, the Research Centre for Socioeconomic and Territorial Studies of Iscte—Instituto Universitário de Lisbon. She was a researcher for Project SPLACH—Spatial Planning for Change, where she was intensively involved in the survey of Lisbon Metropolitan Urban Agriculture and of its Food System. At present, she is developing her doctoral studies research 'Housing in Democracy: Typo-morphological and socio-spatial perspective of housing policies in the Lisbon metropolitan Area'. Her research interests focus on the studies of urban morphology, spatial planning instruments, public policies and the right to the city and to housing.

João Cunha Borges is an Architect and a Research Assistant at DINÂMIA'CET, Iscte—Instituto Universitário de Lisboa. He was a researcher for Project SPLACH—Spatial Planning for Change and was intensely involved in the survey of Lisbon Metropolitan Urban Agriculture and of its Food System. He wrote a dissertation on Aldo Van Eyck and Alison and Peter Smithson, titled 'The dissolution of the modern complex'. At present, he is developing his doctoral studies with the research project 'Social urban forms at the millennial city scene—Workers' housing and landscape (trans)formation from Britain and Portugal'. His research seeks to contribute with a multidisciplinary approach to architectural theory, including Urban Planning History and Sustainability, Aesthetics, Anthropology of Space and Popular Culture.

Patrícia Bento d'Almeida is an Architect and a Researcher at DINÂMIA'CET–IUL, Iscte-Intituto Universitário de Lisboa. She holds a Master and a Ph.D. degree in History of Art from Universidade Nova de Lisboa. She was a researcher for Project SPLACH- Spatial Planning for Change, contributing to the analysis of twentieth century Portuguese Architecture and Urban Planning History and the photographic survey of the food system of Lisbon Metropolitan Area. At present, she is developing a post-doctoral research 'LNEC, the Portuguese Laboratory of Civil Engineering, and the history of research in Arquitectura'. She was the curator of the exhibition Victor Palla and Bento d'Almeida, present at Centro Cultural de Belém in 2017.

Abbreviations

APU	Aliança Povo Unido, United People Alliance
CCDR	Comissão de Coordenação e Desenvolvimento Regional, Regional Development General Commission
CCDRLV	Comissão de Coordenação e Desenvolvimento Regional de Lisboa e Vale Tejo, Regional Development General Commission for Lisbon and Tagus Valley
CIAM	Congrés Internacionaux d'Architecture Moderne
CM Alcochete	Câmara Municipal de Alcochete
CM Almada	Câmara Municipal de Almada
CM Amadora	Câmara Municipal da Amadora
CM Lisboa	Câmara Municipal de Lisboa
CM Loures	Câmara Municipal de Loures
CM Mafra	Câmara Municipal de Mafra
CM Moita	Câmara Municipal da Moita
CM Montijo	Câmara Municipal do Montijo
CM Odivelas	Câmara Municipal de Odivelas
CM Oeiras	Câmara Municipal de Oeiras
CM Seixal	Câmara Municipal do Seixal
CM Sesimbra	Câmara Municipal de Sesimbra
CM Setúbal	Câmara Municipal de Setúbal
CM Sintra	Câmara Municipal de Sintra
CMB	Câmara Municipal do Barreiro
CMC	Câmara Municipal de Cascais
CMP	Câmara Municipal de Palmela
CMVFX	Câma Municipal de Vila Franca de Xira
CPUL	Continuous Productive Urban Landscape
CSOP	Conselho Superior de Obras Públicas, Superior Council of Public Works
DGADR	Direcção-Geral da Agricultura e Desenvolvimento Rural, Portuguese Directory of Agriculture and Rural Development
DGSU	Direcção-Geral dos Serviços de Urbanização, General Directory of Urbanziation services
DGT	Direcção-Geral do Território, General Directory of Territory
FAO	Food and Agriculture Organization of the United Nations
FS	Food System
IAPXX	Inquérito da Arquitectura Portuguesa do Sec. XX, 20th century survey to Portuguese Architecture
IARP	Inquérito à Arquitectura Regional Portuguesa, Survey to the Portuguese Regional Architecture
ISUF	International Seminar on Urban Form
IUA	International Union of Architects
IUUP	Institut d'Urbanisme de l'Université de Paris
JCI	Junta de Colonização Interna, Inland Colonization Commitee

LMA	Lisbon Metropolitan Area
MARL	Mercado Abastecedor da Região de Lisboa, Supply Market of the Lisbon Region
MEMO	'MEMO—Evolution of the Lisbon Metropolitan Area Metabolism. Lessons towards a Sustainable Urban Future' Research Project
OCDE	Organization for the Economic Cooperation and Development
PCP	Partido Comunista Português, Portuguese Comunist party
PDM	Plano Director Municipal, Municipal Master Plan
PDRL	Plano Director da Região de Lisboa, Lisbon Region Masterplan
PGM	Plano Geral de Melhoramentos
PGU	Plano Geral de Urbanização, General Urbanization Plan
PNPOT	Programa Nacional da Política de Ordenamento do Território, National Spatial Plan
PROT	Plano Regional de Ordenamento do Território, Regional Spatial Planning
PROT-AML	Plano Regional de Ordenamento do Território para a Área Metropolitana de Lisboa, Regional Spatial Planning for Lisbon Metropolitan Area
PS	Partido Socialista, Socialist Party
PUCS	Plano de Urbanização da Costa do Sol, Costa do Sol Urbaization Plan
RAN	Reserva Agrícola Nacional, National Agriculture Reserve
REN	Reserva Ecológica Nacional, Nacional Ecological Reserve
RO	Ribatejo e Oeste
SAAL	Serviço de Apoio Ambulatório Local
SFS	Sustainable Food System
SNP	Sociedade Nacional de Propaganda, National Society for Propaganda
SPLACH	'SPLACH—Spatial Planning for Change' Research Project
UM	Urban Morphology
UN	United Nations
WWI	World War I
WWII	World War II

List of Figures

Part I
Spatial Planning and the Territory of the Lisbon Region

It is not uncommon today, at least among scholars, to agree that spatial planning can have a great impact upon the food system, and that its sustainable transition will depend much upon a proper management of resources—notably land. Yet such recognition is fairly recent, and outside academic circles, much remains to be done to achieve an effective inclusion of the food system in the concerns of spatial planning at several scales.

The most recent National Spatial Program (*Programa Nacional da Política de Ordenamento do Território*—PNPOT, 2019) in Portugal has represented a key development in this problem, acknowledging the importance of food security and the environmental impact of current eating habits (Marat-Mendes et al, 2020). But in the Portuguese legal framework, the National Spatial Program presides over sectoral and infrastructural matters, while minute territorial change is presided over by the Municipal Masterplans (*Plano Director Municipal*—PDM), most of which currently do not include any specific initiative regarding food (Marat-Mendes et al, 2020, 2021).

Whether PDMs approved in the years to come will show such sensibility towards the food system or whether the National Spatial Program will become a set of well-intentioned guidelines without materialization remains to be seen. But in the Portuguese context, there is a strong need to emphasize the importance of the challenges advanced at the national scale. First, after more than 100 years of history, Portuguese planning policies remain weak and disconnected from other policies (Campos & Ferrão, 2015). Second, the inception of the food system into public sector policies presupposes a break with long-standing notions regarding both the food system and spatial planning. There are two key strong assumptions that represent a barrier: first, that the food system is a concern of the private sector—and that it is naturally organized and managed to aim for profit—and second, that spatial planning is meant to regulate urbanization processes.

In this first part, we overview the history of Portuguese planning in order to trace how such barriers to the inclusion of the food system in planning policies and instruments have come to solidify. Furthermore, such a historical overview will allow an observation of the key ideas and models that were proposed for the territory of the Lisbon Region during the century of its great transformation.

1 The Emergence of Spatial Planning Policy in Twentieth Century Portugal

1.1 Introduction

A territory successively occupied by Greek, Roman, Celtic, Muslim and Catholic peoples, and with a varying geography, Portugal is moderately diversified in terms of urban form, although in general, as architect Manuel C. Teixeira (b.1951) points out, the typical Portuguese settlement structure has been a mix between orthogonal and organic forms. For centuries in Portugal, most of these settlements developed autochthonously (Teixeira 2016). There are examples of planned villages dating back to the Middle Ages (Albergaria 2007), but while in other countries, formally planned settlements were a common practice, in Portugal such practice was rare. Yet, two Portuguese examples, both from the eighteenth century, stand out, one being the reconstruction of downtown Lisbon after the 1755 earthquake that destroyed the totality of the old settlement (Marat-Mendes 2002), and the other being Vila Real de Santo António, a New Town constructed in the province in the Algarve region.

Once the market centre of the city, a place of trade and housing built by an accumulative and informal process, Lisbon's downtown was shattered by the earthquake, its replicas and the resulting fires. A full reconstruction was necessary to retrieve normalcy in Lisbon, and it was up for Sebastião José de Carvalho e Melo (1699–1782), the Count of Oeiras and Marquis de Pombal, to oversee this reconstruction. Authoritarian but pragmatic, Pombal understood that what was at stake was not only a reconstruction for the present but an opportunity for kickstarting the future of the city. From the five different urban project proposals, the Marquis (by then holding the de facto State power, much more than king Joseph the 1st) selected Manuel da Maya's (1677–1768), a plan that presupposed a new city, a rational and organized downtown catering to the middle classes and replacing the pre-earthquake urban form (Marat-Mendes 2002).

The same craving for rational urban forms and hierarchized infrastructure can be seen in the plan for Vila Real de Santo António, constructed in 1774–1776. When compared with its contemporary new town of Bath, with its highly novel housing types, Vila Real de Santo António's plan is derived from much more rationalistic principles, like those of the new Lisbon downtown, although this did amount, in the Portuguese context, to an important innovation in urban form and in city-building practice.

The reconstruction process of Lisbon's downtown roughly inaugurates a modern approach to settlement construction, in which the plan emerges as an instrument for State control or regularization, but also as an instrument for design. Building-by-building accumulative physical growth is replaced with a more or less integral approach. Moreover, in the reconstruction of Lisbon's downtown, there is for the first time a clearly stated desire to plan the capital city of the kingdom.

However, from the eighteenth century onwards, planning and urban policies were scant. Planning was used as a response to an emergency situation in Lisbon, and seldom elsewhere. The fact that the Portuguese were, until the 1960s, a predominantly rural society, whose urban settlements were few and small, may have contributed to this. Only in the twentieth century did this situation change, concentrating the key paradigmatic changes in planning policy in a rather limited timeframe, as we shall see next.

1.2 The Roots of Spatial Planning in Portugal

Other than sporadic plans for important urban cores, there was a generalized absence of territorial planning policies in Portugal. A key reason for this was that, unlike other nations which started to modernize right after the Industrial Revolution, the country maintained its rural organization. However, the countryside was not always effectively managed. By the late 1860s, an agricultural survey conducted at the national level presented interesting conclusions: from the total agricultural surface of the continental territory, 120.5% was used for cereals, 20.7% for mixed crops, 00.4% for

T. Marat-Mendes et al., *Atlas of the Food System*,
https://doi.org/10.1007/978-3-030-94833-7_1

grazing, 20.2% for vineyards, 70.2% for fruit trees, 20.9% for woods and 70.2% were fallow—however, while 4 642 000 ha were cultivated, 4 138 000 ha were unused and in the regions of Trás-os-Montes, Alentejo and Algarve, barren land exceeded productive land (Branco 2015).

Indeed, in a survey on foreigners who travelled to Portugal from the mid-eighteenth century to the Revolution era in the 1970s, Sociologist Maria Filomena Mónica (2020) notices that many visitors were as impressed with the beauty of the rural landscapes as with the lack of cultivation and proper maintenance that often characterized them, an accusation directed at the aristocracy which did not promote a good use of the territories they associated with, but also at the small land owners who, with the evolution of the political situation, could not properly work their land.

In 1865, the first Portuguese urban planning instrument is legally designated, the Plan of General Improvements (*Plano Geral de Melhoramentos*—PGM), although this political possibility would only come to fruition in the early twentieth century, applied precisely to the Lisbon city (Lôbo 1995).

The plan for the urban extension known as Avenidas Novas, designed by Engineer Frederico Ressano Garcia (1847–1911) from 1863 onwards, was at the basis of the Lisbon PGM, finished only in 1904. Under the PGM, Lisbon urban growth was directed towards the inner city instead of the riverfront (Lôbo 1995). The northwards continuation of the Public Walkway (*Passeio Público*) built during Pombal's regime, which was in reality walled and only accessible for the upper classes—launched a sequence of important avenues and roads as well as public spaces and plots for private-led urbanizing operations (Morais and Roseta 2005). In comparison with the plan for the downtown reconstruction, the Avenidas Novas were mostly structured on circulation and hygiene concerns, although, in accordance with legislation passed for that specific purpose, it also implied an integral vision of the city's territory (Lôbo 1995). At this time, the first efforts towards an urbanism policy in Portugal were developed under the legal figure of the PGM, although they tended to be crushed by political fluctuations.

While the legislation for urban improvements seems to apply only to the main cities, there was a strong emphasis on the problem of agriculture throughout the nineteenth century. From 1850 to 1900, agricultural produce grew at the seemingly impressive rate of 10.5% per year, although this was not due to a real development of agriculture, but rather to the fact that, prior to this era, production is generally very low (Branco 2015). The Portuguese privileged the production of cereals for bread, such as wheat and corn, and of wine, however, at the end of the nineteenth century, Portugal had 30.7 ha of land for each worker, while in France, the value was 50.4 and in England, 90.8 (Branco 2015).

In 1910, the Portuguese monarchy fell and was replaced by a Republic. However, the country that this new regime had to run was challenging to say the least: there was an overwhelming disproportion between rural inhabitants (85% of the population) and urbanites (the remaining 15%), and illiterates represented over 60% of the total population (Branco 2015; Mónica 2020). However, the First Republic (1910–1926), with its fundamentally bourgeois outlook (Varela and Demier 2015) lacked support, as most of the country was divided between supporters of the monarchy and a working-class that did not ally with the bourgeoisie, while Republicans themselves divided into several fractions (Varela and Demier 2015; Abreu 2017). Capitalist organization had been a key policy driver in the nineteenth century, but agriculture had always been one of the sectors offering the greatest resistance to Agrarian Reforms (Branco 2015). A considerable part of the arable territory remained without proper use, with a rural middle class incapable of investing in agriculture, and with large land owners continuously absorbing smaller properties to increase profits while hampering the social mobility of smaller farmers and intensifying the poverty of peasant workers, increasingly prone to proletarianization (Branco 2015; Barreto 2017).

The start of World War One (WW1) did not help the Republic, which had formed six governments since 1910, to stabilize. Discord over whether Portugal should be belligerent further confused an internal scenario of constant political turmoil (Abreu 2017). A direct consequence of this instability was the increase in difficulty to access food, especially by the lower classes, hardly a surprise considering that Portugal had always depended upon foreign trade to ensure its food supply (Abreu 2017; Mónica 2021). Between 1916 and 1917, the urban proletariat was so brutalized by lack of food that several riots, strikes and upheavals rose in the regions of Porto and Lisbon—notably including the looting of warehouses, bakeries and grocery stores (Abreu 2017; Mónica 2021). This situation kept spiralling, with the population rioting against tradesfolk instead of the government, seen as a distant and inaccessible elite. And indeed, the government was so inconsequential in handling this sustenance crisis that it is not an exaggeration to state that "for a few weeks, the State disappeared" (Mónica 2021: 150). On July 1917, Lisbon and its surrounding municipalities were placed under siege (Abreu 2017) although later that year more lootings in food shops were verified.

The key agricultural concern during the First Republic continued to be the immense territory, which remained uncultivated, with propositions drawn up for expropriations. But these attempts ultimately receded to the interest of large land owners (Branco 2015). Despite some mild attempts at mechanization incentives and protectionist laws over cereal production, during the years of WW1, productivity remained low in Portugal—in 1926, the country produced an average of 700 kg of produce per ha, while Denmark produced 3020 kg, Belgium 2450 kg and Holland 2390 kg—even though by then, the national agricultural workforce represented over 60% of the population (Branco 2015).

In 1926, the opposition between the rising bourgeoisie (who supports the Republic) and the decadent aristocracy, on the one hand, and the working classes on the other, was on the verge of civil war, prompting a coup on May 28. A military dictatorship was imposed, and a radical change in regime starts. In this process, the promise of capitalist modernization was paid with the surrendering of political power and civil liberties. António de Oliveira Salazar (1889–1970), the Minister of Finances and a lawyer trained in the Coimbra University, stood out as the preeminent political figure, a "reactionary Catholic who rejected the liberal traditions of the nineteenth and twentieth centuries [and who] hated the Bolsheviks, the Freemasons, the democrats and the liberals" (Mónica 2021: 157). In 1933, he became the head of the New State (*Estado Novo*), a Bonapartist dictatorship, whose key tasks are the regulation of the workforce, the management of ownership concentration in economic conglomerates, the restriction of market competition through protectionist laws, the taming of workers' movements and the optimization of colonial exploitation through mandatory crops and the use of forced labour on a massive scale (Varela and Demier 2015).

From this long-lasting political rollercoaster, the advances made in terms of planning policies were nearly inexistent. The years of the First Republic hardly left any mark in the urbanism of the capital, except on the promotion of a few social housing estates in Lisbon, most notably Arco do Cego and Boa Hora à Ajuda (Agarez 2020).

In general, planning policies remained within the confines of the nineteenth century schism between urbanism policies, on the one hand, and agroforestry planning on the other (Campos and Ferrão 2015). Thus, the scant urban cores existing in Portugal at the time were alienated from the majority of the territory, still rural and managed by widely different rules (Campos and Ferrão 2015). Urbanism policies focused specifically on achieving basic levels of urban hygiene and infrastructuring (Lôbo 1995; Campos and Ferrão 2015).

At the start of the twentieth century, continental Portugal was a small country with a relatively low population—in 1911, there are 5 960 056 inhabitants, with an average family of 40.2 individuals—mostly living on the countryside (DGE, 1913). The backward cultural climate of the country was intimately related to the nearly inexistent industrialization, the lack of access to education by the majority of the population, and the constant upheavals in the political scene, whether motivated by external or internal factors (Mónica 2021). Territorial planning was not considered a priority, as the challenges that usually create greater pressures for spatial planning were generally not felt in the country.

At the time of the chaotic transition from the monarchy to the Republic, a new planning paradigm emerged in the United Kingdom, which would later become the pivotal influence in the modern development of Portuguese cities: Howard's (1902) Garden City. Indeed, the genesis of Howard's vision is not contained within a city, but rather within a region, as the Garden City is really a network of urban settlements functionally connected with agricultural buffers. This is one of the most original yet unacknowledged feature of Howard's paradigmatic vision, making him a pioneer in planning a sustainable urban food system—and one scholarly area where it has been properly acknowledged is precisely that of the food system (Pothukuchi and Kaufman 2000; Steel 2008; Marat-Mendes et al. 2020b).

In Portugal, plans started being designed to program urban expansions, including the partial plan for Porto (1915) designed by British urban planner Barry Parker, since until 1937 there were no Portuguese urban planners. Parker proposes a new City Avenue inspired by Camillo Sitte and by his previous experience as an architect for the Letchworth Garden City and Hampstead projects. He collaborated with celebrated architect Raymond Unwin, who envisioned the Garden Suburb, a variation of the Garden City which applied more smoothly to the design of neighbourhoods, or small urban extensions. His influence in Porto is limited, however, because his plan was not implemented. A purely suggestive and theoretical plan was brought forward by engineer Ezequiel de Campos (1874–1965) in 1935, inspired by the works of Patrick Geddes and conceiving development not a municipal but at a regional scale: Campos establishes a link between the Porto urban centre with its surrounding peri-urban and rural municipalities (Oliveira 2013). In this same line of regional planning, Campos also designed a PGM for Póvoa de Varzim (1920), a coastal town between the cities of Porto and Braga, whose restructuring Campos studied (Lôbo 1995; Marat-Mendes and Oliveira 2013; Oliveira 2013). For the case of Lisbon, urban planning extensions for the city were also prepared. These were commissioned to recognized international architects–urbanists from the Institut d'Urbanisme de l'Université de Paris (IUUP), also followers of Howard's Garden City vision and Geddes pedagogical and methodological approach: survey, diagnosis, and plan, based on direct observation and data gathering (Bardet 1947).

In general, the planning situation of the late nineteenth century and the early decades of the twentieth testify to a relatively disorganized and incoherent planning policy. Urbanism is understood as infrastructuring, precisely the realm into which the State intervened. The construction of buildings and creation of urban functions was carried on by private businesses, having little coordination from the public sector. There is also an absence of intellectual guidance from planning policies: the scale and influences in each plan (including the merely theoretical ones, as Campos') were entirely dependent upon the sensibility and culture of the authors.

Moreover, one has to consider that at the transition to the twentieth century, even the Lisbon region—which included the largest Portuguese city—was fundamentally rural and had not undergone a significant process of modernization or industrialization (Marat-Mendes et al. 2015). Considering that rural and urban spaces were organized by distinct and unrelated policies (Campos and Ferrão 2015), this meant that the scant urbanism laws applied only episodically to a small portion of the territory (Fig. 1.1).

1.3 Agrarian Colonies and the Problem of Rural Territories

One key characteristic of the New State is its exaltation of rurality as the essence of the Portuguese people. Indeed, the country had only two large cities—which were relatively small and underdeveloped in comparison with other European cities—and the large majority of the population lived on the countryside with jobs on the primary sector. Even in the Lisbon Region, rural land uses were predominant and the transition from a solar-based agrarian socio-metabolism to an energy-based industrialized one was only shyly advancing at the dawn of the twentieth century (Niza et al. 2016).

In many senses, Salazar wished to retain this situation, and the New State promoted the vision of the Portuguese as a simple people, living on and from the land, structuring their lives around the unquestionable pillars of God, patriotism, and family. As late as the 1950s, Salazar often highlighted the virtues of traditionalistic peasant life, presenting the Portuguese (to themselves) as immune to the foolish mirages of indefinite (i.e. industrial) wealth and instead seeking a modest but sufficing, healthy and land-bound life (Rosas 2001). Thus, at least for propagandistic purposes, the land and agrarian life were taken to be conditions for social harmony, implicitly fostering a mistrust of technique, of urbanization and of proletarianization—an idea that lied at the basis of another propagandistic key point of the New State, whereby 'honourable poverty' was the price to pay for the righteous absence of unrestrained ambition (Rosas 2001).

Under the surface of this ruralist vision, there are important contradictions. The economic policies of the New State relied heavily upon a corporative system, whose structuring dates to the 1930s, abolishing competition and regulating prices through sectoral corporations ultimately controlled by the State itself, applying also to agriculture (Branco 2015). After the Law of Industrial Development and Reorganization (*Lei de Fomento e Reorganização Industrial*[1]), the New State launched several economic Development Plans[2] (*Planos de Fomento*), which included the support for corporations, aiming to increase productivity and national sustenance. The fact that the basis of support for the regime was largely comprised by the industrial and land-owning bourgeoisie, urbanization (associated with modernization) was, to a certain degree, inevitable to allow the development of capitalist business.

Another important contradiction in the 'myth of rurality' (Rosas 2001: 1035) is that, while the regime and some of its supporters' dwelt on the myth of an honourable, simple and rural country, the real countryside faced severe problems, stemming from underproductive and underexplored land, backward agricultural techniques, loss of rural labour and, in many points of the country, massive migrations towards urban centres in search for industrial jobs (Folgado 2009; Barreto 2017). Indeed, part of the agenda of the New State was to create and regulate a productive workforce (Varela and Demier 2015). Thus, while the primary sector represented a majority of labour in the country, and although most of the population was still rural, the modernization policies of the New State were bound to set forth a crisis on the countryside and an extensive reorganization of its territory.

Within its rural orientation, and aiming for independence from outside trade, the New State promoted several initiatives towards food self-sufficiency. Such was the case of the Campaign of Wheat, which started already in 1929, during the military dictatorship, and which took inspiration from a similar program in Italy, which included a State-determined limit on the price of wheat and the conversion of many agroforestry areas into cornfields. The promotion of Common Land (*baldios*) management, whereby the State appropriated common agricultural land, often promoting its conversion into forestry[3,4] is another example.

Across several policies, it is clear that the New State identified a problem with the productivity of the territory,

[1] Law 2005, published in Diário do Govêrno 54/1945, Série I, 1945-03-14.

[2] The first plan (1953–1958) was ensured through Law 02058 from December 29, 1952; the second plan (1959–1964) through Law 2094 from 25-11-1958; the midterm plan (1965–1967) through Law 2094 November 25, 1958. The third plan (1968–1973) by Law 2133 from December 20, 1967.

[3] This was ensured through Law 1971, published in Diário do Govêrno 136/1938, Série I, 15-6-1938. The 'baldios' were only returned to the local populations after the Revolution, in 1976.

[4] These aims are established in Decree 27.207, published in Diário do Govêrno 269/1936, Série I, November 16, 1936.

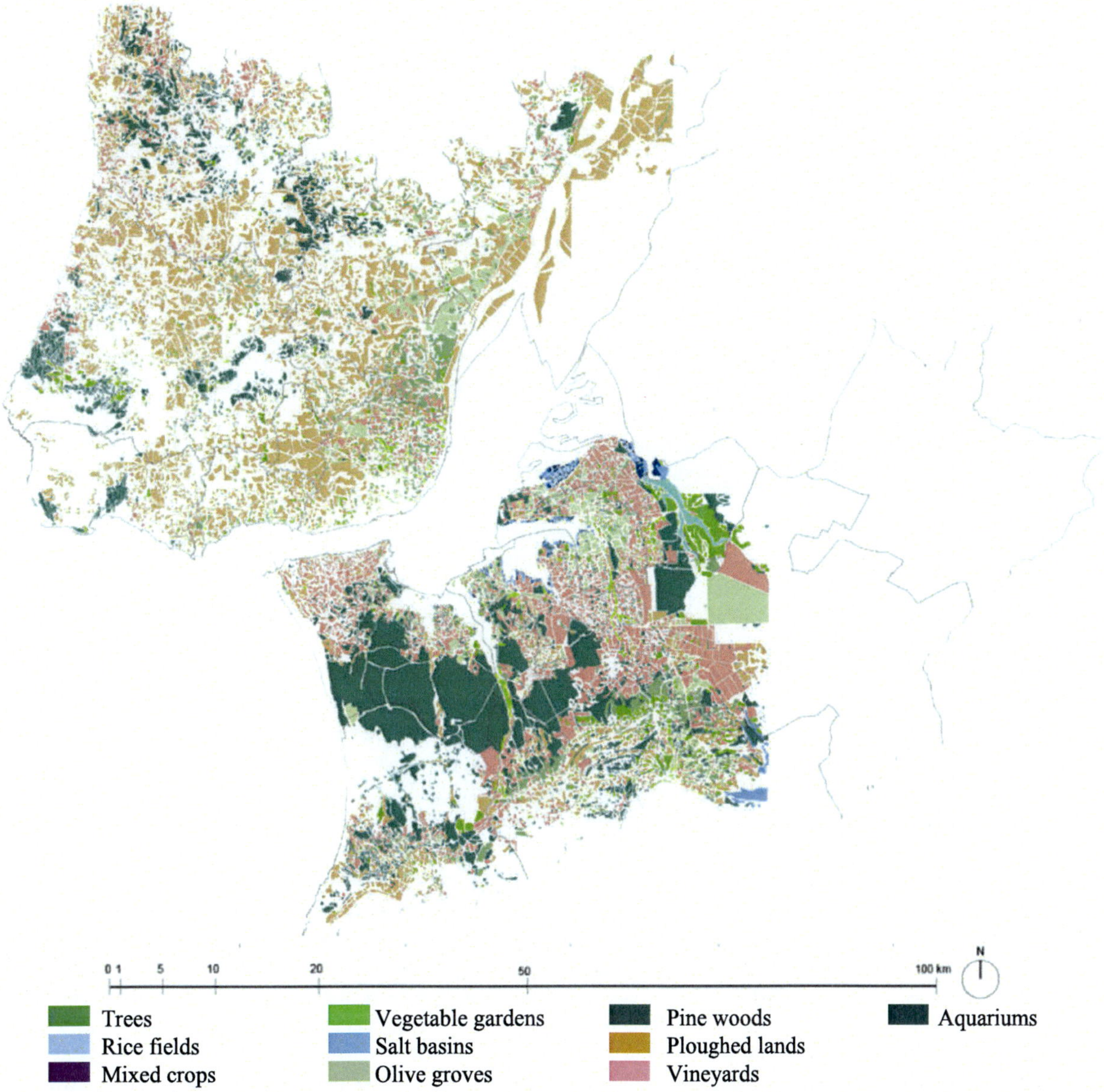

Fig. 1.1 Crops of Lisbon Region in 1900. Source: Marat-Mendes et al. (2015)

and although this ultimately aimed at reducing the dependency of the country from foreign trade, it also responded to a food-production problem which, in Portugal, had several centuries.

In 1939, German Civil Engineer Gottfried Feder (1883–1941), who would become an important political figure under the Nazi regime for his racist views and conservative agrarian theories, proposed the creation of agrarian colonies, in a model physically similar to Howard's Garden City (Feder 1939). Although Feder's plans were not implemented by the Nazis, there were several experiences in European dictatorships with agrarian colonies, including in Spain, Italy and Portugal. This idea was in perfect accordance with the attempts of the New State to boost self-sufficiency and promote a rural lifestyle. As will be seen in Chap. 6, by this time, the diet of the Portuguese was indeed overwhelmingly rural, and traditional cooking—which for the majority of the population was the only accessible one—depended greatly on local products, with a heavy use of vegetables and grain (Graça 2020).

The Inland Colonization Committee (*Junta de Colonização Interna*—JCI) was created in 1936 to prompt solutions for problems identified in rural territories, although in the end its action would—as did other New State initiatives regarding housing—fall short of the original promise. Until 1938, the JCI was directed by the Ministry of Agriculture, but in 1949, it was absorbed by the Ministry of Economy. The key mission of the JCI was the colonization of land for agriculture in both public and private properties, ensuring appropriate infrastructure (Oliveira 2018). Colonies were conceived for a wide array of situations, ranging from colonies for wageworkers and turf farmers to proto-industrial farmsteads with agricultural enterprises (Lima 2016) being perhaps one of the most important attempts to optimize rural productivity. The JCI was tasked with selecting the most appropriate land based on suitability studies and, afterwards, with the works of planning, infrastructuring and implementation of projects.[5]

The JCI joined the National Society for Propaganda (*Secretariado de Propaganda Nacional*—SPN) to promote a romanticized vision of the Portuguese people as naturally rural, thus justifying backwardness and reinforcing the ideology of the regime (Rosas 2001; Guerreiro 2015; Oliveira 2018). If modernity can be seen as a threat to the essence of the Portuguese, then the JCI colonies posed a solution for salvation.

The work of the JCI was in many ways informed by previous works, ranging from the nineteenth and early twentieth century studies of 'spontaneous' rural colonization as well as by several test projects promoted in the countryside—namely in Leiria (region of Beira Litoral) and Sabugal (region of Beira Alta)—by a previous organism, the Direction General of Agrarian Social Action (*Direcção-Geral da Acção Social Agrária*), which had been extinct to make way for the JCI (Guerreiro 2015).

Between 1937 and 1938, the survey for selection of appropriate land was conducted and the results were published in 1939. According to the JCI, most surveyed soils were appropriate for forestry, and a considerable amount of unused land was fit for agriculture, although this would not suffice to overcome the disparity between national food production and expected population growth (Oliveira 2018). In this context, Agrarian Colonies, mostly in the north of Portugal, would promote an increase in agricultural land, but they would also solve the important problem of distributing the population across the territory, avoiding rural abandonment (Oliveira 2018). Inhabitants were granted temporary ownership in the colonies for 3 or 5 years, and from their productivity in this period depended on the continuation of the contract which further stipulated that a sixth of their annual production was to be delivered to the State, a quota later raised to a third of annual production (Lima 2016).

However, while the JCI is today remembered mostly because of this program, it conducted other projects, namely regarding the improvement of physical conditions in existing villages (Guerreiro 2015; Oliveira 2018).

In 1942, the JCI was restructured[6] to improve its functioning, and making it even more similar to foreign experiences, namely those conducted in Spain and Italy. This was a pivotal year for the JCI and the inner colonization program, as the presentation of the project for the Pegões Estate in Montijo, considered a successful work, established the foundations for a methodology of colony development, whose key component was the careful design of the agricultural 'casal' (farmstead), including its cultivation areas and irrigation systems (Guerreiro 2015).

Indeed, in 1946, a legal framework[7] was advanced to clarify design guidelines: each 'casal' was to be constituted by a house and outside areas adequate for rural explorations, particularly for polyculture, to allow constant labour and balance between different productions (Oliveira 2018). The 'casal' was supported by covered areas for livestock and for agricultural equipment (Lima 2016). The JCI conducted the planning and construction of the houses, and in some cases also provided the furniture, agricultural equipment and a battery of seeds to help kickstart production activities (Guerreiro 2015).

In the context of the 1950s, when the role of the JCI was firmly established, many Agrarian Colonies were well accepted by their beneficiaries: in the context of increasing rural unemployment, these colonies offered not only labour but also property—a house and land (Lima 2016). However, intrinsic to this spatial and agricultural program was a social program, as it was also stipulated that each 'casal' was attributed to a male rural worker, under 30, married but unable to provide for his family, and who demonstrated moral suitability.

In some cases, the beneficiaries abandoned the 'casal' out of incapacity to comply with the rules and compromises with the State, which was not specifically due to lack of work, but rather to the fact that some of the land was unable to produce in the volumes expected by JCI technicians (Oliveira 2018). Another important source of conflict in these processes of

[5] These aims are established in Decree 27.207, published in Diário do Govêrno 269/1936, Série I, November 16, 1936.

[6] Decree 32.439, published in Diário do Govêrno 272/1942, Série I, November 24, 1942.

[7] Law 2014, published in Diário do Govêrno 115/1946, Série I, May 27, 1946.

rural colonization was the antagonism by local residents, who resent not only that newcomers were being placed upon land that was once for community use, but was given far better conditions than those existing in traditional villages (Lima 2016; Silva 2020).

The authoritarian overtones of the social program and the indoctrination that was clearly contained in it, as Agarez (2020) recalls, is extensive to all housing promoted directly by the New State, places the JCI work very close to its other European counterparts. Not all built colonies proved successful, as some of them stand today completely abandoned. Yet at the time, the JCI was effective in providing housing for several social strata within rural populations, and the construction of the colonies often used local construction techniques (Oliveira 2018). Preference was also given to individual cottages, although in some cases row houses were employed as a cost-cutting measure (Oliveira 2018). Overall, the architectural typologies and settlement morphologies varied greatly. Moreover, it is important to highlight, with respect to the renewed efforts on internal colonization in the 1950s that

> It its known today that several technicians, foresters and economists sought to go further. In the heat of ministerial offices, they created daring proposals for State intervention, specifically in the property structure. But the pressure of proprietors' groups was stronger in the National Assembly and the Corporative Chamber. The results were reduced to reasonably conservative decisions and good intentions without practical means (Barreto 2017: 65).

The JCI intended to create a group of small landowners to cultivate barren land, however, it had a manipulative side—managing 'from above' the forms of property and the labour relations inherent to rural territories, and imposing specific moral codes. Property, more than urbanization, was the key element of this policy, although from this program resulted in many interesting planned settlements, whose picturesque value is hard to overstress. At the same time, this was really a failed policy: the desired self-sufficiency of the country was not achieved, the program did not go beyond seven new colonies, namely Milagres (1926–1950), Martim Rei (1937–1952), Gafanha (1938–1961), Barroso (1942–1962), Alvão (1942–1953) and Boalhosa (1946–1966), as well as Santo Isidro de Pegões. Although the JCI existed until the end of the regime, its action was rather limited, and after 1960 its activity contemplated only improvement works in already existing settlements (Guerreiro 2015). However, this was the only housing program of the New State on which outside spaces were minutely prepared and sized to foster food production, and where food production was to be the key social and familiar activity. In spite of the undeniable conservative overtones, the inner colonization program has a unique interest for the study of the relationships between housing, planning and food.

1.3.1 The Santo Isidro de Pegões Agrarian Colony

The plan for the Agrarian Colony in the Pegões estate is a hallmark of the JCI, as this was the first project it developed fully and, as stated above, it played a key role in establishing its project methodology and its key component, the agricultural 'casal'. Furthermore, the authors of the colonization plan also intended for it to provide an example of how the rural life and economy of the region should function, aiming at a revitalization of the cultures of the Southern Bank of Tagus (Lima 2016; Silva 2020).

The land is located in the exclave territory of Montijo, near the borderline with Palmela. While today this is one of the key productive territories of the Lisbon Region, in the 1930s, part of it was not cultivated at all, and its scant population had been formed through a process of 'semi-proletarian colonization', whereby houses and small farms were built by rural workers who could not survive solely from their own produce and thus worked for nearby latifundia—notably the Rio Frio Estate in neighbouring Pinhal Novo (Silva 2020). This process marked a first phase of spontaneous colonization of the barren territory of the Pegões estate, which after the intervention of the JCI would become the Santo Isidro de Pegões colony.

The land was a rough and desolate moor, with some spontaneous woods' areas, mangrove patches and cork oaks (Silva 2020). It was crossed by the old Royal Road (*Estrada Real*) from Lisbon to the Spanish borderline—its only point of interest being a fountain built in 1728 (Lima 2016). At the early twentieth century, the estate had nearly 700 ha of arable land, and belonged to José Rovisco Pais (1862–1932), an industrialist and rural land owner, who offered it to the State upon his death. The fact that the land is already nationalized helped prevent the typical local resistance verified in other colonies (Lima 2016; Oliveira 2018; Guerreiro 2015). The preparation studies inquired the first settlers, concluding that out of 800 existing houses, only 50 families were able to provide for themselves from the produce of their land (Silva 2020). In accordance with the New State morality, the authors of the preparatory studies highlighted the family unit as the ideal basis of agricultural production and working on one's own land as a better alternative to wage labour on someone else's land (Silva 2020).

The preparation works included a survey of the territory and the installation of draining and irrigation systems, the removal of sands and the construction of trails.[8] The whole process was very long, as the project started in 1937, public presentation was made in 1942 and construction was

[8] The blueprints can be accessed at Guerreiro, 2016. https://hdl.handle.net/10216/118803.

finished in 1952. The authors of the project were two agronomist engineers, foresters from the JCI, Mário Pereira and Henrique de Barros (1904–2000), with architectural design charged to architects: the homesteads were designed by Henrique Brando Albino (1921–2003) with José de Oliveira Trigo, while public buildings were designed by Eugénio Correia (1897–1985) (Lima 2016). An agrological study was conducted by Joaquim Botelho da Costa (1910–1965) and António Sardinha de Oliveira, while the hydrological study was prepared by Engineer Ricardo Teixeira Duarte (1886–1959) (Silva 2020).

Due to its long-spanning process, there were several changes introduced to the original plan. One of the most striking was that to the two cores originally planned—Pegões Velho with 86 'casais' and Faias with 59—was added a third one—Figueiras, with another 50 'casais' (Silva 2020). These cores are different among themselves, for while in Faias and Pegões Velho housing is dispersed alongside a road or a streamline, in Figueiras the settlement is concentrated (Lima 2016).

The centre of the colony was the Pegões Velhos core, popularly known as Santo Isidro de Pegões (although this name refers to the totality of the three cores). It has a clearly defined centre, a garden and small boulevard where public buildings are grouped: a church, two schools (originally one for each sex), three service houses (for professors or the priest) and a small crescent-shaped commercial gallery. The 'casais' are disposed in oblique position alongside a double street irregularly arranged in the shape of an 8, with the equipment garden occupying the crossroads. A wine Cooperative, which is to ensure the key produce of Santo Isidro de Pegões, was also constructed at the eastern end of the Pegões Velhos core (Figs. 1.2, 1.3 and 1.4).

The 'casais' designed by Albino and Trigo had a front yard with a kitchen garden and a cultivation space in the back, whose dimension varied according to the productive capacity of each family. Notably, the covered porches for cattle occupied nearly half of the housing area. Each 'casal' corresponded to cultivation areas that are not necessarily contiguous with the house itself. These include land for dry

Fig. 1.2 '*Casal*' in the Santo Isidro de Pegões Colony. Source: Authors

Fig. 1.3 Church, in the central avenue of the Santo Isidro de Pegões Colony. Source: Authors

farming, wine and pine, while around the house was a kitchen garden for an orchard and horticultural cultivations for self-consumption, all totalizing about 18 ha for each 'casal' (Lima 2016).

Eugénio Correia designed the remaining public buildings with a clearly modernist taste and using some motifs—such as the arch and the dome—with expressionist tension. All buildings seem like variations upon the key composition of the Church, albeit in a humbler scale and with a more regular structuring (Marat-Mendes et al. 2020a).

Santo Isidro de Pegões is a notable example within the Agrarian Colonies of the JCI, for being its only development promoted south of the Tagus River, but also because it is the only one located within a Metropolitan Region. To be sure, there was no formal definition of the Lisbon metropolis in the 1930s—and when such a delimitation was made legally official in 1959, it did not include the Montijo exclave, where the colony is located. Still, this is the agrarian colony that is closest to a set of urban cores in the whole JCI program: suffice to think that nowadays, it stands about half an hour from Setúbal or from the urban core of Montijo. Santo Isidro de Pegões is also the largest of the JCI colonies (Lima 2016). Another notable exception within the JCI work is that the Colonies of Santo Isidro de Pegões and Gafanha were the only ones constructed in land owned by the State (Silva 2020).

Santo Isidro de Pegões represents a form of spatial planning where food production had the central role. Indeed, the planning process comprised two specific phases: a first one comprising minute studies of the suitability of the land for agriculture and the infrastructure necessary to ensure it, and a second one regarding the colonization structure itself, with the buildings and other physical constructions to support the new productive unit (Silva 2020). With regards to

Fig. 1.4 Home for professor or priest in the Santo Isidro de Pegões Colony. Source: Authors

the latter, Santo Isidro must be highlighted for the quality of its architecture, in the design of both the traditionalist 'casais' and the modernist public buildings and service houses (Lima 2016; Marat-Mendes et al. 2020a). In the second phase, the plan for Santo Isidro de Pegões represents a sort of planning maturity for the JCI, whereby the moralist and psychological justifications of previous colonization experiments are replaced by thorough and detailed scientific studies which approach the territory for its factual productive potentials (Silva 2020).

In 1988, the 'casais' were transferred to the ownership of their inhabitants (Lima 2016). Today, Santo Isidro de Pegões remains inhabited, and while some of the original rural activities are no longer upheld—namely the creation of livestock within the homesteads—the cultivation areas are still in use and the basis of its economy is still agrarian (Lima 2016). With the decline of livestock ownership, the available area in the houses probably increased for most residents. The farm plot in the back is often still in use, while in front yards, cultivation is mostly of ornamental species, discontinuing the horticultural component of the original project. Moreover, the wine Cooperative remains functional, as most equipment originally constructed—and to which little has been added. While individually, most houses show great signs of adaptation and personalization, their serial resemblance is still perfectly perceptible.

In 2008, the process for making Santo Isidro de Pegões graded heritage was kickstarted (Fig. 1.5).[9]

[9] See Edital 142/2008 from 5-09-2008, CMMontijo: 'Deliberação de 20-08-2008 da CM de Montijo a determinar a abertura do procedimento de classificação como de Interesse Municipal' (Deliberation of the Montijo Council determining the opening of the proceedings for classification of Municipal Interest); available in: http://patrimoniocultural.gov.pt/pt/patrimonio/patrimonio-imovel/pesquisa-do-patrimonio/classificado-ou-em-vias-de-classificacao/geral/view/23527781.

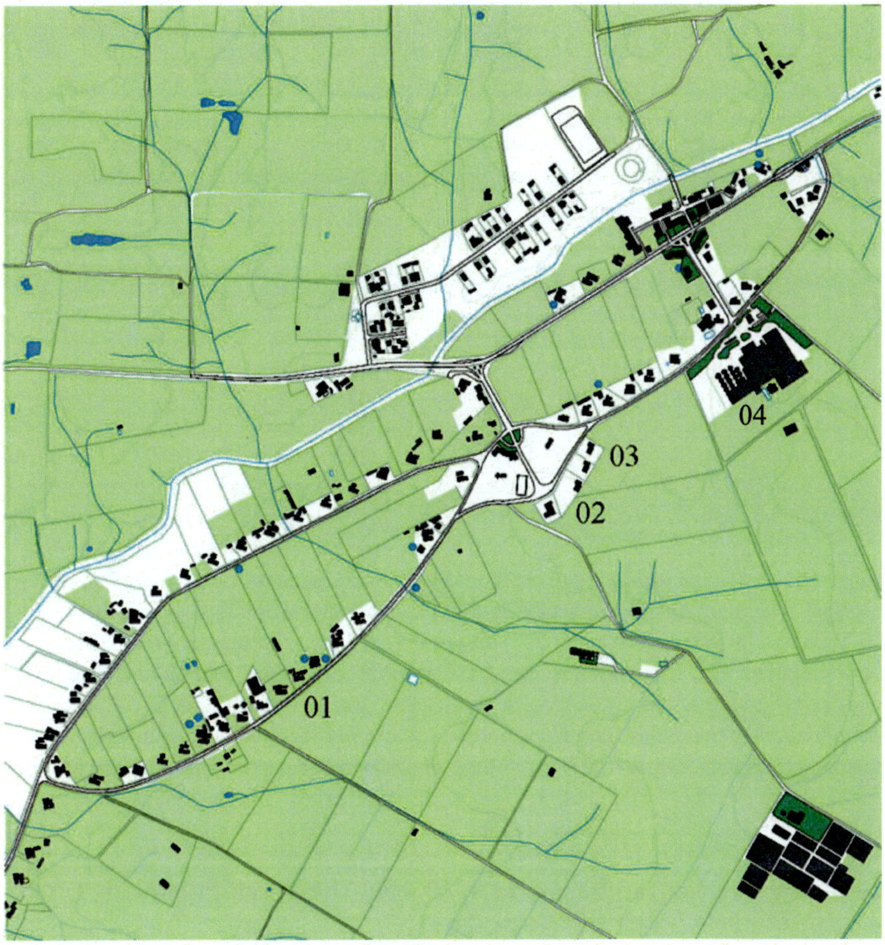

01. 'Casal'; 02. Church; 03. Home for professor or priest; 04. Wine Cooperative

Fig. 1.5 Plan for the Agrarian Colony in the Pegões estate. Source: Authors

1.4 The Rise of the Urbanization Plan

Urbanism policies under the New State dictatorship were especially associated with the figure of engineer Duarte Pacheco (1900–1943), who served two terms as Minister of Public Works, in 1932–1936 and 1938–1943. During his second term, he was simultaneously mayor of the Lisbon City Council (1938–1943). Given his willful and energetic political activity and his modern approach to territorial development, Pacheco cultivated an ambiguous relation within the regime he served, but he is also unmistakably regarded as the founding figure of modern Portuguese planning (Silva 1994; Lôbo 1995; Costa 2012).

Under the leadership of Pacheco, a new legislative framework was kickstarted, applying to all Portuguese municipal centres, as well as towns with 2500 inhabitants or projected to achieve them within ten years (Lôbo 1995). Law 24802[10] was approved in 1934, and its key aims were to infrastructure and modernize existing settlements and to plan or assign extension areas where urbanization was allowed. The privileged legal instrument to ensure such aims was the General Urbanization Plan (*Plano Geral de Urbanização*—PGU), whose key strategy was urban design (Lôbo 1995; Marat-Mendes et al. 2020b). However, this was not the sole instrument to be promoted, notably in the Lisbon Region. To understand this, one must notice the importance of the architecture and urbanism professionals who worked in close collaboration with Pacheco and created the first plans to modernize the capital and its surroundings.

[10] Decree 24802, published in Diário do Govêrno 299/1934, Série I, December 21, 1934.

During Pacheco's first term as a Minister, the severe lack of architects capable of undertaking the urbanization spree launched by the 1934 Law had two responses. The first was the promotion of scholarships for architects to seek education in urbanism abroad, namely at the IUUP. The second is the hiring of foreign architects who could perform such role on short notice, especially with regards to the larger Portuguese cities (Lôbo 1995; Marat-Mendes and Oliveira 2013; d'Almeida 2013, 2015).

Under this second strategy, one of the first professionals to be hired is the French architect Donat-Alfred Agache (1875–1959). At the time of his first invitation to work in Portugal, Agache was already a well-respected planner, having authored planning projects for Dunkirk (1912), Paris (1920), Reims (1921), Creil (1925), Poitiers (1926), Dieppe (1927), Joigny (1927), Tours (1927), Orleans (1927) and Rio de Janeiro (1930). He also had received the third prize in the contest for the Canberra Plan (1912), with some of his solutions being adopted in the city, constructed with basis on the three laureate plans. Although Agache would become most notorious for his work in Brazil, where he also collaborated with a dictatorial regime, his work expressed a vastly modern and even multidisciplinary approach to planning, where the influences of the socioscientific methods of pedagogue Edmond Demolins (1852–1907) and the Musée Social in Paris as well as the sociological ideas of Gabriel Tarde (1843–1904) and Émile Durkheim can be identified (Underwood 1991).

Agache originally comes to Portugal in 1935 in order to restructure the marginal road from Lisbon to Cascais, to accommodate the development of Costa do Sol (meaning the Coast of the Sun), a beach chord from Oeiras to Cascais, where the New State idealized a leisure and summer suburb for the capital's middle class, taking heed of the fact that Cascais, São João do Estoril and São Pedro do Estoril already had this use (Briz 1989). However, Agache's vision differed greatly from this original intent, and upon visiting Costa do Sol, he comes up with a new proposal, where emphasis is laid on the automobile connections between the Lisbon centre and Cascais, not only through the marginal road but also through a highway connecting the city's business centre in Marquês de Pombal to Cascais (Lôbo 1995; Pereira 2009). This implied the construction of a viaduct over the Alcântara Valley, one of Lisbon's most notorious fringe belts. Agache's approach, indebted to the paradigm of the 'city beautiful' (Lôbo 2009), conceives of Costa do Sol as a single territorial unit integrally linked with Lisbon, and while his work, developed in 1934 and 1935, demonstrates his alternation between the regional and the local scale, it was in the former that he focused his attention, designing the infrastructure layout and conceiving a network of free urban spaces articulated with rural buffers.

Agache was also strongly influent in the writing of Law 24802, to which he brought the experience gathered in his previous participation on the French Cornudet Law (1919). With it, Pacheco formalized an interest of the State in urbanization and in the physical growth of cities. But several contradictions spawned from this Law. First, while municipalities were expected to undertake the planning and implementation works, most did not have technical or financial capacity to do so, and it was the central State that ensured such process (Lôbo 1995). Despite its totalitarian aspect, consistent with an authoritarian State, the aggressive laws for expropriation for public interest prompted by Pacheco stand at odds with the regime's ultimate respect for private property rights.

While Pacheco's legislation was the key that presided over the modernization of the Portuguese territory—at least in theory—it also conceived this modernization as a turn against the reality of that very territory. One aspect that was never challenged in Law 24802 and in the planning spawned by it was the fundamental opposition between urban policies and urbanism on the one hand, and rural territory and agroforestry management on the other. While the JCI Agrarian Colonies could be said to represent a sort of 'urbanism in rural settings' (Lima 2016: 19), this experience never influenced urbanization policies, where rural land was acknowledged, but seldom included as part of the total spatial arrangement. From the 1930s to the 1950s, this tacit exclusion of agrarian land from urbanization solutions stood in a discreet contradiction with the paradigmatic tendencies of planning, as will be explained next.

1.4.1 The Era of Étienne de Groër

Architect Margarida Souza Lôbo (b. 1940) has rightly called the period following the approval of Law 24802 the 'era of Duarte Pacheco', when the main cities of the country were finally surveyed and become the object of specific urbanization plans which sought to control and rationalize the process of physical growth (Lôbo 1995). However, in the specific case of the Lisbon Region, this is also the 'era of Étienne de Groër' (1882–1974), as this Russian–Polish architect–urbanist was bound to become one of the key actors in turning the possibilities opened by Pacheco's law into specific plans, at several scales and accounting for different economic and social activities.

Regardless of its ideological overtones, the Pacheco legislation caused a major shift in Portuguese planning culture, as a nearly inexistent policy suddenly was replaced by the demand for over 400 plans, programming the development of numerous towns and cities across the country (Lôbo 1995). As soon as Pacheco returned to the Ministry of Public Works and became Mayor of Lisbon, he invited Agache to return to Portugal and resume the works for the Costa do Sol Urbanization Plan (*Plano de Urbanização da Costa do Sol*—PUCS). However, the urbanist was working in Brazil at the time and declined, suggesting his former collaborator and colleague at the IUUP, Groër who

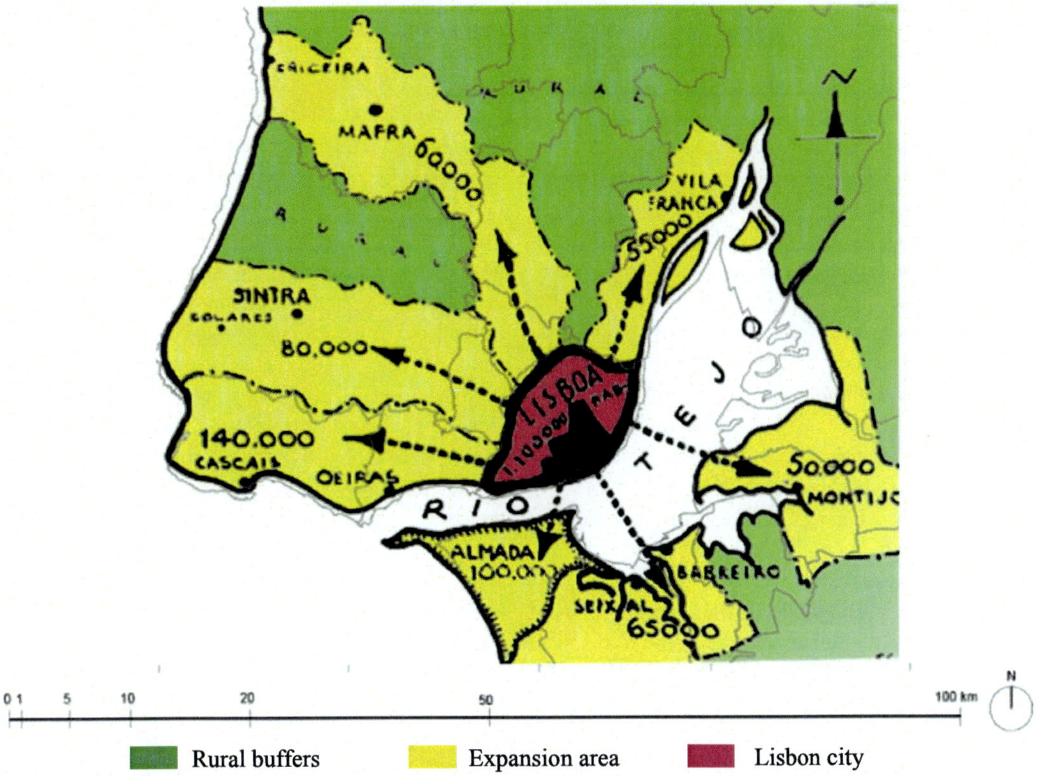

Fig. 1.6 Expansion planning. Source: Authors (Adapted from Gröer, 2006: 58)

would effectively be hired to continue the PUCS. As a professor in IUUP, Groër had become greatly acquainted with the theories of Ebenezer Howard, whom he considered the first truly modern urbanist, as stated in his 'Introduction to Urbanism' (Gröer 1945–1946).

Once in Portugal, Groër works in the improvement or urbanization plans for cities of regional importance, like Coimbra, Braga or Évora, as well as for Luanda, the capital of Angola, a Portuguese colony at the time (Marat-Mendes and Sampayo 2015). In the Lisbon Region, he is originally assigned two tasks: finishing the PUCS and starting the work for the first Lisbon Municipal Masterplan. Perhaps because Groër was tasked with this uniquely important plan for the capital city, he understood very early that to plan what was to happen within the limits of the city itself would not suffice. Thence his scheme of the 'Lisbon Region and its extensions', which he seems to have sketched in 1946, and amended several times (Fig. 1.6). This indicative drawing accompanies the written documents of many of the plans Groër designed for this territory and clearly states the vision that guided the several works he was to develop in the region surrounding the capital, until his departure from Portugal in 1952.

Lacking any legal value, but nonetheless expressing a defined territorial model—centralized, axial, urban–rural—this scheme is indeed the first vision of a true metropolitan region, with a clear notion of what activities must be accommodated and where their development was most adequate. Surely, this vision was highly dependent both upon the social and political conditions of 1940s Portugal, but it also testifies to Groër's interpretation of Howard's Garden City paradigm. Thus, Groër equates a plurality of activities by programming specific development chords for each activity around a well-defined centre and placing a large rural buffer to maintain existing agricultural and forestry activities (Marat-Mendes 2022).

Groër's regional vision had its shortcomings: it segregated economic activities by assigning them to specific (and tendentiously monofunctional) development chords, while keeping industrial activities highly limited, in accordance with the mistrust of the New State in large-scale industrialization. The projected population growth for each of the chords, as well as for the Lisbon city itself, was also severely understated. And finally, considering that Gröer envisioned the development chords not as conurbations—i.e. continuous urbanizations—but rather as urban–rural articulations, the area really made available to overcome housing needs was scant. Groër's preference for low densities would also contribute to intensify the mismatch between the planning vision and the real transformation of the region.

In contrast, the land available for rural activities was comparatively extensive: not only there were fully rural buffers between most chords—where agriculture and forestry were only punctuated by existing villages and thorps—and adding to the protective buffers between each urban core within a chord, as Gröer was adamantly against the merger of different cores into larger settlements. Moreover, even his Lisbon Masterplan, finished in 1948, includes suburban areas classified as rural, although the lack of differentiation between these and the reserves of free space make it likely that these were meant for urban forestry to separate neighbourhoods, and not conceived as productive areas.

It remains disputable whether Gröer's vision of a capital city surrounded by relatively small-scale towns and an extensive rural system was realistic, even considering the New State's anti-urban rhetoric. But while Howard's proposal of a territorial rearrangement was dictated by the need to resolve an existing problem of unsanitary urban mastodons in England—as directly stated in the first chapter of Howard's (1902) book—for Gröer the idea seems to be to avoid the emergence of this problem. Yet he was keenly aware of the importance of technical services (power and water supply, sanitation, waste treatment and mobility infrastructure), considered as one of the foundational basis of modern urbanism—and articulated them with the Garden City ideals expressed in the zoning scheme (Gröer 1945–46). This is at least realistic to the situation of the regional territory at Gröer's time. Suffice to remember that by 1940, the Region had little more than a million inhabitants, mostly living in the Lisbon city. Only in the decade that Gröer arrived in Portugal did the horticultural productions within the Lisbon city—in Lumiar, Campo Grande, Beato, Alto de Pina, Benfica and regular —disappear, while outside city limits, similar cultivations remained (Marat-Mendes et al. 2015). Delayed industrialization, as we will see in Chap. 4, means that a large-scale process of suburbanization will only come to pass long after Gröer's Lisbon masterplan was finished.

The Region was not self-sufficient in its food system, but it is notable that many food items were produced in it, with a natural convergence in Lisbon, where the majority of inhabitants lived. Thus, fish from Setúbal is dried and salted in Coina (Barreiro) and Almada, before reaching the capital; from the Ribatejo through Vila Franca de Xira came products from Alentejo and Santarém, namely olive oil, wine, honey and river fish, and that town itself supplied kale, turnips and peas; from the rural continuum of Oeiras—Cascais—Sintra—Loures came horticultural products and vegetables; while the chord from Alcochete to Caparica (Almada) supplied savoy-cabbages and cauliflower (Marat-Mendes et al. 2015).

The PUCS must be considered a success, as it was effectively implemented and even remained active until the early 1990s, albeit subject to several revisions (Pereira 2009). Although Gröer generally maintained Agache's infrastructural plan, he focused his work on the urban design of the specific settlements to be arranged and extended. Here, he displayed the quality of his delicate design, integrating smoothly the winding street systems of the pre-existing settlements with the planned urban extensions adapted to the territory. The preference for the bourgeois isolate cottage was obvious, and Groër additionally proposed a network of urban green spaces, some of which interplayed with the rural buffer (Marat-Mendes 2009). Although the topography of Costa do Sol was not physically convenient for the installation of large industries, given its drastic valleys perpendicular to the riverside and the danger of casting smoke over residential areas, Groër formally forbade the installation of relevant economic activities, though the maintenance of existing ones is allowed if wind direction is adequate. This turned the PUCS territory into a de facto suburb, functionally dependent upon the city of Lisbon. For Lôbo (1995), this represents a significant departure from Howard's vision, since in this process, the satellite towns (the settlements) did not have any independence or self-reliance ability, always dependent upon the largest city. However, this author underplays the regional vision expressed by Gröer and which is consistent with the system Howard used to articulate cities and towns with their hinterlands.

The family unit constituted a basic issue in Groër's (1946–47) urban planning, determining public space organization according to family needs (creating schools, churches and other equipment within walking distance) as well as private free spaces, including the requirement of a private backyard for growing food and resting. Thus, although Gröer did not include the design or specific assignment of agricultural activities either in the urban settlements or in the rural buffer, he did have familiar food-production scale as an official function for the territory.

Considering the rural-leaning politics of the New State, Gröer found in Lisbon a political climate favourable for exploring his interest in the Garden City. His approach clearly succeeded at planning low-density comfortable neighbourhoods but failed to predict the real changes the city would undergo and design solutions to tackle such problem. However, Gröer's vision is not without merit, as it balanced urbanization with the preservation of rural activities, with positive metabolic impacts.

Regardless of the political implications of Gröer's regional proposal, which he confirmed through plans at the municipal and sub-regional scales, it unmistakably sought a continuity with the situation of the time—instead of promoting urbanization, Gröer limited it to specific development axis, reserving the majority of the territory for rural activities. There can be little doubt, thus, on the preservationist overtones of this strategy, as well as over its *avant-la-lettre* promotion of a short-supply chain of food, which is closer to Howard's theory than most of his other followers in Portugal.

Gröer's vision was coherent with Howard's when we consider the regional scheme he envisioned and which coordinated the different extension chords into an integral metropolitan scheme with a clear distribution of activities, both urban and rural. These activities were dimensioned according to a vision of the country where industrial development was limited and coexisted with rural land and with leisure zones, all favoured by low-rise and low-density neighbourhoods. This vision seemed to collapse in the mid-1950s, when the Lisbon municipality created a new Masterplan and launched the urbanization plans for the eastern end of the city according to very different concepts from Gröer's. These neighbourhoods are rightly credited in architecture historiography for advancing the country's architectural culture towards a more decidedly modern approach, closer to the urbanism principles of CIAM, as will be explained in the following chapter. Additionally, as will be discussed in Chap. 4, by the 1960s, the Lisbon Region territory undergoes a spree of illegal construction, which not only instigates an urbanization pattern very different from that envisioned by urban planners but also has a destructive effect on the extension of productive land.

But the rural-leaning aspect of the Portuguese take on the Garden City must not be overstated. Some of the neighbourhoods and towns created by Gröer or under his influence can seriously be considered as 'villages' within the city, at least considering the state of urban development of the time of their creation. There is also no question that under Gröer were developed comfortable and neat—albeit exclusively middle class—suburban areas that effectively avoided speculative construction or the explosion of illegal housing. Plus, a more modern interpretation of the Garden City and an acceptance of higher residential densities were underway at least since 1938, when the construction of the Encosta da Ajuda was started by the Lisbon Council, and later in 1945 with the Alvalade area, a sort of 'new town in town' whose key housing typology is the four-story block. Such typology became frequent in the development of the suburbs of Lisbon, including in areas built by the private sector. The author of the plans for both Encosta da Ajuda and Alvalade was João Guilherme Faria da Costa (1906–1971), the first Portuguese urbanist, having concluded his studies in the IUUP with a plan for the Portuguese town of Figueira da Foz (Costa 2005; d'Almeida 2013, 2015; Marques 2018).

Because of the idiosyncratic development of Portuguese planning policies, the Lisbon Region was planned unevenly by the State through a plethora of different but often incoherent set of plans at different scales: there were sub-regional units extending beyond municipal borders based on land-use and infrastructure, PGUs guided the urban design of neighbourhoods or towns, and even projects for public buildings and housing typologies were sometimes prepared by the public sector.

It is in the work of Gröer that these planning scales articulated more clearly, as he had the unique opportunity to create the Masterplan with the land-uses of the capital city, as well as to integrate its immediate surroundings.

The legislation—and the larger planning culture—in development from Agache to Gröer collapsed in less than 20 years. The creation of a new legal figure, the Foreplan (*Anteplano*), was key to devoid planning from any ability to succeed. In the early 1940s, the topographical surveys meant to inform PGUs were finished, and municipalities started to submit the resulting plans. Yet the central State processed these PGUs as if they were Foreplans. This was predicted on Law 24802 as a preliminary study meant to guide the general morphology of settlements but lacked legal validity and therefore could not form the basis for land expropriation or even of urbanization authorizations (*licenças*) (Lôbo 1995). This was synthesized in Law 35.031,[11] which formalized the role of the Foreplan, reducing it to the status of a simplified technical study and consequently it undercut much of the work conducted by urbanists in many municipalities.

1.4.2 Vila Franca de Xira Urbanization Plan (1946)

In 1946, Étienne de Gröer and his son Nikita were hired to design the Regional Foreplan Moscavide—Vila Franca de Xira, and the town of Vila Franca de Xira itself receives a PGU, although there is no formal link between these two works.

The authors of the PGU are architects João Guilherme Faria da Costa and Miguel Simões Jacobetty Rosa (1901–1970). These two authors had collaborated on the PGU for Portalegre (1942) and most notably in the Alvalade district in the Lisbon city—Faria da Costa authored the urbanization plan (1945) and Jacobetty was the designer of a significant amount of its housing types (Costa 2005). Both had significant planning experience in the Lisbon Region as well. Faria da Costa oversaw the Alvalde Plan and the PGU for Amadora (1949—not approved), collaborating with Gröer in Almada (1946). Jacobetty was responsible for the plan of Mafra (1946—not approved), he designed the Jamor National Stadium (1939–44) and the S. Miguel

[11] Law 35 031, published in Diário do Govêrno 250/1946, Série I, November 4, 1946.

neighbourhood (1949–51), one of the cells of the Alvalde (Lôbo 1995; Costa 2005; d'Almeida 2013, 2015).

While many settlements on the northern riverside area extending from the eastern end of Lisbon were small-scale villages whose key source of pressure was the nearby installment of industrial facilities, Vila Franca de Xira, whose roots go all the way to the Roman Empire and to the Middle Ages, had acquired by the twentieth century an urban character, and the great barriers to its development were of a geographical and infrastructural nature (CSOP 1955). Indeed, the town was limited on the west by the hill of Monte Gordo, as well as by the passage of the Water Channels of Alviela and Tejo—the latter had been completed under Duarte Pacheco in 1940—which conditioned construction, as did the bridge over the Tagus (inaugurated only in 1951), linking both banks of the river, and the proposed northwards motorway (Jacobetty and Costa 1946).

By the 1940s, the town—which constituted a single parish—counted 9 859 inhabitants (over 9 000 were permanent and the remainder seasonal inhabitants) and 2546 houses—and since population had been growing unevenly since 1900, by this decade, there was already a clear housing shortage, especially for the working class and the lower middle class (Jacobetty and Costa 1946). Despite the physical barriers, Vila Franca has some potential for urbanization, since its terrain is nearly flat and the municipality was already undertaking the infrastructural works to ensure access to clean water (Jacobetty and Costa 1946).

Despite its urban character, the town was still surrounded by important farmsteads, including Quinta do Paraíso, Quinta do Palyart, Quinta do Farrobo, Quinta das Torres and Quinta do Desterro, while a key aspect of its economy and development was the Great Wetland of Tagus (*Lezíria Grande do Tejo*), the territory on the eastern margin of the river, which extended southwards and eastwards, towards the neighbouring town of Samora Correia, outside the Lisbon Region (Jacobetty and Costa 1946). At the time Jacobetty and Faria da Costa designed their PGU, the Great Wetland extended for 68 square miles (109.44 km^2) and produced every year a million and a half litres of cereals, as well as grazing for horses and bulls (Jacobetty and Costa 1946). Bull breeding and bull races were by then—and continue to be—key aspects of the local economy and the local culture.

Other than cereals and grazing, which were the key produce, fishing was also a relevant activity (25 000 shads were sold every year to Lisbon, followed by flounder) as well as fruit production, while among the local industrial activities were rice-mills and wheat grinding—countered by other activities including wool, mosaics, cement and lubricants (Jacobetty and Costa 1946). The commercial activities in the municipality were dominated by fish, cereals—mostly wheat, broad bean, barley, rice and corn—and fruits—mostly cherry, apple, grape, orange and apricot (Jacobetty and Costa 1946). Given this productive economy, it was a key supplier of the capital, as will be seen in Chaps. 6 and 9.

Indeed, be it in the municipality or in the town itself, the economy of Vila Franca de Xira was highly dependent upon food, both its production and transformation. While all this information was accounted for in the descriptive documents of the PGU, little attention was paid to the physical needs for maintaining this rural economy in function. Quite the contrary, and following the general tendency of the age for separating urbanism policy and agroforestry policy, the planners seem to have limited the physical growth of the town, so to hypothetically hamper the conversion of farmland into urban land. While such proposition is not present in the design put forward by Jacobetty and Faria da Costa, they did admit that with the improvement of activity prompted by the Tagus bridge, in the future a new port and a new urban settlement would be created on the other margin of the river (Jacobetty and Costa 1946)—a transformation that has not come to pass.

The PGU proposed extensive works in the existing town, namely a new street system through the demolition of a considerable number of buildings. Such an approach had been taken by Jacobetty in his plan for Mafra, being a key reason for its rejection (Lôbo 1995) and (as we will see next), a similar accusation of disregard with the premodern settlements was brought against the plans designed at this time for Palmela and Setúbal, on the Southern Bank of Tagus. Moreover, as the Gröers also did in their Foreplan, the PGU proposed a new design for the motorways projected by the Portuguese Road Authority, alongside a new knot for accessing the bridge.

To face the lack of space proper for urban extensions, Jacobetty and Faria da Costa propose to limit industrial activities to artisanal production, and the elimination of the existing graveyard and the bull-race arena (both on the southern end of the town). A new area for equipment was proposed on the northern area, erasing the agricultural uses, and replacing them with a new bull-race arena, a sports pavilion and an open-air space for markets and fairs (Jacobetty and Costa 1946) (Figs. 1.7 and 1.8).

Jacobetty and Faria da Costa furthermore propose, again through demolitions, the creation of a new civic centre, a large square around that were gathered buildings of public importance, including the Town Council, a library, a museum banks and hotels (Jacobetty and Costa 1946). Outside this civic centre, equipment included several schools (including technical) and a hospital. All these stipulations were made aiming for a population of, at most, 20 000 inhabitants (Jacobetty and Costa 1946), about twice the population existing at the time of the PGU.

Although the plan regulations allowed for buildings up to three floors, this was mostly for a reason of economy—and

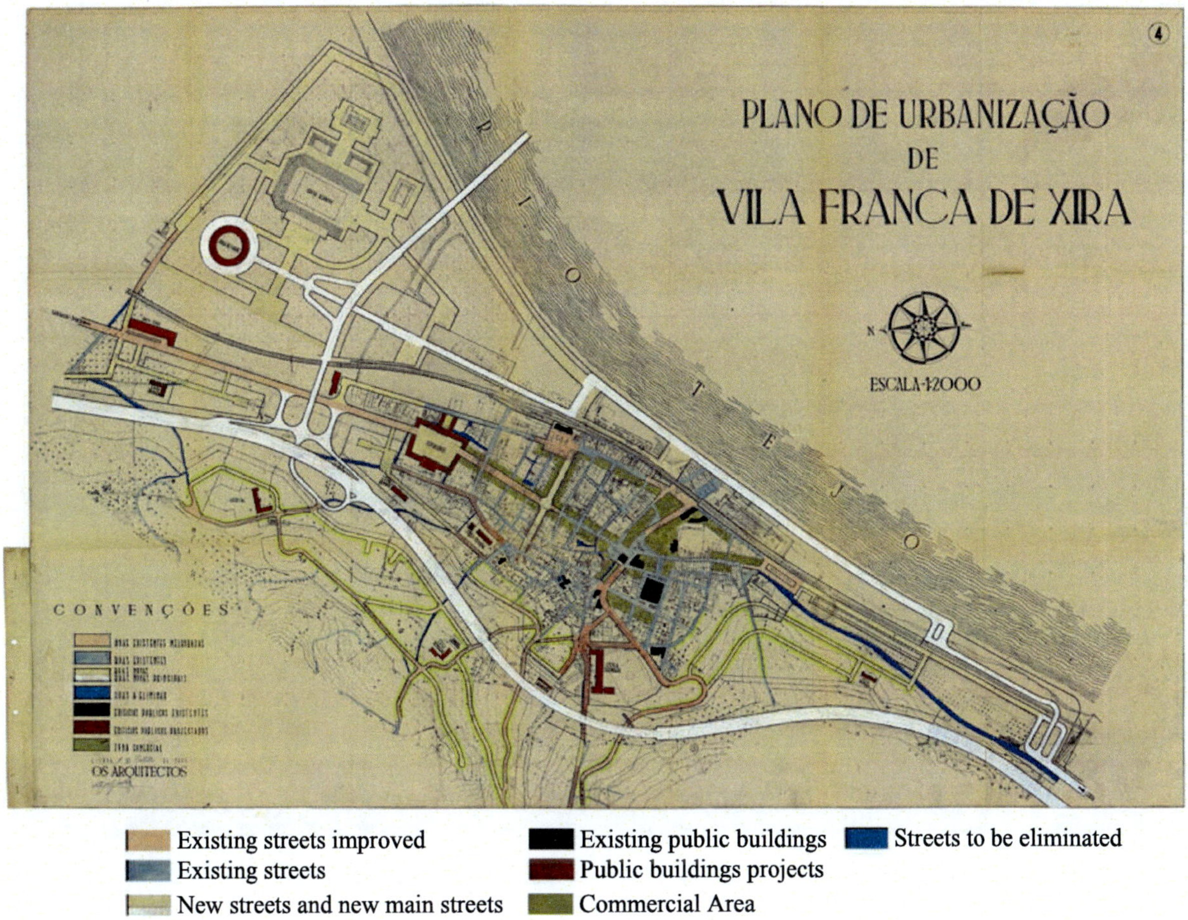

Fig. 1.7 The Vila Franca de Xira Urbanization Plan. Source: PT GFC-01-002/2 DES.00028, Guilherme Faria da Costa, 1946 - Arquivo DGPC/SIPA

implicitly for lack of available space—but throughout the plan regulation, there was an unmistakable defence of the individual house as the ideal housing typology, allowing familiar independence as well as space for a vegetable garden or an orchard (Jacobetty and Costa 1946).

While there were several zones predicted for urban expansion, Jacobetty and Faria da Costa did present a detailed plan highlighting the area for the civic centre and for one of the extension areas, westwards in the lower slope of the Monte Gordo (Fig. 1.9). According to the zoning scheme, this area was assigned for housing for the poor and the middle class. Accordingly, 86 plots for individual houses with front- and backyards were created in two loosely-formed groups whose morphology was adapted to the terrain and centred on equipment—in one case, the hospital and in another, the school. This urban extension was located westwards of the new motorway, while on the opposite side, was proposed a new housing area composed of twin semidetached houses, it seems likely that the neighbourhoods on this side of the motorway were expected to be inhabited by the middle-class—an aim that was intensified by the immediate presence of the hospital. In general, what this detail demonstrates is that the territorial model proposed by Jacobetty and Faria da Costa treated Vila Franca as a properly urban centre, justifying a suburban development, slightly segregated, and seeking, through low-density and generous garden space (both public and private) to integrate in the rural area that surrounded the town. Faria da Costa used a similar strategy in Encosta da Ajuda, back then a satellite of Lisbon, surrounded by the Monsanto Forestry Park (d'Almeida 2013). However, the plan excessively segregated different segments of the population according to their social class (CSOP 1947).

In 1947, the Superior Council for Public Works (*Conselho Superior de Obras Públicas*—CSOP) gathered to evaluate the work of Jacobetty Rosa and Faria da Costa, considered as a foreplan. The resulting technical report criticized the value attributed to statistical information, the

Fig. 1.8 Overview of the northern extension area of the PGU, Vila Franca de Xira. Source: Authors

lack of accurate accounts of climatic conditions, the alterations to the motorway projects of the JAE, the distribution of housing densities across the different zones, the location of the industrial zone on the riverside which eliminated the public space in existence there, the placement of the green spaces and some equipment, the predicted number of schools, and the generally expensive costs of implementation of the proposed territorial model (CSOP 1947). While advancing several aspects in which the presented plan could be improved, the definitive version is recommended to be designed only after the conclusion of the Regional Foreplan in development by Gröer and his son, to ensure the integration of the development of Vila Franca de Xira in the wider industrial zone it belonged to.

The technical report by the Council of Vila Franca de Xira also pointed out several problems, including the lengthy planning process, during which buildings were approved in zones the plan assigned for other uses, the demolitions proposed in the old core and mostly, the costly implementation that the plan predicts and which, the Council predicted would spawn negative results. Other negative appreciations included the halting of development tendencies, the refusal of collective buildings that were considered necessary to ensure proper housing for everyone, the creation of a new sports area in flooding land. Moreover, the Council criticized the plan for proposing a civic centre while maintaining residential density as low as possible when, in reality, what is necessary was the construction of houses and apartment buildings as close to the centre as possible. The proposed new sports area was in contradiction with business interests and with tradition, since the existing bull-race arena was favourable for 'esperas de toiros'.[12] Finally, the plan restricted construction to a degree that would allegedly hamper a true modernization of the town (CMVFX 1948). While many of the critiques are justifiable, and between the CSOP and the Town Council, important flaws were pointed to the proposal of Jacobetty and Faria da Costa, it seems clear in many of the criticisms, especially from the Council, that the key problem of this plan was that it was far beyond the financial—and legal—possibilities of this municipality.

[12] Also known as 'Largadas de Toiros', these are feasts where bulls are freed to run through the streets of the town, being provoked and confronted by attendants. It is still celebrated today.

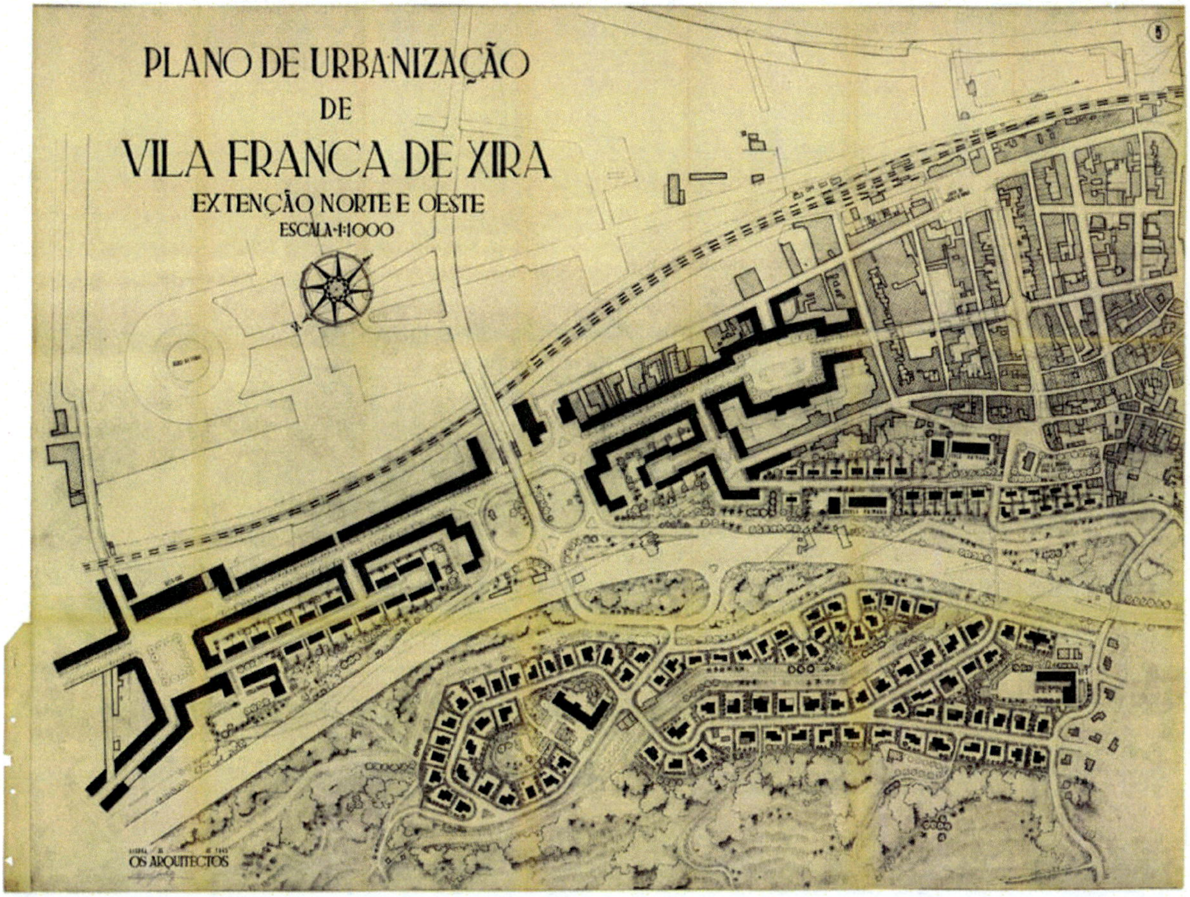

Fig. 1.9 The Vila Franca de Xira Urbanization Plan. Source: PT GFC-01-002/2 DES.00029, Guilherme Faria da Costa, 1946 - Arquivo DGPC/SIPA

Interestingly, there are no mentions in the technical report of the negligence in accounting for the rural character of the town and its economy. The proposed PGU is quite similar to other PGUs designed for areas where the agrarian aspect was far from relevant. Indeed, the practice of PGUs only seems to further impose a distinction between urban settlements and rural space, whereby the latter is the negative of the former, providing it with a physical and functional limit. The different technical reports highlighted the unrealistic relation between housing density and the defence of individual housing (CSOP 1947) and the key proposition of the PGU, which was the creation of a new area for public activities, which seems out of tune with both the ongoing development of the town and with the traditions and aspirations of the local population (CMVFX 1948). Despite the influence of the Garden City principles, which Faria da Costa shared with Gröer in their several collaborations, this did not help the planners to conceive a solution whereby the low-density housing strategy would be met with a more creative use of the agricultural surroundings of Vila Franca de Xira (Fig. 1.10).

The PGU was never revised, and its original proposal cannot be confirmed in the territory. The physical limitations to the growth of Vila Franca de Xira are still clear today, and while the settlement remains visibly compact, its modernization was ensured through the construction of buildings of considerable density. The area that received the detailed plan in the PGU is curiously occupied with a street system very similar to the one proposed by Jacobetty and Faria da Costa, although it does not correspond to any preexisting trail or road.[13] However, in this new neighbourhood, no equipment was built—while the PGU proses a hospital—and construction favoured high-rise construction instead of single-family homes. Collective slabs in this area vary between 5 and 9 floors, while towers reach 14 floors. In the road proposed in the PGU linking the new neighbourhood to

[13] This is testified by the plan of the existing settlement included in the PGU process. In this location, there is a single road—it is only in the PGU that such road is transformed to get the lasso shape it currently has.

Fig. 1.10 Extension area assigned in the PGU for low-density housing, Vila Franca de Xira

the rural area towards Povos and Castanheira do Ribatejo, another neighbourhood was built in 1976: the Pedra Furada Estate, designed by architect Margarida Vieira Neidercorn (Bandeirinha 2017). Built to house working-class slum-dwellers in the aftermath of the Revolution, this estate is built under the Local Service Support (*Serviço de Apoio Ambulatório Local*—SAAL), it is constituted by rows of two-story single-family homes, perhaps the only setting in this area that vaguely meets the proposal of the 1940s PGU.

1.4.3 The General Urbanization Foreplan of Palmela (1948–1954)

By the mid-twentieth century, Palmela was a relatively small-scale town on the northern slope of a mountaintop around a walled Castle (Fig. 1.11), construction of which dates to the Roman Empire. This Castle had ensured that Palmela was a strategic place for military and religious functions, particularly after the kingdom of D. João I (1385–1433), who ordered a considerable extension to the already existing building. Its urban structure, which remains very perceptible today, confirms the tendency identified by architectural historian Spiro Kostof (1936–1991), whereby a fortified palace gives rise to an unplanned precinct of artisan, craftsmen and military officials, as well administrative personnel, all attracted by the concentration of wealth presupposed by the Castle (Kostof 1991).

However, by the nineteenth century, Palmela had lost all geopolitical importance and the 1855 Administrative Reform deprives it of the statute of municipality, integrating it in Setúbal. In 1926, the military dictatorship agrees to reinstate it as an autonomous municipality. From the beginning, the municipality consisted of small-scale settlements punctuating an extensive rural territory, notable for its production of olive oil, apples, wine and livestock. Indeed, its productive intensity has stood the test of time, as will be seen in Chap. 6, since it remains today a key territory on the rural aspect of the Lisbon Region.

Fig. 1.11 View of the Palmela town and Castle. Source: Authors

Under Pacheco's Law, this vast rural territory required only one Urbanization Plan, although some secondary settlements, like Pinhal Novo, had recently been urbanized. Beyond the legal obligation, the Palmela town was indeed insufficient for its population and even more for the projected tenfold population increase, achieving 4728 inhabitants by 1970.[14]

Architect João António Aguiar (1906–1974) was hired to design the PGU. With a voluminous experience, which included collaborations with Luís Cristino da Silva (1896–1976), Ernesto Korrodi (1870–1944) and Agache (Lôbo 1995), it is little surprising that Aguiar was one of the most requested urban planners of the 1940s, pairing with Étienne de Gröer and Faria da Costa, and also that in 1947, he became the director of the newly constituted Office for Colonial Urbanization (*Gabinete de Urbanização Colonial*).

Aguiar would be, throughout his life, responsible for 22 plans in the Portuguese territory.[15] Among these was the General Urbanization Plan of Setúbal, whose first version had been presented in 1944, generating great public discontent and accusations of lack of knowledge about the city (Lôbo 1995). The extensive demolitions proposed by Aguiar displeased the locals and the municipality itself and led to several revisions in 1947, 1955, 1962 and 1972 (Pereira 2007). As such, Aguiar started to work more or less at the same time on the plans for Palmela and Setúbal. This situation with Setúbal is relevant, because it expresses a tendency for large interventions in premodern cores, common not only in Aguiar's work (including in Palmela) but also in the planning wave of the 1940s and 1950s in general (Lôbo 1995).

[14] This information, as well as others pertaining to this case study, are drawn from the typescript written documents available at the Palmela Municipal Archive, in two files with the title 'Ante-Plano Geral de Urbanização de Palmela'.

[15] Specifically for the following localities: Albergaria-a-Velha, Alto da Vila (Alenquer), Arraiolos, Belmonte, Cantanhede, Castelo Branco, Covilhã, Faro, Guarda, Monfortinho, Olhão, Palmela, Penhas da Saúde, Santarém, S. Miguel da Guarda, Setúbal, Viana do Castelo, Vila Real, Vila Viçosa, Vimieiro, Viseu and Vouzela—cf. Lôbo (1995).

For Palmela, Aguiar designed a PGU (which in the end was only approved as a Foreplan) where he proposed interventions in the old core and an urban extension northward. In total, he stipulated that the town should achieve a density of 125 inhabitants/ha. The pre-existing street system was generally maintained, with punctual widening of street dimensions to accommodate traffic. A civic centre, a town hall, a postal office, a cinema, a marketplace, a bus station and elementary schools were proposed as new equipment. Green parks and sightseeing spots were also projected. Commerce and industry were not expected to grow significantly but were proposed in continuity with pre-existing commercial areas. While basic sanitary infrastructure and public lightning were included, expropriation processes were avoided to ensure a relatively cheap implementation.

Despite the desired economic restraint, Aguiar suggested that both public and private investment should concur to the implementation process. Accordingly, he pointed out some priority public works, including the equipment core (concentrated in a square, the Praça de São João), the pavement works in the old core and the construction of the urban extension and links to the old centre. To ensure a clearly defined intraurban road, Aguiar proposed the demolition of existing hovels and of buildings of considerable value, justified by the creation of easy links between buildings of public interest (Aguiar 1948) (Fig. 1.12).

The land-use scheme was based on four types of zones: housing; commerce, services and housing; green areas; and reserved plots (Fig. 1.12). A first housing zone (R1) was the old core, acknowledged for its picturesque interest and for having an ambience of its own, even if its buildings were not remarkable. Several demolitions were predicted, for both sanitary reasons and for infrastructure facilitation. A second zone (R2) concerned the northern section of the urban expansion, with detached individual houses and row individual houses or rental buildings (*prédios de rendimento*). The plan imposed a height limitation, with one floor for individual houses and two floors or 70.5 m height for rental buildings.

Furthermore, two zones for commerce, services and housing (C1 and C2) were planned nearby public buildings. Commerce was to be placed at the ground floor of buildings, whereas services may occupy other floors or even the entirety of buildings. According to the regulation, buildings in all the town had a maximum height of two floors, although for non-residential buildings, three floors could be allowed.

With the exception of the old urban core, zones with residential use were subjected to a construction limitation of 20% of the total plot area, with a minimum of 40% of the area used for trees or vegetation, recommended for the mediation between street and building. There was an obvious sanitary motivation in this rule, to avoid the excessive compaction verified in the old core. However, the reservation of a considerable area for vegetation was most likely also motivated, as was common in Portugal residential planning at this time, by the belief that families needed a cultivation space, at least for some horticultural crops for self-consumption.

Distributed in these areas were ten blocks destined for public interest buildings, some indicated by Aguiar, others remaining unspecified. These blocks were in strategically prominent places, expected to stand out for good architectural design, increasing the value of the new urbanizations. The three green zones did not have a regulation, but rather specific indications for each. A first zone (V1) was a tree boulevard, the second zone (V2) the main town park, and the third zone (V3) was located on the northern slope of the Palmela Castle, the largest leisure area of the town and a protection buffer for the Castle.

A reservation area of 600 m was placed around the urban limit for ensuring panoramic views, although this was not represented on the plans. Allowed uses within this reservation included agriculture and supporting facilities, as well as the construction of detached houses whose construction should not surpass 1% of the total plot area.

Among other equipment, the construction of a slaughterhouse is recommended, although its location was not in the town, but suggested to be midway towards Quinta do Anjo, a town that would be urbanized in the late 1970s, with a plan advanced by the private sector. This way, the municipality would easily provide veterinary assistance for both locations. Quinta do Anjo was probably not an arbitrary reference, considering that since the early twentieth century it had an economy strongly focused on cheese production, wineries and bakeries, thus being a strong pole for rural economy (JFQA, s/d). Finally, a marketplace was indicated to include a covered as well as an open-air area, the latter supposed to accommodate the parking of vehicles and livestock. This very specific recommendation demonstrates that Aguiar had a notion that the commerce in Palmela would be of a strong rural character. Indeed, cartographical information from 1900 and 1940 (Marat-Mendes et al. 2015) demonstrates that, although there were some shifts and fragmentations in agricultural estates around the town, it continues to be surrounded by agriculture and agroforestry, with some tree areas even being replaced by productive species (Fig. 1.13).

There is no indication that Aguiar surveyed or accounted directly for the agrarian aspect of Palmela. Indeed, this dimension is awkwardly absent from his plan. However, the food system is directly present in the planning process. This can be highlighted on the survey mapping existing buildings of 'Commerce and industry' (Fig. 1.14). From this map, winehouses show such predominance that they were the object of a separate map. Although Aguiar made no

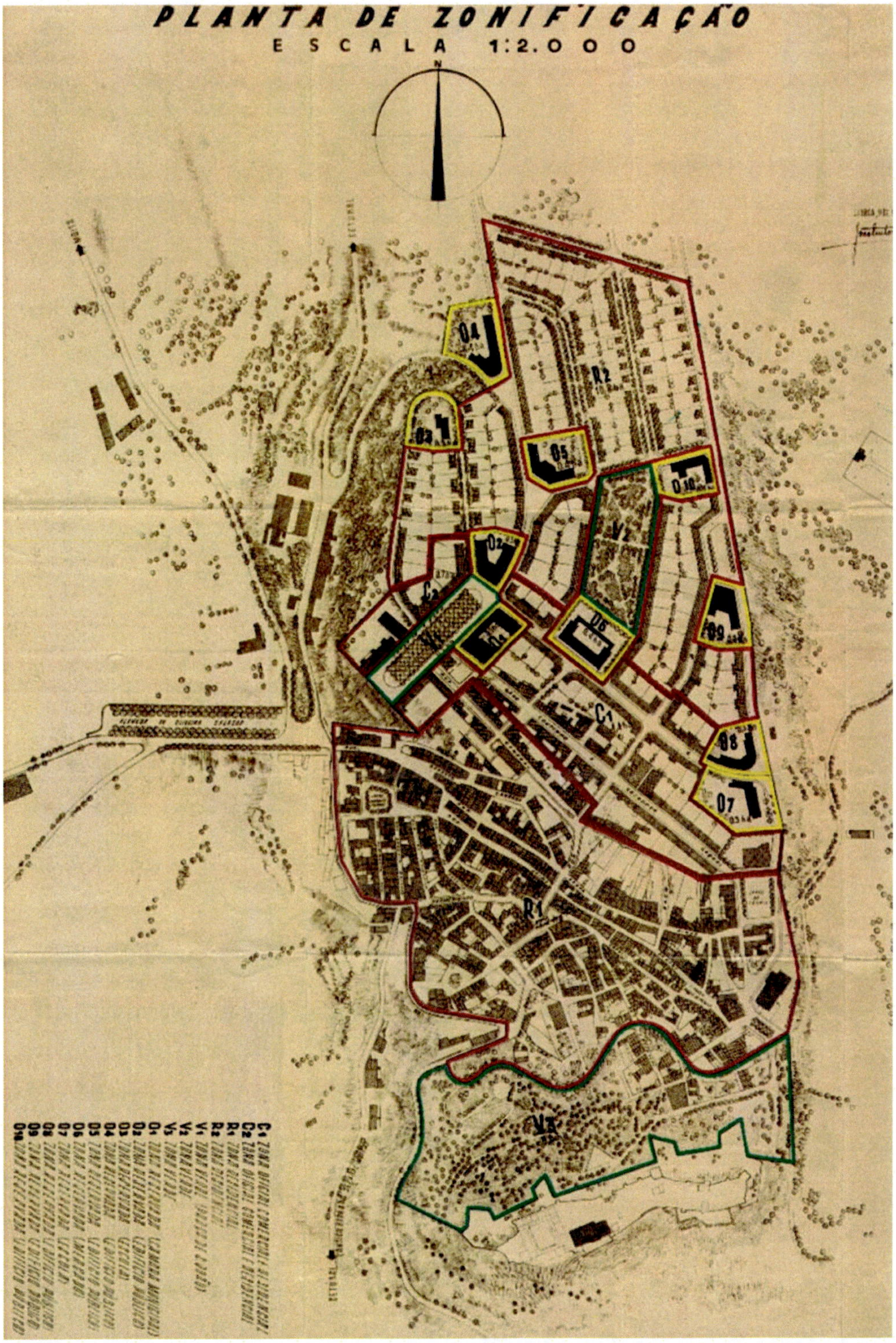

C1 and 2 - Official residential commercial zone
R1 and 2 - Residential zone
V1 to 3 – Green spaces zones
O 1 to 10 - Area reserved for public buildings

Fig. 1.12 The General Urbanization Foreplan of Palmela. Source: Arquivo Municipal de Palmela

1900

1940

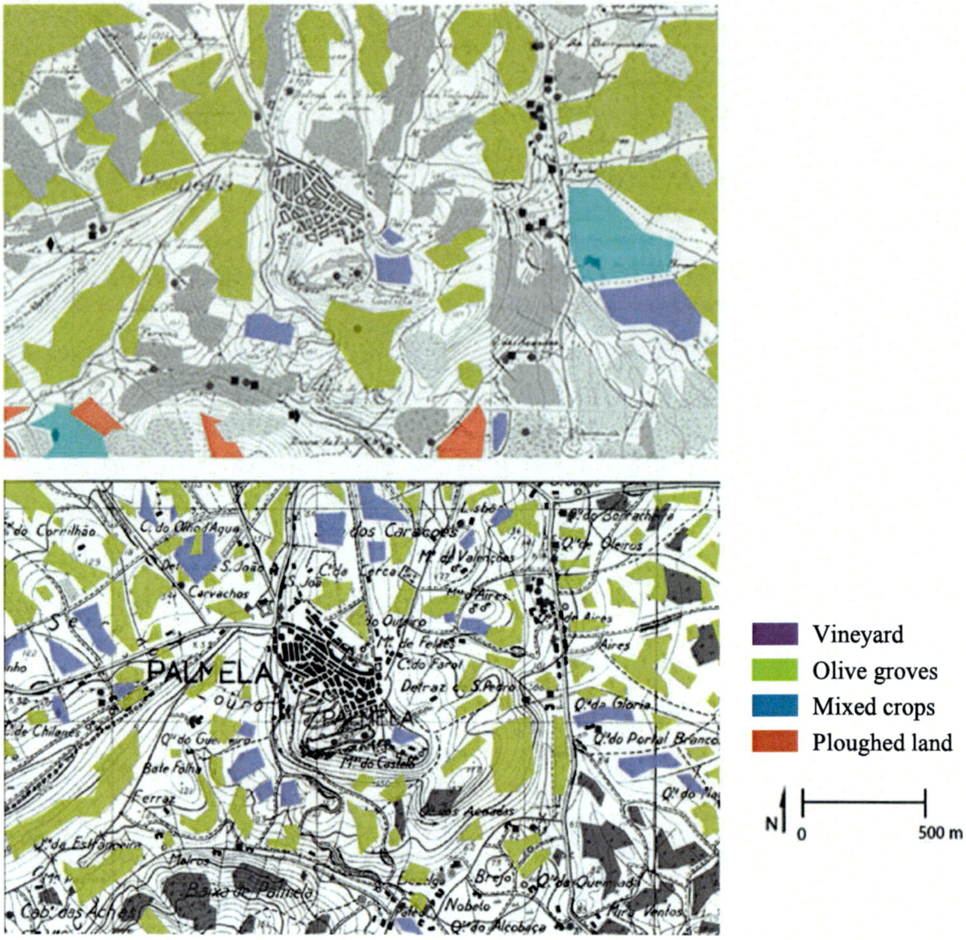

Fig. 1.13 Key cultivations in the vicinities of the Palmela town. Source: Authors (Adapted from Marat-Mendes, 2015)

stipulations particular to this form of commerce, it is noteworthy that he included this survey in the process, and it probably demonstrates what he had in mind when considering the market and the emergence of new commerce. Moreover, not only in the municipal territory, but specifically around the Palmela town, olive groves and vineyards were a very relevant land use. Thus, the suggestive inclusion of the winehouses survey may demonstrate that, by this time already, the commercial aspect of the food system was clearly identified with the urban realm, while the productive aspect, overridden in Aguiar's process, was excluded.

Beyond the winehouses, commercial activities on the map include taverns, grocery shops, bakeries, butchers, dairies, olive oil mills and restaurants, besides other non-food related activities, such as cappellists, carpenters, car workshops, construction materials storage, shoe stores, inns and drugstores.[16] It is thus clear that food-related commerce was a key component of the Palmela town, even prior to its modern planning.

In 1954, and after an evaluation by the CSOP and access to several population surveys conducted in 1950, Aguiar presented a 'Remodelling'[17] of the first plan. In this revision, it was mentioned that some of the indications of the original version had already been implemented, and most of the changes regarded the street system. A new land-use scheme was advanced, more detailed than that of 1948, introducing an industrial zone meant for the relocation of stables, barns and mills eliminated from the old urban core. A second

[16] The map further enlists the Philharmonic Society, a football field, public buildings, garages and fountains, which are not commercial activities.

[17] In the Arquivo Municipal de Palmela, only the written documents of this remodelling are deposited. As such, we had no access to the corresponding graphic documents. However, it is possible to have some clues to how some of the remodelling aspects would be, by comparison with the graphic documents of the 1948 version.

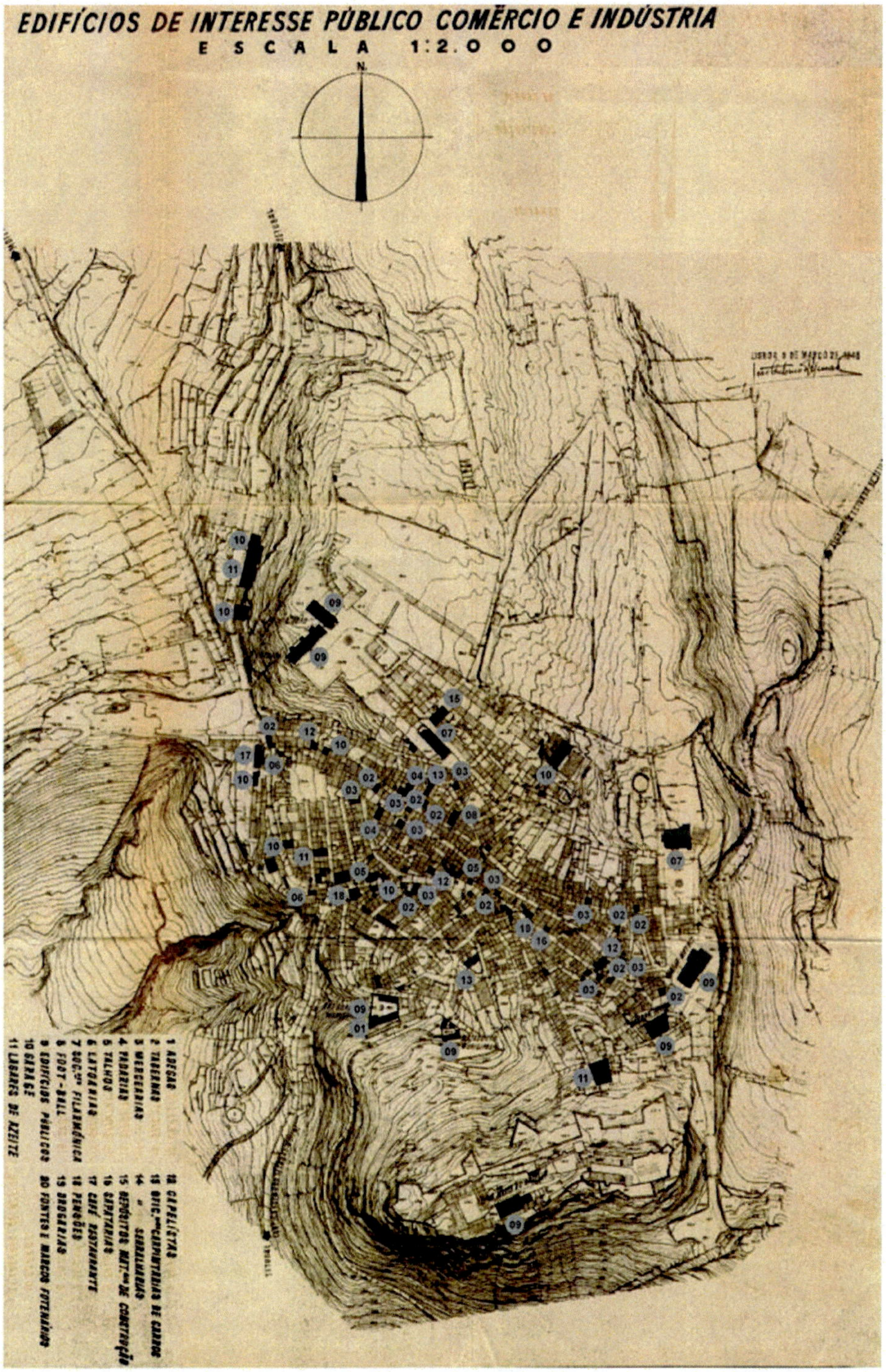

01 – Winery
02 – Taven
03 – Grocerystore
04 – Bakerystore
05 – Butchery
06 - Lotteries
07 - Philharmonic society
08 – Soccer
09 – Public buildings
10 - Parking garage
11 - Oil mills
12 - Chaplains
13 - Workshops
14 - Workshops
15 - storehouse of materials
16 - Choe store
17 – Coffees and Restaurants
18 - Hotels
19 – Stores
20 - Water sources

Fig. 1.14 The General Urbanization Foreplan of Palmela: buildings of public interest, commerce and industry. Source: Arquivo Municipal de Palmela

Fig. 1.15 São Braz, in the vicinity of the Palmela town. Source: Authors

industrial zone was located nearby the Largo do Chafariz, where several garages already existed.

The rural buffer was now called a 'Protection rural zone' and the content of this classification changed. Although agriculture and rural buildings were still allowed, this was extended to public buildings and industries precluded on the previously indicated industrial zone. Furthermore, the 600 m offset was extended to 1 km,[18] nearly double the original area. Significantly, this buffer was also meant for future urban expansion, a direct contradiction with the dispositions of the 1948 version.

In 1955, this new version of Aguiar's plan was approved, although not as a PGU, but instead as a Foreplan. Thus, it is little surprising that it was generally not implemented. Quite the contrary. The old core generally retained its original features, although the transition to modern areas was urbanized more or less accordingly to Aguiar's design. Yet the urban extension itself, while located where Aguiar had planned, did not follow his design closely.

Indeed, there is a vague similarity between the street system designed by Aguiar for the extension area and the one in existence today. This may result from the fact that the architect chose the most favourable site for urbanization and sought to adapt it to its geography. However, the envisioned model departed significantly from reality, and many areas were urbanized long after the Foreplan was approved. The low residential density planned by Aguiar proved to be in sharp contradiction with the goals of local social agents. Thus, a four-floor height was the rule in the areas more immediately close to the old core—in line with the stipulations of the Foreplan remodelling—but as one moves north, five-floor buildings start to appear, and at the northern end of the town extension, one finds four tower blocks, each with six floors, in the area planned for detached houses and two-floor rows (Fig. 1.15).

[18] A dimension which, by Aguiar's own standards might have been somewhat excessive, as it is the same he had proposed in 1945 in his plan for the much larger city of Faro, in Algarve. Also noteworthy is that the regulations for the rural buffer in the remodelling are the same as those in the Faro plan—cf. Lôbo (1995: 179).

The rural protection buffer could be confirmed in the immediate surroundings of the town, where only a few buildings appeared and where agricultural fields of different dimensions remained an important land-use, including allotment gardens. However, this buffer does not extend for a 1 km offset around the town edge. Within this distance, some low-density suburbs have developed, with typologies closer to those Aguiar probably had in mind in the 1940s. These include Volta da Pedra, Aires, Baixa de Palmela, Pegarias, Carvalhos and São Braz. The latter has developed north of the Windmills Route, photographed three times in the Foreplan regulations, although excluded from the planned area—making it likely that Aguiar saw them as an important landscape element. Currently, these windmills serve an aesthetic and symbolic role.

References

Abreu L (2017) O Serviço de Aviação do Corpo Expedicionário Português na Grande Guerra: Da Vontade à Realidade. MSc dissertation, Faculdade de Letras da Universidade de Lisboa, Lisboa

Agarez R (ed) (2020) A habitação apoiada em Portugal. Fundação Francisco Manuel dos Santos, Lisboa

Aguiar JA (1948) Anteplano geral de urbanização de Palmela. Arquivo Municipal de Palmela (typescript)

Albergaria H (ed) (2007) Vilas medievais planeadas de Portugal. IERU, Coimbra

Bandeirinha JA (2017) O Processo SAAL e a Arquitectura no 25 de Abril de 1974. Imprensa da Universidade de Coimbra, Coimbra

Bardet G (1947) L'Urbanisme. Presses Universitaires de France, Paris

Barreto A (2017) Anatomia de uma revolução – a Reforma Agrária em Portugal 1974–1976. D. Quixote, Lisbon

Branco JPF (2015) Políticas agrárias e florestais em Portugal: da regeneração à Política Agrícola Comum, uma abordagem Histórico-Geográfica. MSc dissertation. Universidade de Coimbra, Coimbra

Briz MGG (1989) A arquitectura de veraneio: os Estoris 1880–1930. PhD thesis, FCSH-UNL, Lisbon

Campos V, Ferrão J (2015) O Ordenamento do território: uma perspetiva genealógica, ICS Working Papers 1, 2015. Instituto de Ciências Sociais da Universidade de Lisboa, Lisboa. http://www.ics.ul.pt/flipping/wp2015_1/index.html

CMVFX - Câmara Municipal de Vila Franca de Xira (1948) Apreciação do Ante-plano de urbanização de Vila Franca de Xira. Arquivo Municipal de Vila Franca de Xira, Atas das reuniões da Câmara Municipal, Ata de 7 de setembro de 1948 (PT/MVFX-ARQ/AAL/CMVFX/B-A/003)

Costa JP (2005) Bairro de Alvalade: um paradigma no urbanismo português. Livros Horizonte, Lisboa

Costa SV (2012) O País a régua e esquadro. Urbanismo, Arquitectura e Memória na obra de Duarte Pacheco. IST Press, Lisbon

CSOP – Conselho Superior de Obras Públicas (1947) Parecer 1893. Arquivo Municipal de Vila Franca de Xira, Planos diretores e de urbanização, Parecer n.º 1893 do Conselho Superior de Obras Públicas referente ao anteplano de urbanização de Vila Franca de Xira (L/A.002)

CSOP – Conselho Superior de Obras Públicas (1955) Parecer dirigido ao Ministro das Obras Públicas. Arquivo Municipal de Loures, Processo 2116, pp 22–75

d'Almeida PB (2013) Bairro(s) do Restelo. PhD thesis, FCSH-UNL, Lisbon

DGE (Direcção-Geral de Estatística) (1913) Censo da população de Portugal em 1911. https://www.ine.pt/xportal/xmain%3Fxpid%3DINE%26xpgid%3Dine_publicacoes%26PUBLICACOESpub_boui%3D73210860%26PUBLICACOESmodo%3D2%26xlang%3Dpt+&cd=3&hl=pt-PT&ct=clnk&gl=pt

d'Almeida PB (2015) Bairro(s) do Restelo. Panorama Urbanístico e Arquitectónico. Caleidoscópio, Casal de Cambra

Feder G (1939) Die neue Stadt. Versuch der Begründung einer neuen Stadtplanungskunst aus der sozialen Struktur der Bevölkerung. Julius Springer, Berlin

Folgado D (2009) A nova ordem industrial. Da fábrica ao território de Lisboa: 1933–1968. PhD thesis, Universidade de Lisboa - Faculdade de Letras, Lisbon

Graça P (2020) Como comem os portugueses. Fundação Francisco Manuel dos Santos, Lisbon

Groër É (1945–1946) Introdução ao Urbanismo. Boletim da Direcção Geral dos Serviços de Urbanização 1:17–86

Guerreiro FC (2015) Colónias Agrícolas Portuguesas construídas pela Junta de Colonização Interna entre 1936 e 1960. A casa, o assentamento, o território. PhD thesis, FAUP, Porto

Howard E (1902) Garden cities of tomorrow. Swan Sonnenschein & Co, London

Jacobetty M, Faria da Costa JG (1946) Plano de urbanização de Vila Franca de Xira – Peças Escritas. Archive Forte de Sacavém, SIPA - PT GFC-01–002/2, Guilherme Faria da Costa

JFQA - Junta de Freguesia da Quinta do Anjo. s/d. História. https://www.freguesiadequintadoanjo.pt/resenha-historica. Accessed 12 May 2021

Kostof S (1991) The city sahaped. Urban patterns and meanings through history. Thames & Hudson, London

Lima P (2016) A Colónia Agrícola de Santo Isidro de Pegões (Montijo). Câmara Municipal do Montijo, Montijo

Lôbo MS (1995) Planos de urbanização – a época de Duarte Pacheco. FAUP Edições, Porto

Lôbo MS (2009) O Plano de Urbanização da Costa do Sol face ao paradigma da Cidade-Jardim. In: Pereira M (ed) O Plano de Urbanização da Costa do Sol - uma visão inovadora para o território. Câmara Municipal de Oeiras, Oeiras, 42–63.

Marat-Mendes T (2002) The sustainable urban form. A comparative study in Lisbon, Edinburgh and Barcelona. Ph.D. thesis, University of Nottingham, Nottingham UK

Marat-Mendes T (2009) O PUCS e os vazios planeados. Novas oportunidades para o ordenamento sustentado da Costa do Sol. In: Pereira M (ed) O Plano de Urbanização da Costa do Sol - Uma visão inovadora para o território. Câmara Municipal de Oeiras, Oeiras, pp 92–122

Marat-Mendes T, Oliveira V (2013) Urban planners in Portugal in the middle of the 20th century: Étienne de Gröer and Antão Almeida Garrett. Plan Perspect 28(1):91–111

Marat-Mendes T, Sampayo M (2015) The Plano de Urbanização da Cidade de Luanda by Étienne de Gröer and David Moreira da Silva (1941–1943). In: Silva CN (ed) Urban planning in Lusophone African Countries, Routledge, London, pp 57–77

Marat-Mendes T (cord), Mourão J, d'Almeida PB, Niza S (2015) Atlas da Água e da Agricultura. Região de Lisboa 1900–1940. Water and Agriculture Atlas: Lisbon Region in 1900–1940. Iscte Instituto Universitário de Lisboa/ DINAMIA'CET-IUL, Lisboa

Marat-Mendes T, Borges JC, Lopes SS (2020a) Villages and colonies: food and prayer in rural Lisbon. Paper presented at the 3rd International Seminar Architectures of the Soul. Iscte Instituto Universitário de Lisboa, Lisbon, March 2020

Marat-Mendes T, Borges JC, Lopes SS, Pereira MM (2020b) Where the fields have no name: urban-rural transitions in the Lisbon Region planning history. Cidades, Comunidades & Territórios 41:105–132

Marques B (2018) Faria da Costa e o(s) Plano(s) de Urbanização da Figueria da Foz. Caleidoscópio, Casal de Cambra
Mónica MF (2020) O olhar do outro – estrangeiros em Portugal do século XVIII ao século XX. Relógio d'Água, Lisboa
Mónica MF (2021) O meu país – notas sobre o nacionalismo. Relógio d'Água, Lisboa
Morais JS, Roseta F (2005) Os Planos da Avenida da Liberdade e seu prolongamento. Livros Horizonte, Lisboa
Niza S, Ferreira D, Mourão J, d'Almeida PB, Marat-Mendes T (2016) Lisbon's womb: an approach to the city metabolism in the turn to the twentieth century. Reg Environ Change 16(6):1725–1737
Oliveira V (2013) A evolução das formas urbanas de Lisboa e Porto nos séculos XIX e XX. Edições UP, Porto
Oliveira AM (2018) Colónias agrícolas da Junta de Colonização Interna no concelho de Montalegre – Modos de habitar a ruralidade. MSc dissertation, FAUP, Porto
Pereira M (ed) (2007) Setúbal – a cidade e o rio. eGeo, Setúbal
Pereira M (2009) O Plano de Urbanização da Costa do Sol – o pioneirismo de um plano sub-regional. In: Pereira M (ed) O Plano de Urbanização da Costa do Sol - Uma visão inovadora para o território. Câmara Municipal de Oeiras, Oeiras, pp 24–41
Pothukuchi K, Kaufman JL (2000) The food system. J Am Plann Assoc 66(2):113–124. https://doi.org/10.1080/01944360008976093
Rosas F (2001) O salazarismo e o homem novo: ensaio sobre o Estado Novo e a questão do totalitarismo. Análise Social 35(157):1031–1054
Silva CN (1994) Políticas urbanas em Lisboa 1926–1974. Livros Horizonte, Lisboa
Silva ME (2020) Estado, território, população: As ideias, as políticas e as técnicas de colonização interna no Estado Novo. PhD Thesis, Universidade De Lisboa, Iscte-Instituto Universitário de Lisboa, Universidade Católica Portuguesa, Universidade De Évora, Lisbon and Évora
Steel C (2008[2013]) Hungry city—how food shapes our lives. Vintage, London
Teixeira MC (2016) As formas da cidade portuguesa. In: Oliveira V, Marat-Mendes T, Pinho P (eds) O estudo da forma urbana em Portugal. Universidade do Porto, Porto, pp 25–54
Underwood DK (1991) Alfred Agache, French Sociology and Modern Urbanism in France and Brazil. J Soc Archit Hist 50(2):130–166
Varela R, Demier FA (2015) The development of Portuguese capitalism (XIX–XX): economics and Politics. Critique 43(3–4):421–437

2 Change of Paradigm: A Modern Programme

2.1 Introduction

Portugal kept politically neutral during WW2, and thus it did not suffer the destruction of its built heritage. However, it still faced pressures to modernize cities, as population increasingly abandoned the countryside seeking better living and working conditions around the Lisbon and Porto industrial belts (Pereira 1963). Yet with the recession caused in Europe by the effects of the war, Portugal was affected too. Despite the propagandistic promotion of rural life by the New State and the policies for contention of industry and of cities, by the mid-1950s, Portuguese agriculture is still far from thriving. The majority of agricultural land was cultivated by small peasants—50% for imperfect familiar holdings—i.e. insufficient to ensure self-sustenance, forcing the owners to seek additional labour—32% for perfect familiar holdings and 17.7% for wage–labour explorations (Branco 2015). The majority of the population was indeed rural, but it was also poor and its agriculture lagged.

In 1946, the LNEC is founded for investigations and experimental studies on civil engineering. In the 1960s, architecture and urbanism were introduced as research areas at LNEC, originating the Construction and Housing Division (CHD) within the Buildings and Bridges Department (d'Almeida and Marat-Mendes 2019; Marat-Mendes et al. 2021; d'Almeida et al. 2020).

Investigation interests at LNEC included the housing shortage, which lingered in Portuguese cities, resembling European post-WW2 conditions studied at other international research labs such as the 'Land Use and Built Form Studies' (Steadman 2016) and the 'Building Research Establishment' (Atkinson 1996) in the UK; the 'National Swedish Institute for Building Research' (Kemeny 1997), the 'National Board of Public Building', the 'Kungliga Tekniska Högskolan Royal Institute of Technology' and the Lund University, in Sweden (Rosenberg 2012); the 'Centre Scientifique et Technique du Bâtiment', 'Centre d'Études des Groupes Sociaux' and the 'Centre Nacional de la Recherche Scientifique' in France (Lengereau 2018) and, to an extent, the 'Centro de Estudos de Urbanismo e Habitação Engenheiro Duarte Pacheco' in Portugal (d'Almeida and Marat-Mendes 2019; d'Almeida et al. 2020; Marat-Mendes et al. 2021).

Indeed, there were several research topics and interests that were shared by such European research labs and LNEC. Knowledge exchange and official visitations between different labs were common, and in the case of LNEC, there are abundant reports noticing experiences shared at international congresses and meetings. Particularly important is the 8th International Architects' Union (Union Internationale des Architectes, UIA) Congress in Paris 1965, where architect Nuno Portas listed and exposed the topics under study in international centres, as well as those conducted at LNEC regarding housing and urban space (Portas 1965).

Partly energized by these advancements and international contacts at LNEC, from the late 1950s to the 1960s, a shift emerged in Portuguese planning practice and construction. No territory shows this more clearly than that of the Lisbon Region, in both officially planned urbanization and in operations carried in spite of the existing legal framework. With respect to urbanizations promoted by the State, there remained a strong design component, which allowed architects to direct research conclusions to the testing of specific urban form solutions, thus contributing to modernize architecture and the city itself (Grande 2012; d'Almeida 2015; Marat-Mendes and Cabrita 2015; Borges and Marat-Mendes 2019; Borges et al. 2022).

The 1960s represent a moment of particular tension for the ideological contradictions of the New State. The regime sought to integrate the single national market in the European economies, while at the same time being forced to acknowledge the incompatibility between industrial conditioning and the development of the national industrial fabric demanded by the industrial bourgeoisie (Barreto 2017; Louçã 2020). With the start of the Colonial War in 1965, there was a mass mobilization of male workers to the overseas battlegrounds, while a large amount of men escaped the country towards France, Switzerland, Belgium,

T. Marat-Mendes et al., *Atlas of the Food System*,
https://doi.org/10.1007/978-3-030-94833-7_2

Germany and Luxembourg, among others. By 1960, 43% of the population was still working in agriculture, but the soil surface really occupied with agriculture was in decline—the State support for growing wheat had extended to land unsuited for it, and new policies were created to promote forestry instead, as well as livestock breeding and technical mechanization (Branco 2015).

The policies for forestry promotion also catered to the relaxation of industrial conditioning, since in agrarian communities, especially in the mountains, the conversion of common land (*baldios*) into forestry implied the replacement of a source of communitarian wealth—chiefly constituted by grazing areas and of self-consumption cultivation—with forestry, whose wealth proved useful for the industrial bourgeoisie (Branco 2015). And in practice, such policies gave rise to opposition from local populations, but largely benefited industrial owners, particularly in the sectors of wood and cellulose production (Branco 2015).

Another effect of these forestation policies was the increased abandonment of small-scale agriculture and livestock, making dispossessed peasants more willing to become wage workers (Barreto 2017). Salazar's original policies meant to limit industrial expansion—which may have contributed to undercut the planning efforts developed by planners towards the modernization of industrial towns. But this approach relaxed in the 1960s. Indeed, although the dictatorship officially promoted the vision of Portugal as a rural country, by the 1950s, it was already clear that it also supported the interests of a number of industrial developers, with increasing investments made available for industry in all the Development Plans. With the exit of Salazar as a chief of State in 1968, Marcello Caetano (1906–1980), his replacement, was more willing to open the 'single market' of Portugal to Europe and to further promote the interests of industrial owners (Louçã 2020). Such a transformation could not have been achieved without the physical growth of cities, and although the two metropolitan regions often urbanized through illegal operations (Pinto and Guerra 2019), there was also, especially in Lisbon, a larger presence of the public sector as an agent of urbanization, even promoting planning strategies and practices sometimes followed by the private sector itself. Outside the Lisbon city, illegal construction and shanty-towns took a much larger role and, as will be discussed in Chaps. 4 and 6, this has a significant impact on rural settlements and on the extension of the Lisbon hinterland, increasingly opening the door to a less and less local-based food system.

A key policy that influenced the planning paradigm of the late 1950s and the 1960s was the approval of Law 42454 (1959),[1] which established rent categories for social housing, allowing the construction of large neighbourhoods in expropriated rural land on the eastern end of the Lisbon city. Three key urbanization plans were then developed: Olivais Norte (1959),[2] Olivais Sul (1960–62)[3] and Chelas (1960–1964).[4] To bring this Law to reality, the Lisbon Council forms the Technical Office for Housing (*Gabinete Técnico de Habitação*—GTH) which gathered a set of energetic urban planners and urbanists, some of them informed by the research being conducted at LNEC and also acutely aware of urban and architectural achievements in foreign countries. The Plan for Olivais Norte employed all of the principles of the Chatre d'Athénes (1933)—albeit applied to a single neighbourhood—and from Olivais Sul to Chelas, the key experiences of architecture and planning in the second postwar, from the Italian social housing programme INA-Casa to the experimental approach of Team 10, became important references, even in spite of the conservative regime (Borges and Marat-Mendes 2019, 2020; Borges et al. 2020).

As architects and planners, LNEC researchers left an important mark on this phase of development of the Lisbon Region. In the Lisbon city, Bartolomeu Costa Cabral (b.1929) and Nuno Portas designed housing complexes for Cell C of Olivais Sul; Francisco da Silva Dias coordinated the Urban Plan for Chelas and for the Vale Escuro—Alto da Eira Plan (1970); Gonçalo Byrne (b.1941) and António Reis Cabrita (b.1942), designed the housing estate in Chelas commonly known as 'Pink Panther' (1971–1979). Silva Dias also coordinated the plan for the Zambujal neighbourhood (1973–1974), then in Oeiras, now in Amadora. In Almada, Costa Cabral collaborated on the White Housing Estate (1976), and Maria da Luz Valente Pereira on the Urban Plan for Monte da Caparica (1972–1983), and Silva Dias collaborated with planner José Rafael Botelho (b.1923) in the infrastructural works for the inception of the Tagus Bridge (then Salazar Bridge, now 25th April Bridge). In Setúbal, Silva Dias co-authored a Masterplan designed in 1963 and Byrne designed the rehousing neighbourhood of Casal das Figueiras (1974–1978).

[1] Law 42454 published in Diário do Govêrno 188/1959, Série I, 18-08-1959.

[2] Urbanization Plan conducted by the GTH predecessor, the Urban Studies Office (*Gabinete de Estudos Urbanísticos*—GEU). Coordinators were architect Pedro Falcão e Cunha and engineer Sommer Ribeiro, landscape architecture by Álvaro Ponce Dentinho and housing buildings by Pedro Cid, Fernando Torres, João Braula Reis, João Matoso, António Pinto Freitas, Artur Pires Martins e Cândido Palma de Melo; elementary school by Victor Palla and Bento d'Almeida.

[3] Urbanization Plan coordinated by José Rafael Botelho, Carlos Duarte, Mário Bruxelas, Celestino de Castro and António Pinto Freitas, from the GTH.

[4] Urbanization Plan originally coordinated by José Rafael Botelho (1960–62) and later by Francisco Silva Dias (1962–64) with collaboration from architects João Reis Machado, Carlos Worm, Alfredo Silva Gomes and Luís Vassalo Rosa from the GTH.

In general, from the 1950s onwards, Portugal seemed to go through a process of paradigmatic shift with regards to the sensibilities and ideals of architecture and planning professionals. Since the death of Duarte Pacheco there had been a frank waning of planning policies and a generalized lack of efficiency of designed plans, but in the 1950s, without any large legal transformation, architects and planners change their attitudes towards the territory and towards the city. New ideas and new forms of habitat were prompted in the Lisbon Region, some catering to more orthodox modernist theories—such as Olivais Norte or the neighbourhood of Nova Oeiras—but others already *au-pair* with radical new ideas of the 1960s and 1970s—such as Olivais Sul, Chelas, the Integral Plans of Almada, Setúbal and Zambujal, the different projects for the Almada city (Baptista and Melâneo 2020a), or the continuation of the Restelo Urbanization in Lisbon (d'Almeida 2013, 2015) (Figs. 2.1 and 2.2).

This process testifies to the massive impact of housing provision in transforming planning policies and, through them, the territory. In Lisbon, the GTH promoted the urbanization of Olivais Sul and Chelas as integral urban areas, with a strong emphasis on housing (often with high-density solutions) but also with a minute planning of infrastructure, public space, equipment, commercial areas and gardens. The concentration of these different aspects into a single plan established a standard for urban planning that would be solidified by the Integral Plans, following a similar methodology (Vilaça and Ferreira 2018). The Housing Development Fund (*Fundo Fomento à Habitação*—FFH), which promoted the Integral Plans, as well as other social housing projects, exemplifies the generalization of the methodologies defined in the GTH.

This process was incited by updated planning methodologies and also by the increasing ability of architects and planners to base their work on references from other Western countries, most of which were not subjugated to dictatorships as Portugal was. Indeed, such references often have to be negotiated with the cultural and political limitations of the country, and, as will be seen in Chap. 4, the role of rural

Fig. 2.1 View of Cell B of Olivais Sul. Source: Authors

Fig. 2.2 View of Zone I of Chelas. Source: Authors

architecture was a fundamental key to unlock the ability to design, in this context, an architecture that was simultaneously international and Portuguese. However, it must be noted that New State officials often lacked the proper intellectual and critical instruments to correctly understand the implications of urban and architectural design, as was abundantly clear with the work of the GTH in the 1960s (Borges and Marat-Mendes 2020).

At the same time, most of the international references that became central for Portuguese architects and planners were based on the assumption—seldom challenged in the CIAM meetings and in their aftermath—that planning applied to urban, but not to rural, settings. There were exceptions, however, one such case being that of architect Octávio Lixa Filgueiras (1922–1996), who finished his Architectural Degree in Porto with a dissertation and a project on rural planning, provocatively titled 'Urbanism: a rural theme' (Filgueiras 1953). Soon afterwards, a Portuguese team presented a project for a rural housing estate at the 10th CIAM in Dubrovnik 1956 (Marat-Mendes and Borges 2017), while in Team 10 meetings, projects for suburban and even rural settlements are often presented (Van Den Heuvel and Risselada 2005).

Regardless of the effects of such projects in other contexts, in Portugal, the sense of a long-lasting delay in comparison with other European countries probably intensified the rush with which architects and planners seek to install a properly urban landscape in the growing cities of the second half of the century. At the same time, the abandonment of the Garden City paradigm that had guided Portuguese spatial planning in the 1940s and 1950s contributed to complete the abandonment of the role of rural land in urbanization. With the arrival of high-density construction and collective housing estates, open spaces started to be conceived mostly from the perspective of public leisure space, with the productive dimension abandoned. What is interesting is that these paradigmatic changes developed without a change in the general planning regulation, whose keys remain in the Law 24802 (the PGU Law) and Law 35-031 (the Foreplan Law). Only in 1971 was this framework fundamentally transformed with the publication of Law 560/71 and the Decree 561/71,[5] which redefined the hierarchy of Portuguese plans and their implementation.

These laws applied to the same locations as Pacheco's Law had, but clearly excluded rural land, which would only be considered in the immediate surroundings of urban cores, and only in Setting Plans (*Planos de Conjunto*) which were elaborated by the central State (Pereira 1997), while the municipalities were charged with creating Detailed Plans (*Planos de Pormenor*) applied to specific urban areas included in the Setting Plans. Indeed, this Law did not fundamentally change the planning practice that was already in existence—rather, its key role was the elimination of the Foreplan ambiguity, by reestablishing a clear set of planning scales and responsibilities (Pereira 1997).

According to the new legislation, the Setting Plan included urban and surrounding rural areas and was assigned to the central State, whereas Urbanization Plans were created by municipalities and approved by the State, and finally, the Detailed Plan frames in the plans at larger scales were prepared and approved by the municipalities (Pereira 1997).

In general, after the end of WW2, the policies of the New State increasingly disregarded agriculture, although it represented over 30% of the national GDP and 40% of the labour force (Branco 2015). The movement clearly indicates that cities were growing and that the countryside was being abandoned, increasing the pressure to create adequate planning instruments to preside over the spatial changes that announce themselves. This migration to the city was slow, disorganized and left behind a landscape of neglect, where unused land was invaded by pine trees and bush, with private owners ensuring only a minimal maintenance and the State not taking part in the problem (ISA 2006). The 1970s planning regulation resulted in ambiguous responses to the problem of illegal construction, which, in the two metropolitan areas, had given rise to a spree of land-use conversions. The growing urbanization applies also to the food system, as the 1960s are a decade of drastic change in the key dietary habits of the Portuguese, losing its rural and local basis (Graça 2020), as will be discussed in Chap. 6.

2.2 The Lisbon Region Masterplan (1959–1964)

The Lisbon Region Masterplan (*Plano Director da Região de Lisboa*—PDRL) was the first legally sustained planning attempt of the Lisbon Region as a whole. Its necessity and utility are often underlined by planners and technicians at the municipal and national scale, since plans at lower scales, either local, municipal or sub-regional, lacked coordination and an integral programme for the organization of economic activities around the obvious centre constituted by the capital city (CSOP 1947, 1955; CMVFX 1948).

In 1959, the authorization for creating the PDRL, as well as the specific delimitation of the Metropolitan Region, was published.[6] In 1960, works for the plan were already advancing. Its coordinator was Miguel Rezende, an engineer, and its design counted with the participation of engineers Mário de Azevedo, Marcolina Azevedo and Acácio Durão, as well as architects Borges de Campos,[7] Filipe Lopes, Geraldes Cardoso, Jaime Azevedo, Hipólito Raposo, Cecília Eloy and Varandas Monteiro. It was the first official planning instrument to conceive the region in its totality, and to attribute a basic land-use to all its components. In terms of development, it proposes five suburban cores around Lisbon, supposed to relax its population growth and organize construction activities, which were destroying rural soils with urbanizations that often would lack proper conditions.

The Foreplan phase of the PDRL was finished in 1964, and while it received praise, it remained without approval from the Central State. Initially, it did function as an unofficial orientation for municipal decisions, but lacking legal approval, it eventually fell into oblivion, its proposals blatantly overridden by the spree of both legal and illegal urbanization that stems from the 1960s boost of the construction sector (Gonçalves 1978; Pereira and Nunes da Silva 2008; Ferreira 2010) (Fig. 2.3).

The PDRL proposed a modern urban concentration, intensifying the Lisbon settlement within the city itself and through contiguous urban settlements, while promoting suburbanization through a restrained expansion of existing settlements, ignoring the already existing disperse areas of illegal construction (Pereira and Nunes da Silva 2008).

While the implementation of the infrastructure was made impossible the lack of State approval, the contrast between the planned urban extension and the existing one is due also to a lack of will to either frame or halt the existing trends.

An original feature of the PDRL was its classification of green fields through a complex system ranging from public parks and woods to the specification of soils with high agriculture potential. Generally, the rural soils of the Southern Bank of Tagus seemed to be more prone to agroforestry, while some similar spaces were to be found in Sintra and Mafra. However, the PDRL did not include the Montijo exclave, which remains today highly productive in

[5] Law 560/71 and Law 561/71, both published in Diário do Govêrno 294/1971, Série I, 17-12-1971.

[6] Decree 2099, published in Diário do Govêrno 186/1959, Série I, 14-8-1959.

[7] Fernando Lorenzini Borges Campos is the architect charged with revising the industrial zoning of the northern extension of Lisbon, as will be seen in Chap. 9.

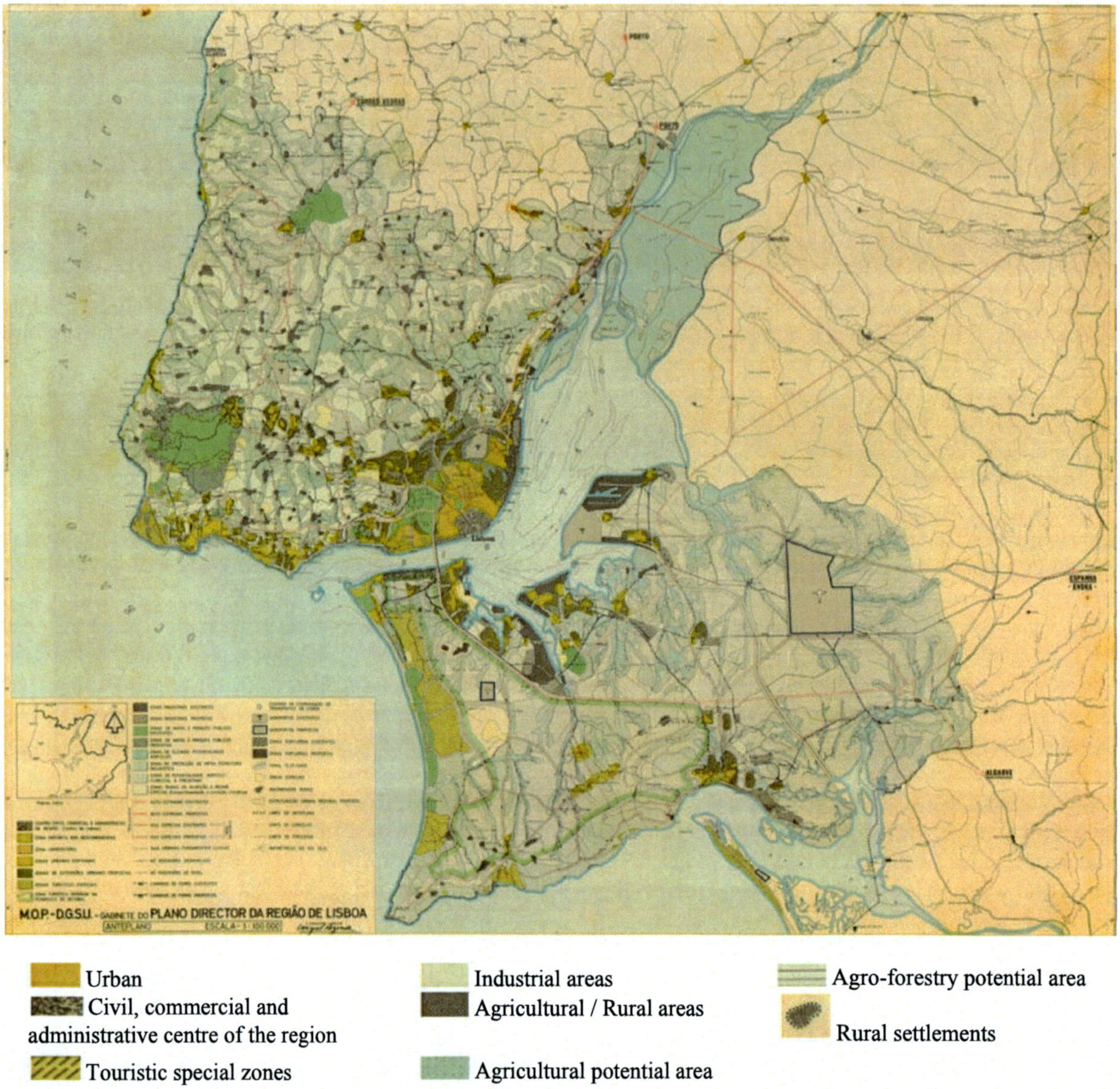

Fig. 2.3 The Lisbon Region Masterplan. Source: Direcção Geral do Território, License number nº 170/20

agroforestry—namely with olive groves, vineyards and cork production. In a sense, despite the important step of integrating rural soils in the same planning instrument as urban land, the primary goal of the PDRL was to regulate urbanizing activities, hence excluding a more complex planning strategy to important agricultural territories within the region, whose contribution to its economy was not dispensable (Fig. 2.4).

With respect to soils with high agricultural potential, these disseminated throughout most of the territory, denser in Sintra, Loures and Mafra, as well as Moita. However, the territory with the most agricultural potential was clearly that of Vila Franca de Xira, which, at the time, included the land of the Great Wetland of Tagus (the largest continuous agricultural terrain in the PDRL) and the three 'mouchões'.

In the 1980s, the integration of Portugal in the European Union created a set of conditions under which the implementation of the motorway infrastructure network predicted in the PDRL finally became possible. Community funding, an intensification of foreign capital and the increase of financial capacity of families through credit easement and privatization of banks would stimulate a rise of consumption in goods, including housing (Pereira and Nunes Silva 2008; Mourão and Marat-Mendes 2015).

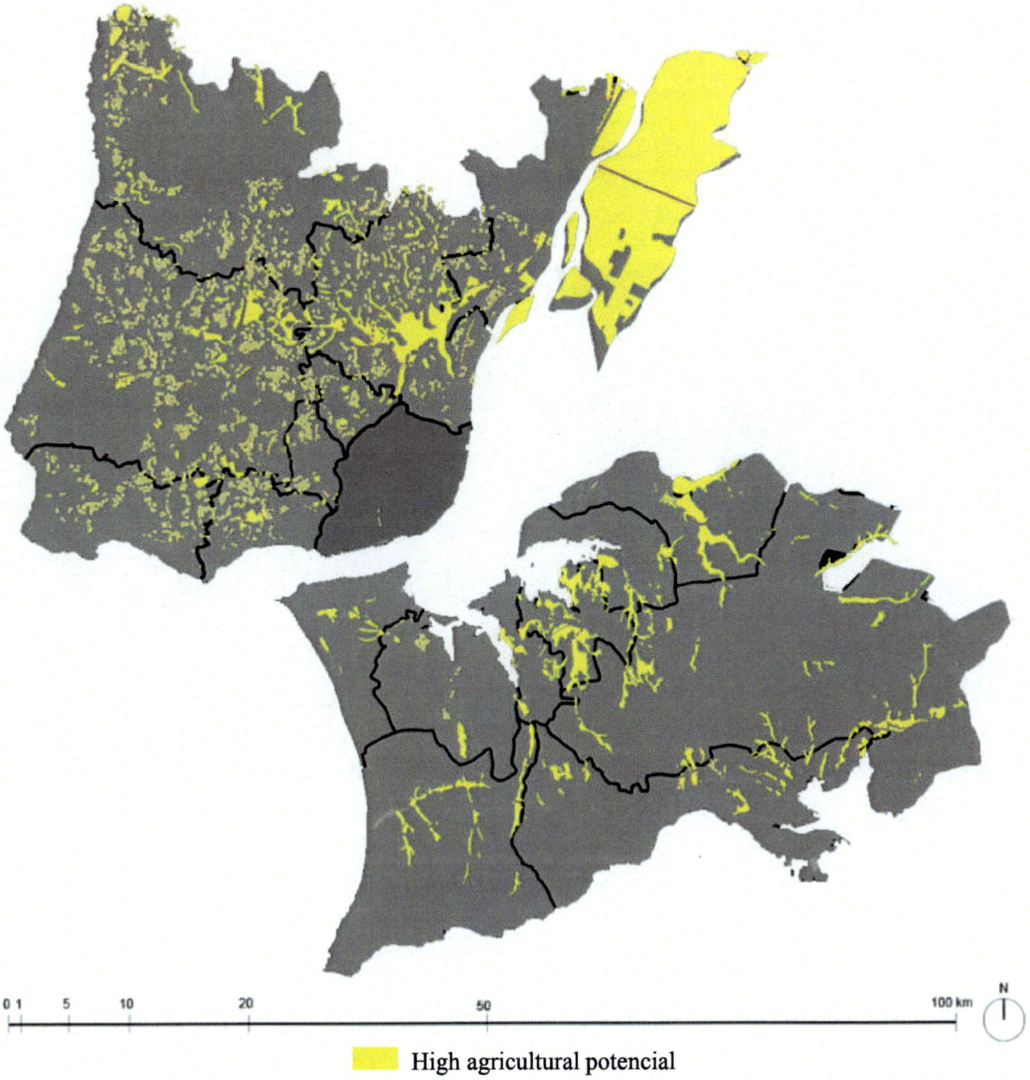

Fig. 2.4 Soils with high agricultural potential from the PDRL. Source: Authors (Adapted from MOP, 1964)

2.3 The Almada Urbanization Office (1955–1965)

Despite its location in the Southern Bank of Tagus—from which Lisbon remained physically disconnected until the construction of the first bridge in 1962—Almada was always a key territory for the suburbanization of the capital. Its key uses were, until the early twentieth century, military and rural, with several farmsteads and thorps, a pattern of land-use that was going to be revulsed by industrialization (Baptista and Melâneo 2020a). The recognition of the strategic importance of Almada happened in the early 1940s, and in 1946, when planners Étienne de Gröer and João Guilherme Faria da Costa are hired to design a plan for the municipality. Many important parts of this plan are now lost, but the ones that remain suffice to clearly understand the vision of this first plan (Galvão 2019; Baptista and Melâneo 2020a). It was meant for a group of towns, the 'Almada group', which is equated in Gröer's (1946) regulation with the sub-regional chords of Costa do Sol or Moscavide—Vila Franca de Xira (Fig. 2.5).

Gröer's (1946) report for the plan regulation constituted an important description of the conditions of the Almada territory prior to the great transformations that were to be operated in it in the mid-twentieth century. At that time, 41.7% of the Almada population worked in agriculture, with the remainder of the population working in industrial or otherwise urban jobs, many of which were really located in Lisbon (Gröer 1946). Most of the local industry was for ship

Fig. 2.5 Postcard from the late nineteenth century, showing the Pragal area. Source: Centro de Documentação do Centro de Arqueologia de Almada

construction, cork and brick production—only one of the main factories referred in Gröer's (1946) description was food-related, the mill in Caramujo. Agriculture was an important part of the territory: the predominant form was the mid-sized property, and thus there were no large-scale agricultural explorations. However, certain cultures stood out, namely wine, wheat, potatos and vegetables, running on traditional techniques (Gröer 1946). Livestock was another important rural activity, with breeding of sheep, bovine, horse, hinnies, goat and swine, the latter of which was brought from the Alentejo region for both consumption and breeding (Gröer 1946). There were further mentions to poultry, including the common Portuguese species, namely, partridge, rabbit, hare and quail, as well the breeding of chickens and roosters in rural homesteads (Gröer 1946). Wholesale trade of foodstuffs was referred as an important economic activity, including the storage and exportation of wine, olive oil, cereals and codfish, with refrigeration facilities on the industrial areas to intermediate the sales to retail commerce (Gröer 1946).

While there is no full chart of the land-use for the whole municipality, plans were designed for specific settlements—the Almada centre by Gröer and the towns of Costa da Caparica and Trafaria by Faria da Costa. The remainder of the territory was considered rural, including the central area of the municipal territory almost until its southern borderline, starting in Porto Brandão with its preserve factory facing the Tagus, and extending south to Casas Velhas, Arieiro, Monserrate, Palhais, Marco Cabaço, Vale de Cavalos, Charneca, Costas de Cão, Murfacém, Estrelinha, Vila Nova, Pilotos and Sobreda (Gröer 1946). This classification as rural land precluded construction and sought to maintain the agrarian aspect of the local economy functioning, and only a special regulation was attributed to the village settlements themselves, establishing a protective limit and regulating the allowed construction, following the regulation created for working-class housing areas where plots may vary from 300 to 400 m^2, a small dimension in comparison with other housing areas in the plan (Gröer 1946). Considering the remainder of Gröer's plans in Portugal, the one for Almada stands out for having more industry and less urban green spaces than usual—although this was justified, in principle, with the semi-rural character of many settlements and the large rural area (Marques 2020).

As there is no general plan where the totality of the land-use scheme is demonstrated, there can be little certainty on how such zoning applied, but the territorial model is clear enough: urbanization is placed at the eastern end of the

Fig. 2.6 Almada centre. Source: Authors

municipality, touristic towns are placed at the Western coastal area and villages are preserved (Figs. 2.6 and 2.7).

Gröer's plan was not approved, although some of its proposals can be confirmed in the territory (Baptista and Melâneo 2020a). However, not only was it necessary for Gröer to revise his plan in 1950, but the problems with speculative development continued to increase to the point where the original plan ceased to make sense. In 1955, the Almada Council acknowledged this problem and tried to respond to it by forming the Urbanization Office (*Gabinete de Urbanização da Câmara Municipal de Almada*—GU-CMA), whose chief-architect was José Rafael Botelho, appointed by Faria da Costa for that effect (Cavaco 2010). As mentioned earlier, Botelho was simultaneously one of the coordinators of the Urbanization Plan for Olivais Sul in Lisbon, one of the most important examples of the new methods of municipal neighbourhood planning, in this case inspired not only by the principles of modernism but also by Botelho's own in-loco studies of English New Towns of the second postwar (Cavaco 2010; Borges and Marat-Mendes 2020). Because of this influence, Botelho did not fully reject Howard's Garden City theory, and accepted satellite towns as a feasible solution for the problem of urban congestion in larger cities. However, in Botelho's planning work, this reference was matched with influences from CIAM modernism and from the planning experiences of Patrick Abercrombie (1879–1957) in London (Cavaco 2010).

The work of the GU-CMA covered all the municipal territory. For some years, Botelho was mostly focused on designing Partial Plans (*Planos Parciais*): this was due to the urgent need to stipulate rules for the extension of existing settlements, which was already taking place since the 1940s (Baptista and Melâneo 2020a). Although Botelho followed through with some of the suggestions of this prior plan, namely, the concentration of urban areas in the eastern end of the municipal territory, there were significant departures from both the aims and the preferential morphology of Gröer and Faria da Costa. For starters, Botelho challenged the assumption that Almada should be conceived as a suburb of Lisbon, and instead proposed a satellite city with a life of its

Fig. 2.7 Cova da Piedade, Almada. Source: Authors

own, defending a rescaling of services, commerce, equipment and public space (Marques 2020). Botelho was critical of the former regional organization, whereby all municipalities around the Tagus were functionally dependent upon Lisbon (Cavaco 2010).

The preference for low-density development with cottages and yards was also abandoned in favour of a blatantly modern approach, with higher densities, collective housing and large-scale urban parks and equipment (Baptista and Melânio 2020a). The backyards and interiors of blocks were eliminated, and all façades of the buildings became elements of public space (Marques 2020). In these Partial Plans, Botelho proposes a zoning scheme that matched the CIAM tradition, based on the distinctions between housing, labour, mobility and leisure (Cavaco 2010).

The work of the Partial Plans was finished in 1962, but meanwhile, Botelho launched other planning instruments with great importance for Almada. The first is a Masterplan contemplating nearly all the municipal territory. Predicting a city for 90 000 inhabitants, Botelho advances an urban super structure of public equipment to create jobs and use workforce locally (Cavaco 2010; Marques 2020).

To ensure the implementation of the plan, Botelho suggested several administrative and juridical initiatives. These included optimizing expropriation laws, easement of bank loans for municipal works and the redistribution of properties and obligations among owners of land necessary for plan implementation (Cavaco 2010). Moreover, Botelho suggested that Law 42454 becomes extended to the whole of the Lisbon Region to ensure the effective implementation of planned solutions (Cavaco 2010) (Fig. 2.8).

Simultaneously, Botelho developed in GU-CMA the Plan for the National Park of the Setúbal Peninsula (1959–1964), which included territories from Almada, Seixal, Barreiro, Sesimbra, Setúbal and Palmela, as well as the Tróia Peninsula, already a part of the Alentejo Region. Botelho understood this plan, which he designed with little supervision from the Almada Council, as an instrument to articulate the development of urbanization with landscape conservation of natural resources within a truly regional perspective (Cavaco

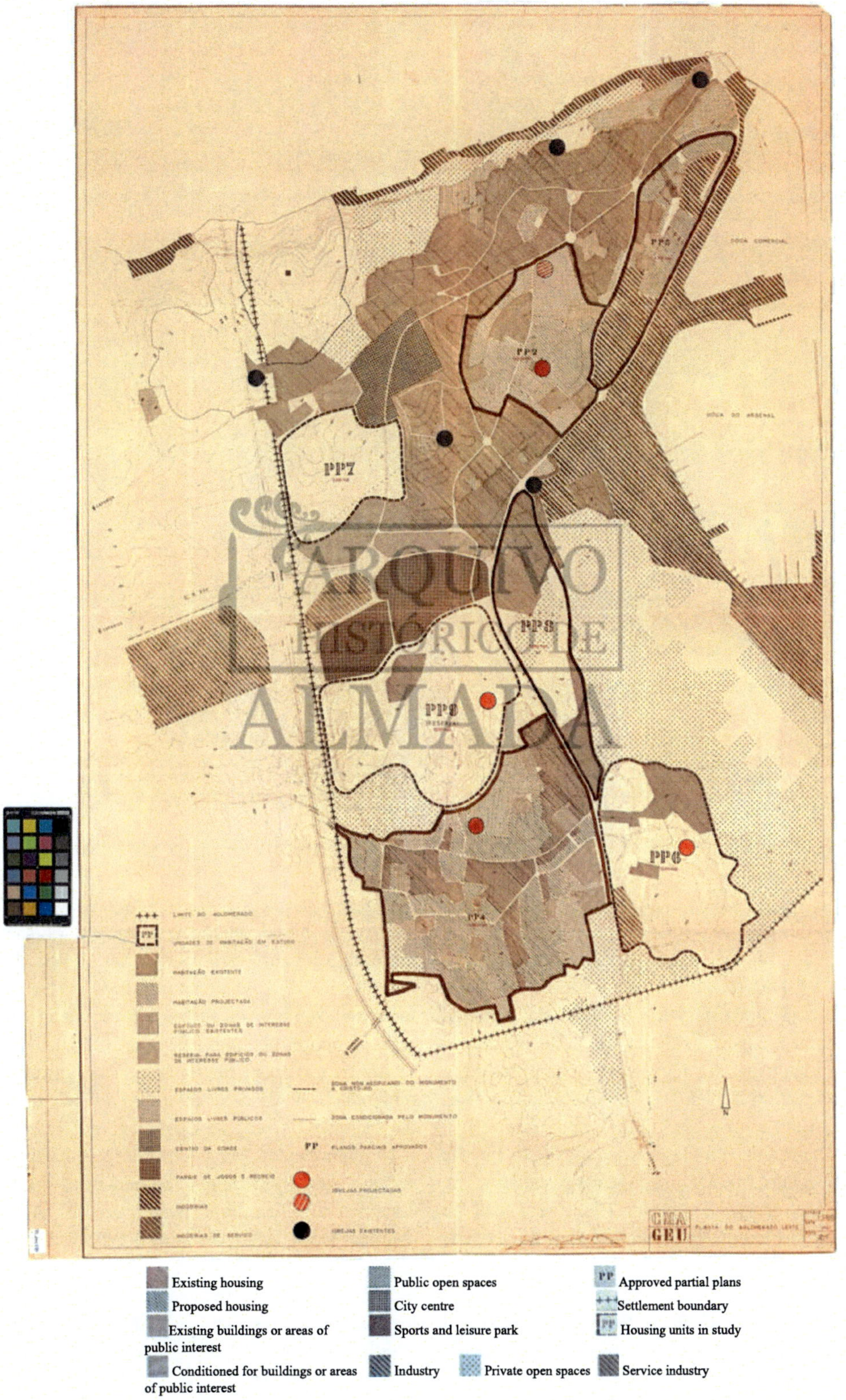

Fig. 2.8 Partial plans of the eastern agglomeration (1955–1975). Source: CM Almada

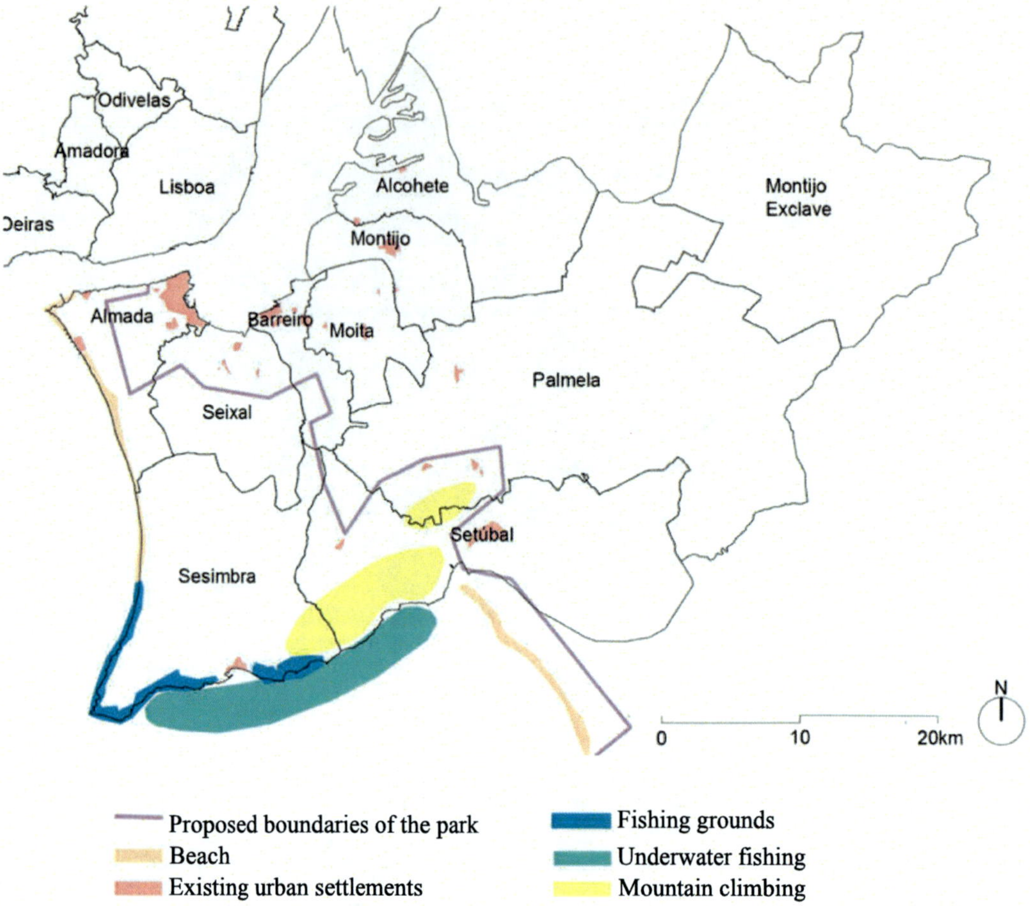

Fig. 2.9 Scheme of the Plan for the National Park of the Setúbal Peninsula. Source: Authors, adapted from Baptista and Melâneo (2020a)

2010). Although this plan paid little attention to agricultural land, it safeguarded all the coastal area of the Sesimbra municipality as a fishing area and of the maritime area from Praia da Baleeira (Sesimbra) to the Portinho da Arrábida (Setúbal) for submarine fishing. The coast is thus conceived as an area with importance for food access, complementing the prior planning efforts, which had focused on leisure uses on the coast of Almada. Considering the importance of fish in the dietary habits of the Region—and which can be confirmed in accounts of that time (Marat-Mendes et al. 2015) and in local recipes (Modesto 1982), and will be further discussed in Chap. 6, the importance of this coastal safeguarding for the regional food system can hardly be overestimated (Fig. 2.9).

From Gröer to Faria da Costa to Botelho, Almada had plans with great quality. Even if subjected to different strategies, all sought to take hold of the rapid urban development the city faced. Indeed, Almada was in this process of speculative transformation even before the first plan, as Gröer (1946) had already noted. Often, this urban growth was really based on the construction of shacks, and the unsanitary lugubrious houses Gröer (1946: 185) called 'human stables' and which were indeed a characteristic of the industrial suburbs of Lisbon, acknowledged even in Portuguese fiction literature (Borges 2019).

Speculative construction was a key motivation of planning efforts in Almada, and it was not halted by them. While the majority of Faria da Costa's plan for Costa da Caparica was implemented, as was Gröer's for the centre of the Almada city, from Botelho's plans nearly nothing has been constructed. Thus, the development of Almada was not determined by urban planning—instead, private developers kept constructing without the necessary authorizations from the Council and, after Botelho leaves the GU-CMA, many of his plans were revised precisely to legalize buildings that had been constructed illegally (Cavaco 2010).

Thus, the pessimistic accounts advanced by both Gröer and Botelho are confirmed: profit-seeking from private developers and the laxness of the Council ended up being the driving forces of the city's modernization.

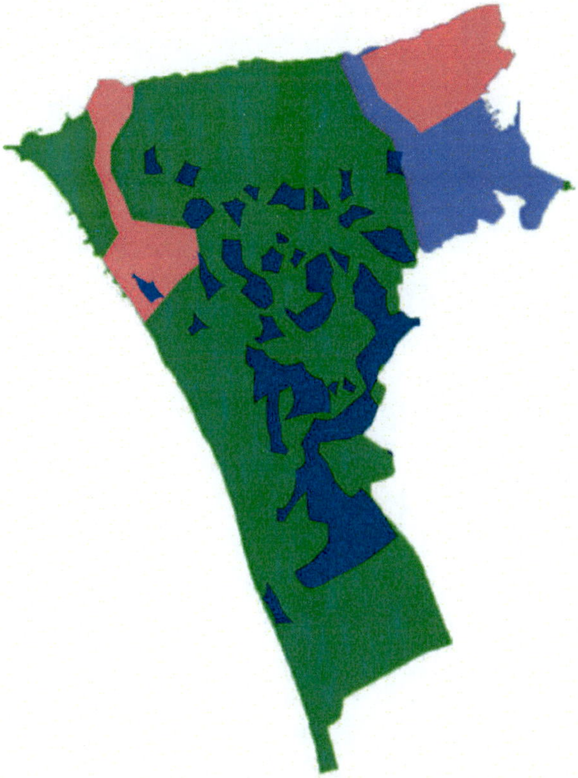

Rural zone and villages from the plan of Étienne de Gröer (1946)

Urban areas from the plan of Étienne de Gröer (1946)

Urban areas from the plan of José Rafael Botelho (1959)

Illegal housing areas developed in rural land

Fig. 2.10 Relation of illegal housing to planned urban and rural areas in Almada. Source: Adapted from Baptista and Melâneo (2020a) and Soares (1987)

The plans from the GU-CMA represent an important attempt of conceiving a modern urbanism and architecture in the context of metropolitan suburbanization—which makes it likely that, if Botelho's designs had been fulfilled, they would have become hallmarks of Portuguese history of architecture—as Olivais Sul undoubtedly has. Yet, the concern—however justified—over the spree of speculative urbanization may have prompted an increased lack of concern over the particularities of rural land, which, in the GU-CMA, is seldom mentioned. If Gröer had sought to preserve the farmsteads and agricultural explorations in the inland of the municipality, this extensive part of Almada became a sort of negative space, a limit of the urban, more than a space properly defined.

This assumption turned out to be a problem, since the lack of a proper land-use regulation from the Council made it easier for many of these rural estates to be converted into housing areas (Soares 1987), some of which were and remained illegal for decades (Fig. 2.10).

This means a significant decrease in food-production area and, arguably, a displacement of certain supply chains that had been in place until the early twentieth century, as described by Gröer (1946). In a press article of the 1960s about Almada, journalist Tomás Ribas (quoted in Domingues 2020: 52) refers with amazement to the reemergence of Almada as a legitimate satellite town—and among his many notes, there are curious mentions of this problem: "Lordly manors were brought down. The farms have disappeared. The men of real estate partitioned the land: they opened streets and squares, raised houses and neighbourhoods. […] Cafés, pastry shops and restaurants have spread their awnings on the shade of arcades [as well as] modern delicatessen and grocery shops". This description, however unintentionally, demonstrates a key understanding of twentieth century cities and the role that the food system was supposed to have in them. While commerce and consumption are an integral and emblematic aspect of urban life, other activities were reserved for an abstract outside, as if food appears on the plate out of magic (Steel 2008).

While many rural settlements and farmsteads are still in existence today in Almada, the rural zone predicted by Gröer is no longer there, although it is particularly interesting to note that most illegal urbanizations on that very territory were carried on during the democratic period. Moreover, it must be noted that the municipality finished its first democratic Director Plan in 1997, which has been in revision process for some years now.

The years of fluctuating planning local policies and of speculative development have left an unmistakable mark in the landscape of Almada. A pivotal surrounding area of Lisbon, and a key territory of the Southern Bank of Tagus—for three decades, it was its only direct link to the Lisbon city—it is as chaotic a territory as it is diverse. In that sense, it is a particularly telling example within the planning history of the Lisbon Region (perhaps the country):

> While Almada is one of the most planned municipal territories of the country, with works produced by a very significative set of architects and urbanists, reality does not seem to reflect this, manifesting the distance between planning and its implementation (Baptista and Melâneo 2020b: 9).

2.4 The Santo António dos Cavaleiros Urbanization (1964–1989)

Both within and around the Lisbon city, large-scale collective housing estates had a fundamental importance in the 1960s and 1970s. While the action of the State was mostly

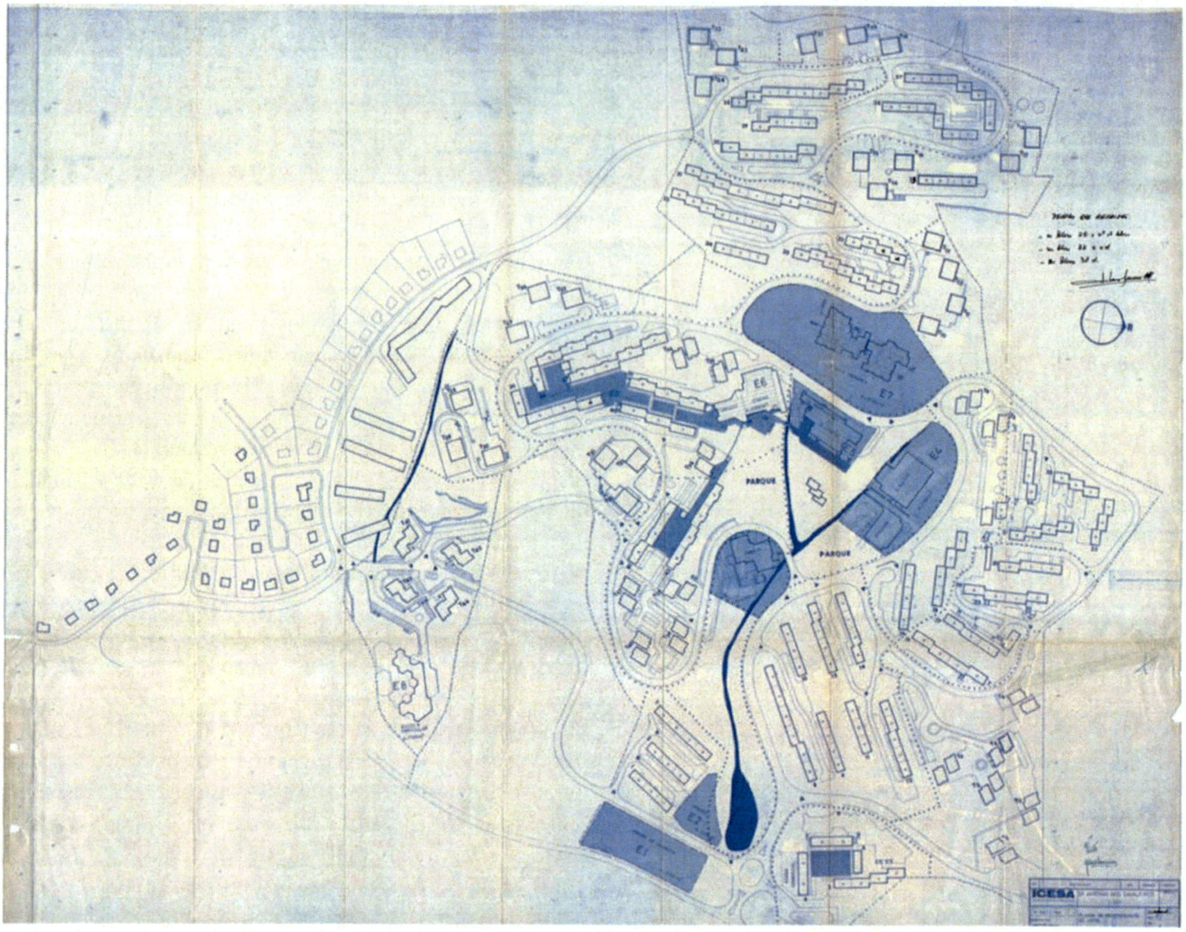

Fig. 2.11 The Santo António dos Cavaleiros Urbanization Plan. Source: SAC - Planta de referenciação de lotes - Processo 22 128 - Alvará de Loteamento nº 8/1976

confined within Lisbon, with the council urbanizations of the eastern end of the city, in its metropolitan region, private companies such as URBANCO, SIURBE, J. Pimenta, EMACO, Torralta or Grão-Pará stand out, with operations in municipalities such as Oeiras, Cascais, Sintra and Loures, with António Xavier de Lima (1926–2009) standing out in the Southern Bank of Tagus, although many of his developments were not fully legalized—as was the case with Quinta do Conde and Fernão Ferro in Sesimbra. Many of their operations sought rural land to conduct urbanization processes but often consisted of monofunctional urban areas limited to residential buildings and immediate commerce (Fig. 2.11).

ICESA (*Indústrias de Construção e Empreendimentos SARL*) is another of these private companies rising in the 1960s as a response to a severe housing shortage that afflicted the Lisbon Metropolitan Region. Its factory was located on Póvoa de Santa Iria, a town in the southern border of Vila Franca de Xira, and in it were developed techniques of prefabricated construction, namely through the use of precast concrete panels in FIORIO process (Fernandes 2009). Architect Alberto Reaes Pinto (b.1932) had a pivotal role in the ICESA developments: having studied mass-construction techniques at the Societé Fiorio (1964–67) at Linoux, France, Reaes Pinto introduced this technique to ICESA and soon started to put it to practice. In 1965, ICESA asked LNEC for a technical study on a series of housing projects to ascertain the level of satisfaction in terms of functional demands and the preconized dimensions (Portas and Gomes 1965), further exposing the rationalized and highly organized form of production promoted by ICESA.

The opportunity was provided by Santo António dos Cavaleiros, whose first planning phase ran from 1964 to 1966. Located 15 min away from the Loures town, and in a north–south-oriented slope facing the Loures Floodplain, this was to be a sort of new town of fully modern design. It rose over the thorp of Casal dos Cavaleiros, an agricultural and leisure location, with a few farmsteads and thorps, as well as the Monastery of the Holy Spirit, built in the sixteenth century. Once, from this slope, one could oversee the sailable waters of the Trancão River, a route for bringing agricultural products to the capital (AMLoures 2020). At the 1960s, it stood in a strategic place for ICESA, for it was relatively close to both their facilities in Póvoa de Santa Iria and to the Lisbon city (Fernandes 2009).

Santo António dos Cavaleiros was then created, according to a plan for 42 ha which, despite being developed in the private sector, shares many characteristics with the planning practices of the public sector in Lisbon, namely through the inclusion of a careful and diversified equipment scheme and a project for landscape and public spaces (Fernandes 2009; Borges and Marat-Mendes 2019; Borges et al. 2020). Significantly, part of the 3000 houses to be built for 12 000 people were promoted by the Portuguese Mutual Assurance Fund (*Fundo das Caixas de Previdência*) for its programme of Economic Houses (Ferreira 2010; AMLoures 2020; d'Almeida and Marat-Mendes 2021).

ICESA created three interlinked offices for Santo António dos Cavaleiros: an urbanization service, a department for architecture and pre-fabrication, and a department for calculations, measurements and budgets (Fernandes 2009; Ferreira 2010). Architects Reaes Pinto and Fernando Ressano Garcia (1927–2016) were joined by landscape architect Gonçalo Ribeiro Telles (1922–2020), topographer António de Oliveira Barroso and Engineer Pereira Gomes to design the plan (AMLoures 2020). Through these offices and multitude of professionals, there was a minute control of the design process, from the urbanization plan to the mass-produced pieces used for assembling the buildings (Pinto 2013).

In 1966, construction started. The street system was adapted to topographical level curves to ease the passage of cranes, while the width of streets was also determined by this factor (Fernandes 2009). Pedestrian pathways were created separate from the roads, to cut across the places with better landscape (Fernandes 2009). The territory was modelled at some points, through retaining concrete walls, whose formwork was also designed at ICESA. As construction moved forward, less rigid compositions were attempted.

However, a key change to the original project has to do precisely with the equipment scheme. A Civic Centre with commercial activities was planned to be the central area of the plan. This would include municipal and administrative services, a fireman station, a post office, a bus station, a church, a swimming pool complex, a gym, a cinema and theatre, a gallery, a library, kindergartens, schools, sports arenas and a bull race arena (Ferreira 2010). However, the Civic Centre was relocated in order to support the construction of the first phase of construction—which implied a reduction of its dimension (Fernandes 2009). By 1975, the plan was revised and most of the equipment planned earlier was eliminated (Figs. 2.12 and 2.13).

Buildings were constructed through modular pieces, and had great internal diversity, including flats with two, three and four bedrooms while maintaining an expressly regular external design. This was managed through the combination of different types of flat on each floor (two-bedroom + three-bedrooms or three-bedroom + four-bedroom). Overall, ICESA created five different standard buildings for Santo António dos Cavaleiros, all consisting either of 11 floor towers (with a non-residential ground floor) or five floor slabs.

Much about Santo António dos Cavaleiros failed to meet the expectation, although it was always far from a failure. Its community was greatly marked by the processes of decolonization after the 1974 Revolution and by the arrival to Lisbon of proletarianized peasants (AMLoures 2020). Moreover, most of the equipment ended up being constructed by the Loures Council in the 1990s, which means for years the town remained without proper services. Some of the plots originally reserved for public equipment remain vacant until today. In 1997, a couple of abandoned towers were imploded (Ferreira 2010).

Its architecture, strongly committed to modern industrialized techniques, resonates with the architectural aesthetics of late modernism and particularly of British Brutalism, in its use of prefabricated pieces as a construction technique and as a design strategy. Curiously, the buildings designed for economic housing were considered a successful development and would be replicated by ICESA, albeit at smaller scales, in two other neighbourhoods: Quinta do Morgado in Lisbon and Olival de Fora in Vialonga (Vila Franca de Xira).

Today, the town is surrounded by other urban neighbourhoods, such as Quinta da Caldeira, the Cidade Nova urbanization and the Bela Vista towers. However, Santo António dos Cavaleiros remains a somewhat central area for this agglomeration. For all its modern outlook, it presented itself as a 'garden city at the entrance of Lisbon' (AMLoures 2020), which is not an exaggeration of the landscape project by Ribeiro Telles. Indeed, while all buildings in the town give off a clear sense of high density and urbanity, there is a

Fig. 2.12 Main entrance to Santo António dos Cavaleiros, Loures. Source: Authors

balance between them and the lush gardens spreading over the slope. Moreover, the cultivated fields on the lower end retain something of its former pastoral landscape.

However, this was not contemplated directly in the plan. Although the Loures Floodplain is a highly productive territory and one of the most dazzling landscapes of the Lisbon Region, it has no relation with Santo António dos Cavaleiros, and it was never considered a territory that could have any functional link to the facing urbanization. Rural land, even when it was in use, as is the case here, became increasingly estranged from the urban realm and from planning, which may control all elements from the plan "to the screw" (Fernandes 2009: 117) but does not ask how people are to be fed in a new urbanization.

Overlooking Santo António dos Cavaleiros is the Southern Coastline (*Costeira*) of Odivelas, a steep hill, which from the 1960s onwards, became a hotspot for illegal construction. Between such illegal neighbourhoods are several olive groves: together with the vegetable gardens and small farms of the Trancão Floodplain, at the bottom of Santo António dos Cavaleiros, this is a landscape that displays the disordered way in which urbanization took over the capital's hinterland (a process which will be discussed on Chap. 4). But it also bears witness to the way in which planning, even when effective from the perspective of urban space, failed to create productive and imaginative relationships with rurality.

Fig. 2.13 Olive trees at the gardens of Santo António dos Cavaleiros town, Loures. Source: Authors

References

AMLoures (Arquivo Municipal de Loures) (2020) Em Loures o passado tem futuro: Santo António dos Cavaleiros. Arquivo Municipal de Loures, Loures

Atkinson G (1996) Thoughts during the building research establishment's 75th anniversary. Constr Hist 12:101–107

Baptista LS, Melâneo P (2020a) Cidade consolidada. In: Baptista LS, Melâneo P (eds) Almada: um território em seis ecologias. Câmara Municipal de Almada, Almada, pp 11–16

Baptista LS, Melâneo P (2020b) Almada: um território em seis ecologias. In: Baptista LS, Melâneo P (eds) Almada: um território em seis ecologias. Câmara Municipal de Almada, Almada, pp 7–9

Barreto A (2017) Anatomia de uma revolução – a Reforma Agrária em Portugal 1974–1976. D. Quixote, Lisbon

Borges JC (2019) Babilónia superlotada – representações de Lisboa na obra de Luísa Dacosta. Revista do CESP 39(62):233–254

Borges JC, Marat-Mendes T (2019) When Lisbon met the team 10 cluster city. Joelho 10:86–99. https://doi.org/10.14195/1647-8681_10_5

Borges JC, Marat-Mendes T (2020) Conversa com Francisco Silva Dias sobre o Plano de Urbanização de Chelas. Cidades Comunidades Territórios 41:1-23

Borges JC, Marat-Mendes T, Lopes SS (2020) Chelas Zone J revisited: urban morphology and change in a recovering neighbourhood. In: Strappa G, Carolitti P, Leva M (eds) Urban substrata and city regeneration. Proceedings of the 5th ISUFitaly international conference, Rome, 19–22 Feb 2020, pp 751–760

Borges JC, Lopes SS, Fernandes RDP, Marat-Mendes T (2022) Planning at the edge: urbanism and socio-political transition in Chelas, Lisbon. Planning Perspectives https://doi.org/10.1080/02665433.2021.2001364

Branco JPF (2015) Políticas agrárias e florestais em Portugal: da regeneração à Política Agrícola Comum, uma abordagem Histórico-Geográfica. MSc Dissertation. Universidade de Coimbra, Coimbra

Cavaco CSRG (2010) Formas de habitat suburbano. Tipologias e modelos na área metropolitana de Lisboa. PhD Thesis. FAUL, Lisbon

CMVFX - Câmara Municipal de Vila Franca de Xira (1948) Apreciação do Ante-plano de urbanização de Vila Franca de Xira. Arquivo Municipal de Vila Franca de Xira, Atas das reuniões da Câmara Municipal, Ata de 7 de setembro de 1948 (PT/MVFX-ARQ/AAL/CMVFX/B-A/003)

CSOP – Conselho Superior de Obras Públicas (1947) Parecer 1893. Arquivo Municipal de Vila Franca de Xira, Planos diretores e de urbanização, Parecer n.º 1893 do Conselho Superior de Obras

Públicas referente ao anteplano de urbanização de Vila Franca de Xira (L/A.002)

CSOP – Conselho Superior de Obras Públicas (1955) Parecer dirigido ao Ministro das Obras Públicas. Arquivo Municipal de Loures, Processo 2116, pp 75–22

d'Almeida PB (2013) Bairro(s) do Restelo. PhD Thesis, FCSH-UNL, Lisbon

d'Almeida PB (2015) Bairro(s) do Restelo. Panorama Urbanístico e Arquitectónico. Caleidoscópio, Casal de Cambra

d'Almeida P, Marat-Mendes T (2019) Nuno Portas and the research on urban morphology at the Portuguese National Laboratory for Civil Engineering: retrieving a possible glossary of urban form. In: Carlotti P, Ficarelli L, Ieva M (eds) Cities in the making and future urban form, 4th ISUFItaly conference READING BUILT SPACES. U+D Editions, Italy, pp 213–222. https://www.urbanform.it/books/

d'Almeida PB, Marat-Mendes T (2021) A atualidade da investigação científica em arquitetura e urbanismo desenvolvida no LNEC (1961–1979) face ao desafio da sustentabilidade. Cidades Comunidades Territórios 15–40

d'Almeida P, Marat-Mendes T, Toussaint M (2020) Portugal's rising research in architecture and urbanism: the influence of international research centers and authors. J Urban Hist 1–28. https://doi.org/10.1177/0096144220968078

Domingues A (2020) Desapareceram as quintas. In: Baptista LS, Melâneo P (eds) Almada: um território em seis ecologias. Câmara Municipal de Almada, Almada, pp 47–51

Fernandes APS (2009) Habitação (colectiva) modular pré-fabricada; considerações, origens e desenvolvimento. MSc disserattion. FAUTL, Lisbon

Ferreira, BM (2010) (In)Formar a cidade contemporânea. MSc dissertation, Iscte Instituto Universitário de Lisboa, Lisbon

Filgueiras OL (1953) Urbanismo: tema rural. Concurso para obtenção do diploma de arquitecto, FAUP, Porto. https://repositorio-tematico.up.pt/handle/10405/48394

Galvão MT (2019) Étienne De Gröer e o Plano de Almada. Unpublished MSc Thesis, Universidade de Coimbra. https://estudogeral.sib.uc.pt/handle/10316/90208

Gonçalves F (1978) A mitologia da habitação social – o caso português. Cidade Campo 1:21–83

Graça P (2020) Como comem os portugueses. Fundação Francisco Manuel dos Santos, Lisbon

Grande N (2012) O Ser Urbano nos Caminhos de Nuno Portas/The urban being: on the trails of Nuno Portas. Imprensa Nacional Casa da Moeda, Guimarães

Groër É (1946 [2006]) Urbanização do concelho de Almada (1946): Análise e programa – relatório. Anais de Almada 7–8:151–236

ISA – Instituto Superior de Agronomia (2006) Proposta Técnica para o plano Nacional de Defesa da Floresta contra incêndios – Anexo 1. http://www.isa.utl.pt/pndfci/A1_Perspectiva_Historica.pdf

Kemeny J (1997) The abolition of the National Swedish Institute for Building Research: social change & political compromise. Scand Hous Plan Res 14(3):133–147. https://doi.org/10.1080/02815739708730430.

Lengereau E (2018) Aux Origines de la Recherche Architecturale en France. In: Cohen J (ed) L'Architecture et Connaissance Scientifique. Éditions du Patrimoine, Centre des Monuments Nationaux, Paris, pp 20–29

Louçã F (2020) A jaula oligárquica: a modernização conservadora ao longo do século XX. In: Rosas F, Louçã F, Lopes JT, Peniche A, Trindade L, Cardina M (eds) O século XX Português. Tinta da China, Lisbon, pp 117–204

Marat-Mendes T, Borges JC (2017) A integração da sustentabilidade no ensino da arquitetura: a sua contribuição na consolidação do papel social do arquitecto. Paper presented at the international conference 'Projetar a cidade com a comunidade: reflexões sobre processos participados', FAUTL, Lisbon, 8–9 June 2017

Marat-Mendes T, Cabrita MA (2015) Morfologia Urbana e Arquitectura em Portugal. Notas sobre uma Abordagem Tipo-Morfológica. In: Oliveira V, Marat-Mendes T, Pinho P (eds) O Estudo da Morfologia Urbana em Portugal. FEUP, Porto, pp 65–94

Marat-Mendes T (coord), Mourão J, d'Almeida PB, Niza S (2015) Water and agriculture atlas: Lisbon region in 1900–1940. Atlas da Água e da Agricultura. Região de Lisboa em 1900–1940. Iscte - Instituto Universitário de Lisboa, DINÂMIA'CET-IUL, Lisboa

Marat-Mendes T, d'Almeida PB, Borges JC (2021) Concepts and definitions for a sustainable planning transition: lessons from moments of change. Eur Plan Stud. https://doi.org/10.1080/09654313.2021.1894095

Marques E (2020) O planeamento urbano em Almada. In: Baptista LS, Melâneo P (eds) Almada: um território em seis ecologias. Câmara Municipal de Almada, Almada, pp 227–234

Modesto ML (1982 [1990]) Cozinha tradicional Portuguesa. Verbo, Lisbon

Mourão J, Marat-Mendes T (2015) Urban Planning and Territorial Management in Portugal. Antecedents and Impacts of the 2008 Financial and Economic Crisis. In: by Knieling J, Othengraffen F (eds) Cities in Crisis – Socio-Spatial Impacts of the Economic Crisis in Southern European Countries. Routledge, London, p 157–171. https://doi.org/10.4324/9781315725048

Pereira R (1963) Problemática da Habitação em Portugal. Análise Soc 1:33–66

Pereira M, Nunes da Silva F (2008) Modelos de ordenamento em confronto na área metropolitana de Lisboa: cidade alargada ou recentragem metropolitana? Cadernos Metrópole 20(2):107–123

Pereira MIBR (1997) P.M.O.T. – A regulação quantitativa e a qualidade do ambiente urbano. MSc dissertation, Universidade Técnica de Lisboa, Lisbon

Pinto AR (2013) Bairro de Santo António dos Cavaleiros. Passagens – Paisagens Distantes a CRIL uma avenida Pós-moderna. 1:204–209

Pinto TC, Guerra I (2019) Housing policies, market and home ownership in Portugal: beyond the cultural model. Cidades Comunidades Territórios 39:101–114

Portas N (1965) Estudos Sobre Habitação: Relato Sucinto dos Contactos Estabelecidos por Ocasião do Congresso UIA. LNEC, Lisboa

Portas N, Gomes N (1965) Estudo analítico de projectos de habitação: Projectos "ICESA". LNEC, Lisboa

Rosenberg F (2012) Science for architecture. Designing architectural research in post-war Sweden. Footprint. Delft Arch Theory J 10/11:99–112. https://doi.org/10.7480/footprint.6.1-2.752

Soares LB (1987) Transformação informal do território. Situação na Área Metropolitana de Lisboa. A Cidade em Portugal, Povos e Culturas 2/3:345–351

Steadman P (2016) Research in architecture and urban studies at Cambridge in the 1960s and 1970s: what really happened. J Archit 21(2):291–306. https://doi.org/10.1080/13602365.2016.1165911

Steel C (2008[2013]) Hungry city—how food shapes our lives. Vintage, London

Van Den Heuvel D, Risselada M (eds) (2005) Team 10—in search of a utopia of the present. nai010 publishers, Amsterdam

Vilaça E, Ferreira T (2018) Os anos de crescimento (1969–2002). In: Agarez R (ed) Habitação – cem anos de políticas públicas. IHRU, Lisbon, pp 317–364

3 The Democratic Planning Policy

3.1 Introduction

On the dawn of April 25, 1974, through a secret operation, the Portuguese Armed Forces barricaded the head of the New State Marcello Caetano at the Carmo Headquarters in Lisbon and forced him to surrender his position. The people were told to stay at home, but they ran in mass to the streets, although bloodshed was effectively avoided. In the barrels of military guns, bullets were replaced by carnations—hence the epithet of Carnation Revolution. After 41 years, the dictatorship was overthrown. While April 25 represents the fall of the old regime, it stands between a set of revolutionary struggles in the former African colonies, which originally undercut the power of the imperialist State, and a continental revolutionary period that would last until November 25, 1975. In 1976, a Constitution of the Portuguese Republic was approved,[1] thus finalizing definitively the process. In practice, Portugal went through a lively revolutionary period, where workers took over workplaces (including the rural fields), but in over a year the regime stabilized without changing key production relations, and reinstating the figure of the State, albeit in the form of a democracy (Varela, 2019).

While the Revolution was, to an extent, defeated, the country would not be the same afterwards. Political and administrative power was redistributed, and this naturally affected the control over the territory and its management instruments. While during the dictatorship, municipal power tended to take a more passive role, writing technical reports regarding urbanization plans developed by the Central State, the democracy soon acknowledged that control over the territory was to be attributed to the local community. The practical empowerment of municipalities was considerable, reflecting a replacement of faith in the State in local development—the resulting new paradigm of 'Municipal State' was in tune with the European economic crisis of the late 1970s and early 1980s (Lopes, 1990) and the wave of neoliberal privatization that these crises spawned.

Social and spatial specificities, small-scale interventions and territorial integration become pressing problems, which municipalities were more equipped to understand since they were in principle closer to the local communities they represented. Thus, the central element of the democratic policy would be the Municipal Masterplan (*Plano Director Municipal*—PDM), applied to the totality of each municipal territory.

LNEC researchers actively contributed to support the outlining of the PDM, as attested in a number of collaborations and contributions to specific PDMs across the country (Portas 2005, 2005a; Gonçalves 1974; Grande 2012). Indeed, Nuno Portas left LNEC to become the Secretary of State for Housing and Urbanism (1974–1975) and the initiative for creating the PDM was launched by Portas and by Fernando Gonçalves (b.1946), another LNEC researcher (Gonçalves, 1974, 1978a; Grande, 2012; d'Almeida & Marat-Mendes, 2021). Thus, many research themes and approaches taken by LNEC researchers had already anticipated the approach of the democratic PDM, whereby the social and economic development of municipalities was associated with urban management rules regarding the occupation, use and transformation of the territory, and with municipal works and investments (Lopes, 1990; Pereira & Silva, 2008, d'Almeida et al, 2020).

[1] Decree of 794/76, published in Diário da República n.º 259/1976, Série I de 1976–11-05 10–4-1976.

T. Marat-Mendes et al., *Atlas of the Food System*,
https://doi.org/10.1007/978-3-030-94833-7_3

However, the first decades of the democratic era also gave rise to other important spatial policies, namely the legislation over cultural heritage,[2] on the environment,[3] on regional planning,[4] on municipal spatial plans[5] and on special spatial plans[6] (Cabral & Portas, 2011). Additionally, there were three policies which impacted spatial planning greatly, namely the establishment of the National Agricultural Reserve (*Reserva Agrícola Nacional*—RAN),[7] the National Ecological Reserve (*Reserva Ecológica Nacional*—REN),[8] as well as of a National Network of Protected Areas (*Rede Nacional de Áreas Protegidas*),[9] developed under conservative governments, under the orientation of Gonçalo Ribeiro Telles, a landscape architect who served shortly as Minister of State and of Quality of Life (1981–1983).

Over time, the regime of the Reserves suffered alterations, but the principle of establishing, at a national scale, a precise map of soils whose agricultural or ecological importance was officially recognized and protected was an important step of the democratic State towards ensuring the fulfilment of its obligations recognized in the Constitution. Namely, the State was mandated to promote a proper development of all the national territory as well as to protect and value the environment and natural resources in its Article 9, section c. Interestingly, the territory and its resources were equated, in the Constitution, with the cultural heritage of the Portuguese people.

After some revisions, the RAN is currently considered as the lands which, for its agro-climatic, geomorphological and pedological characteristics, are most suitable for agricultural activity. The RAN is a land management instrument, taking the form of a restriction of public utility, by establishing a set of conditions for the non-agricultural use of the soil, and playing a fundamental role in the preservation of the soil resource and its allocation to agriculture (DGADR, 2015).

A key problem in establishing the RAN was how to define its areas, demanding a transitional regime based on capacity charts for soil categorization. This included not only class A and B soils but also low alluvial and colluvial soils and other soils whose integration was considered convenient. Subsequently, all activities that could diminish or destroy the agricultural potential of RAN soils were prohibited.

In 2009, the RAN legislation was updated[10] and the methodology for land classification was compatibilized with recommendations from the FAO, although it remained ultimately shaped and designed by the Portuguese Directory of Agriculture and Rural Development (*Direção-Geral da Agricultura e do Desenvolvimento Rural*—DGADR). In this new version, the management of the RAN was divided between the national entity of the RAN and the Regional Development Commissions (*Comissão de Coordenação e Desenvolvimento Regional*—CCDR), with different scopes of action. In the past, the CCDRs had more legal power including those for imposing fines for the transgression of regulations and replacement of the respective soils, and competences it ceased to hold (Marat-Mendes et al, 2020).

In the most recent version, only land units with a high/moderate aptitude for agricultural activities integrate the RAN. However, a particular area can be included (or excluded) for being subject to a major investment to increase productive capacity, when its use is critical for the economic viability of pre-existing agricultural holdings, or if it holds strategic, pedogenetic or patrimonial interest. Conversely, land from urban perimeters classified in the PDM as urbanized soil, soil for urbanization or ecologically relevant soil, cannot integrate the RAN. Moreover, urban lands in intermunicipal or municipal territorial plans are also excluded.[11]

All actions that weaken or destroy the agricultural potential of lands and soils, including urbanization, construction or expansion works remain forbidden but may be approved in the absence of a viable alternative outside the RAN. Exceptions are allowed for rural tourism, residential tourism and nature tourism, complementary to agricultural activity, for recreational and leisure facilities complementary to agricultural activity and rural areas and sports facilities. The option to make regulations flexible to cater to tourism, albeit very common, has proved, at least in the Lisbon Region, to be highly problematic, since touristic development has often been used as a cover-up for the development

[2] Law 13/85, published in Diário da República 153/1985, Série I, 6–7-1985.
[3] Law 11/87, published in Diário da República 81/1987, Série I, 7–4-1987.
[4] Decree 176-A/88, published in Diário da República 115/1988, Série I—Suplemento, 18–5-1988.
[5] Decree 69/90, published in Diário da República 51/1990, Série I, 2–3-1990.
[6] Decree 151/95, published in Diário da República 144/1995, Série I-A, 24–6-1995.
[7] Originally established by Decree 451/82, published in Diário da República 265/1982, Série I, 16–11-1982.
[8] Originally established by Decree 321/83, published in Diário da República 152/1983, Série I, 5–7-1983.
[9] Originally established by Decree 19/93, published in Diário da República n.º 19/1993, Série I-A, 23–1-1993.
[10] Decree 73/2009, published in Diário da República n.º 63/2009, Série I, 31–3-2009.
[11] Decree 199/2005, published in Diário da República n.º 216/2005, Série I-A, 10–11-2005.

of real estate which, once finished, will be used as a regular residence (Pereira & Silva, 2008). That such a threat may be posed to the RAN is obviously a cause for concern, since it undermines the very principle of the reserve, which is the protection of food production activities.

As for the REN, it constitutes a biophysical structure with ecological value and sensitivity or exposed and susceptible to natural risks. It is a public utility constraint that conditions the occupation, use and transformation of the soil to uses and actions compatible with its objectives (CCDRLVT, 2008).

Promoting the symbiosis between human beings and nature, the REN meant to ensure that exploration of natural resources and territorial use do not threaten natural value. Hence, the REN includes coastal and riverside areas, inland and maximum infiltration waters and sloping areas. Actions or activities that may jeopardize or destroy the biophysical potential of these areas are precluded—except for mining, quarries and other infrastructure, which are dependent on authorization from two ministries.[12] While the original regime did not mention the criteria or methodology for classifying land as belonging to the REN, in 1990, a transitional regime was approved, aiming to preserve national ecosystems and biodiversity.

In 2006, the REN was revised[13] to ensure its powers as a management instrument. A key difference concerned the entities that acted in the delimitation of REN: while originally, this competence was shared between different ministries and by the REN Commission, after 2006, delimitation proposals were advanced by the CCDRs, which were empowered to develop studies and to consider the exclusion of previously built areas, or areas necessary for housing, economic activities or infrastructure. Other revisions of REN took place, culminating in 2019 when the Portuguese government reviewed it to consider the problems of climate change, the compatibility with the Paris Agreement and the safeguarding of water resources to prevent water scarcity.[14]

In the case of the Lisbon Region, democratic municipal powers, as defined by the 1976 Constitution, encountered a territory with rising problems and visible contrasts. On the one hand, there was a legal city, made of suburban neighbourhoods and urbanizations isolated on rural land, already with considerable density but lacking in infrastructure, equipment and urban life; while on the other hand, the 'illegal' city, fed through the land market, was chiefly made of low- or mid-density housing and lacked proper urban features (Pereira & Silva, 2008). The economic and productive basis of the Lisbon city transformed rapidly, with the intensification of suburban expansion and the rise of a complex network of mobility, and while the democratic planning policy would favour the municipal scale, the area of influence of the Lisbon city increasingly spilled to the surrounding territories (Ferreira, 2010). This growing periphery remained dependent upon Lisbon, where third sector jobs were strongly concentrated, as were large equipment and specialized commerce and services (Pereira & Silva, 2008).

Thus, from the late 1960s to the late 1980s, there were important urban operations privileging the periphery, from those which extended (and fundamentally altered) pre-existing settlements like Odivelas, Linda-a-Velha, Nova Carnaxide, Santo António dos Cavaleiros and Vale da Amoreira, to what were in reality (albeit not formally) new towns in plain rural land, like Cidade Nova, Sassoeiros, Portela de Sacavém, Quinta da Piedade, Miraflores, Alfornelos and Setúbal's Bela Vista (Ferreira, 2010; Rafacho, 2011; Lopes et al, 2021), not to mention the countless operations of illegal urban parcellment promoted by private landowners, like Vila Amélia, Ferão Ferro and Quinta do Conde (Gonçalves, 1978; Portas et al, 2007).

The 1970s were, however, marked by complex phenome such as an industrial recession caused by the energetic crisis, the hundreds of residents of the former African colonies who returned or came to Portugal after the retreat of the Portuguese, while conflicting notions of freedom and the disorganization of the public sector both give a new breadth to illegal construction (Gonçalves, 1978; Pereira & Silva, 2008). Slowly, Lisbon saw its population rise with the growth of the third sector economy, while in the Southern Bank of Tagus, especially in coastal areas, there was a growth of second residence homes (Pereira & Silva, 2008).

In 1982, the PDM Law was published, followed in 1988 by the Regional Planning Law, but the situation in the Lisbon Region was unfavourable in terms of effective planning. Integral Plans approved prior to the Revolution were concluded in Monte da Caparica (Almada), Bela Vista (Setúbal) and Zambujal (Amadora, then Oeiras). The short-lived SAAL programme promoted the construction of rehousing neighbourhoods for slum-dwellers, with many examples in both banks of the Tagus. Some municipalities, like Setúbal (1977) and Moita (1983), created PDMs, but their effectiveness was variant. After the failure of the Lisbon Masterplan by the GEU in 1959, another Masterplan had been started by French Engineer Georges Meyer-Heine (1905–1984), who completed it in 1967. Oddly, this

---

[12] These are the Ministry for Quality of Life (*Ministério da Qualidade de Vida*) and the Ministry of Industry, Energy and Outside Trade (*Ministério da Indústria, da Energia e da Exportação*)—both of which are extinct by now.

[13] Decree 180/2006, published in Diário da República n.º 172/2006, Série I, 6–9-2006.

[14] Decree 124/2019, published in Diário da República n.º 164/2019, Série I—28-8-2019.

Masterplan was indeed approved 10 years later, already during the democratic era.

Only in 1990, with the publication of the Law of Municipal Spatial Plans,[15] which specified planning and implementation procedures, but most importantly made the PDM mandatory, did municipalities start to prepare and approve their PDMs. By the late 1990s, it seemed like the planning situation was regularized, all over the country and even within the Lisbon Region, but problems with implementation and revision started to emerge. Municipalities often lacked the power, organization or even the will to implement the territorial models proposed in the PDMs, while many, expected to be updated or redesigned every decade, remain in activity today.

Throughout its evolution, the PDM brought important advantages in comparison with the former planning legislation of the New State. Namely, it ensured the management of all the national territory, not just urban cores; it brought autonomy for municipalities to decrease their dependence in face of institutions of national scope and it allowed for coordinated spatial strategies where rural land ceased to be the negative of urban spaces and was allowed (at least theoretically) to be considered an integral part of the municipal spatial system.

However, it also had its shortcomings. First, over time, the PDM created a polarized dynamic where municipalities set a strategy through abstract land-uses, but de facto development and design was ensured by the private sector. This translated into a policy whose real emphasis was on management, rather than on planning: while the PDM might condition the process of land-use change, it was powerless to propose a strategy to guide the private sector, especially with regards to agricultural explorations, which remain fully dependent upon the interests of proprietors. Associated with this problem was another one, as cultural as it is legal: the contradictions between established land-uses and rights inherent to private property. Indeed, while PDMs were binding in principle to both private and public actors, there was a tendency to consider that private property allowed owners to do as they please within their land, which over time has led to a widespread lack of compliance with regulations that hit the metropolitan regions especially hard (Ferreira, 2010; Mourão & Marat-Mendes, 2015; Abrantes et al, 2016). Finally, the paradigmatic vision of the PDM was sustained, perhaps abusively, in the idea that more developed cities were an example to be followed by all types of habitats, and that planning at the municipal scale was enough to resolve inequalities between different realities within the same country (Baptista & Henriques, 1985). This determinism contributed to undermining the efficiency and even the accuracy of many planning instruments approved during the democratic era.

3.2 The First Attempts at an Integral Planning Structure

In 1982, with the first democratic PDM Act,[16] the central State empowered municipalities to plan the local territory, mobilizing towards this end the PDM. The term Director Plan had been employed occasionally during the New State, namely for the Masterplans of Lisbon (by Gröer in 1938–1948, by the GEU in 1959 and by Georges Meyer-Heine in 1967) as well as the plan for Porto by Antão de Almeida Garrett from 1952 and for the PDRL of 1964. What distinguished these plans from PGUs was that they applied to the totality of the municipal territory (usually through a land-use scheme), and not just a particular settlement. After 1982, the expression gets a new reach, as it no longer applied merely to the spatial or territorial model that was being proposed. Indeed, the PDM was also meant to coordinate strategies for spatial, social and economic local development (Lopes 1990).

Perhaps because of the large number of dimensions and involved actors, the process of elaboration of the PDM tended to be complex and time-consuming. Until 1990 its elaboration was not mandatory, thus many municipalities failed to create a PDM, instead used urbanization plans—sometimes created during the New State—or studies of refurbishing for illegal urbanizations, thus promoting sparse and site-specific interventions and decisions (Pereira & Nunes da Silva, 2008).

The situation changes in 1990 when the Law for Municipal Spatial Planning made the PDM mandatory and established it for coordination of plans at a smaller scale, namely the Urbanization Plans and the Detailed Plans. Both of these were supposed to include urban design, while the PDM would focus on land-use macro-zoning. A key strategy for prompting the elaboration of PDMs was making both expropriation process and the access to Structural Funds for Community Support dependent upon having an approved PDM.

[15] The above-mentioned Decree 69/90.

[16] Decree 208/82, published in Diário da República 119/1982, Série I, from 28-5-1982.

The new planning scheme implied a demise of urban design, which had been the chief planning strategy since the legislation created by Duarte Pacheco. From the more lenient land-use scheme inherent to the PDM, a liberal and fair articulation between public and private interest was expected to emerge (Lopes, 1990; Marat-Mendes et al, 2020). Programmes for public construction, especially to tackle housing shortage and eradication of slums in the two metropolitan areas continued to exist, but their mechanisms were more directed towards financing, and less towards design (Tulumello et al, 2018).

In the Lisbon Region, most PDMs were approved during the 1990s, in response to the new demands of the Law for Municipal Spatial Planning. However, these had some serious problems: oversized areas for urbanization supported by unrealistic perspectives of growth, scarce attention to the pre-existing city and the generic integration of illegal construction areas in urban areas (Pereira & Nunes da Silva, 2008). Furthermore, three key frailties already announced the problems for the food system that were to arise within the new planning scheme, namely the merely incidental interest on rural space with clear inability to interpret or coordinate its fringes with urban space, the allowance of building in rural space for several activities only limiting the parcel occupation and the straightforward rendering of agricultural and forestry areas as they were, without valuing them as productive spaces (Pereira & Nunes da Silva, 2008).

Briefly, while democratic legislation did include rural spaces within the same planning structure as urban spaces, a certain inanity in dealing with the countryside still permeated in the first generation of PDMs. In a sense, rural spaces were merely catalogued and mapped, but no strategic vision was promoted for them and no intervention or coordination by the public sector was advanced. Private property did triumph over the management of the territory as a public good—suggested by the 1976 Constitution—and thus, in Portugal, was installed a tendency verified in other contexts, whereby the food system is understood as being fully within the responsibility of the private sector and thus outside the intervention ability of the State (Pothukuchi & Kaufman, 2000). With the successive responses to the EU Common Agricultural Policy, the erasure of the former diversity of local food systems in Portugal was bound to become certain.

The same triumph of private property lies behind the oversizing of areas for urbanization since the State was incapable of controlling the phenomenon, which remained dependent upon landowners with their goals, means, timing and strategies (Pereira & Nunes da Silva, 2008).

While the 1990 Law proves effective in prompting municipalities to approve their PDMs by the late 1990s, in 1998, a new Planning Law was published, rendering many options followed in the first generation of PDMs obsolete. Another demonstration of the hardships of Portuguese planning policies.

3.2.1 The Amadora PDM (1994)

Approved and published in 1994, as the result of the exhortation of the 1982 Planning Act, the Amadora PDM remains to be revised today, implicitly remaining active (CMAmadora, 1994). At the time, the city council was governed by the United People Alliance (*Aliança Povo Unido*—APU),[17] a coalition of the Portuguese socialist (*Partido Comunista Português*—PCP) and other smaller parties of a communist or socialist orientation.

Amadora results from a complicated administrative process, having been a civil parish of the Oeiras municipality until 1979. Only then did it become an independent municipality, of an overwhelmingly suburban character, being intrinsically linked with the Lisbon city through Benfica, a large zone in the northwest sector of the capital city. Amadora results from the expansion of the small village of Porcalhota which was surrounded by farmsteads until the twentieth century when it became a much sought-after territory for the instalment of industries. It had an urbanization plan in 1949 designed by Faria da Costa, but this was not approved (Silva, 2018) and in the year prior to the Revolution, one of its areas, Zambujal, received an Integral Plan, implemented in its housing dimension, but less so with regards to public space. Amidst the lack of planning stipulations, the Amadora territory was especially prone to illegal construction, ranging from unauthorized apartment buildings to shanty towns, like Cova da Moura or the bygone Fontaínhas and Ribeira da Falagueira.

In the Amadora PDM, and undoubtedly because of its time of elaboration, the food system is never directly mentioned. There is also no land classified or categorized as agricultural, although some of its lands are included in the RAN. Industrial spaces are numerous in Amadora, and there are no specificities regarding food-related industries. The PDM is not structured around a clearly stated ideological vision, although it has some relevant regulations concerning the environmental quality of the municipality. Not only there is careful stipulation as to natural spaces (and equivalent ones), but there is also a whole chapter on pollution, including dispositions towards preventing it in the air, water and the soil. This is even more important considering the heavy industrial character of the municipal economy.

[17] A coalition of the Portuguese Communist Party (*Partido Comunista Português*—PCP) and other smaller parties of a communist or socialist orientation, which include the Ecological Green Party (*Partido Ecologista Os Verdes*—PEV) and the Portuguese Democratic Movement/Democratic Electoral Commission (*Movimento Democrático Português/Comissão Democrática Eleitoral*—MDP/CDE).

Fig. 3.1 An overview of Amadora from the Falagueira civil parish. Source: Authors

The PDM is, like most of its generation, supported by the land-use system, even including a chapter with many definitions that help interpret the functioning of such a system. However, once it is established, it merely provides regulations on scale, lining and height. Land-use is further distributed in each Planning Units, adding specific interventions according to the needs of the zone. However, there are no interventions of a rural character, nor specifications for agricultural space, as there is no land classified or categorized thus. All specifications relating to agriculture are to be found in the regulations for RAN.

The regulation for green protection spaces states that these may be used for equipment for the satisfaction of population needs, and while this could serve as framework for the creation of horticultural parks, a decree from 2004 has suspended it (Fig. 3.1).

Although the Amadora PDM is clearly structured and does present a system of land-uses and a realistic division of its territory in Planning Units, it does suffer from trying to cater to the installed tendencies, favouring urbanization in nearly all the territory. While there are spaces assigned for the construction of new industries, in general, the PDM does

Fig. 3.2 Examples of rural remains in the Amadora centre. Source: Authors

not necessarily envision a multifunctional space: it basically presupposes urban settlements with industrial parks and natural spaces (or equivalent) between them. Another important aspect that undercuts the possibilities of Amadora is the lack of clear goals towards articulation with the Lisbon municipality—other than through transportation (Fig. 3.2).

Indeed, the pressure for urbanization in Amadora has been strong for a long time, especially considering that it provided easy links with the northwestern end of Lisbon city and that it lacked properly approved plans capable of regulating its growth. Thus, throughout the years, Amadora grew under the guise of the private sector, and its territory is marked by a wide spectrum of different residential spaces, ranging from high-density estates for the upper-middle-class such as the Alfragide Towers designed by the studio of architect Francisco Conceição Silva (1922–1982) (Leite, 2007; Fiúza, 2013) to social housing estates like Alto do Moinho designed by Francisco Silva Dias for SAAL (Bordalo, 2007) to illegal neighbourhoods with single-family and apartment buildings like Brandoa, to former industrial neighbourhoods like Venda Nova, to shanty towns such as Cova da Moura.

The strength of urbanization in Amadora, as well as its diversity and complexity, makes it a particular challenge to rebalance the resulting spatial system. However, it is undeniable that the approach to green spaces is narrow-minded, by considering they all are of urban character—and therefore destined for leisure. Such approach does not even translate the reality of unbuilt space in Amadora, where some agricultural explorations remain, scattered among urban cores and classified as 'Green spaces for protection' (as in Venda Nova and Reboleira), not to mention the considerable amount of leftover urban space occupied with urban agriculture, which has been partially acknowledged by the City Council. Moreover, despite the continuous pressure for urbanization, there remain in Amadora today many vestiges from its rural past, from farmstead walls to premodern houses in the rural ('saloio') style characteristic of the Lisbon Region (Fig. 3.2). Their protection is also not properly safeguarded in the PDM.

Currently, Amadora is also challenged by the decline of its industrial estates, many of which are not derelict or obsolete, suggesting important questions about their future use (Figs. 3.3 and 3.4).

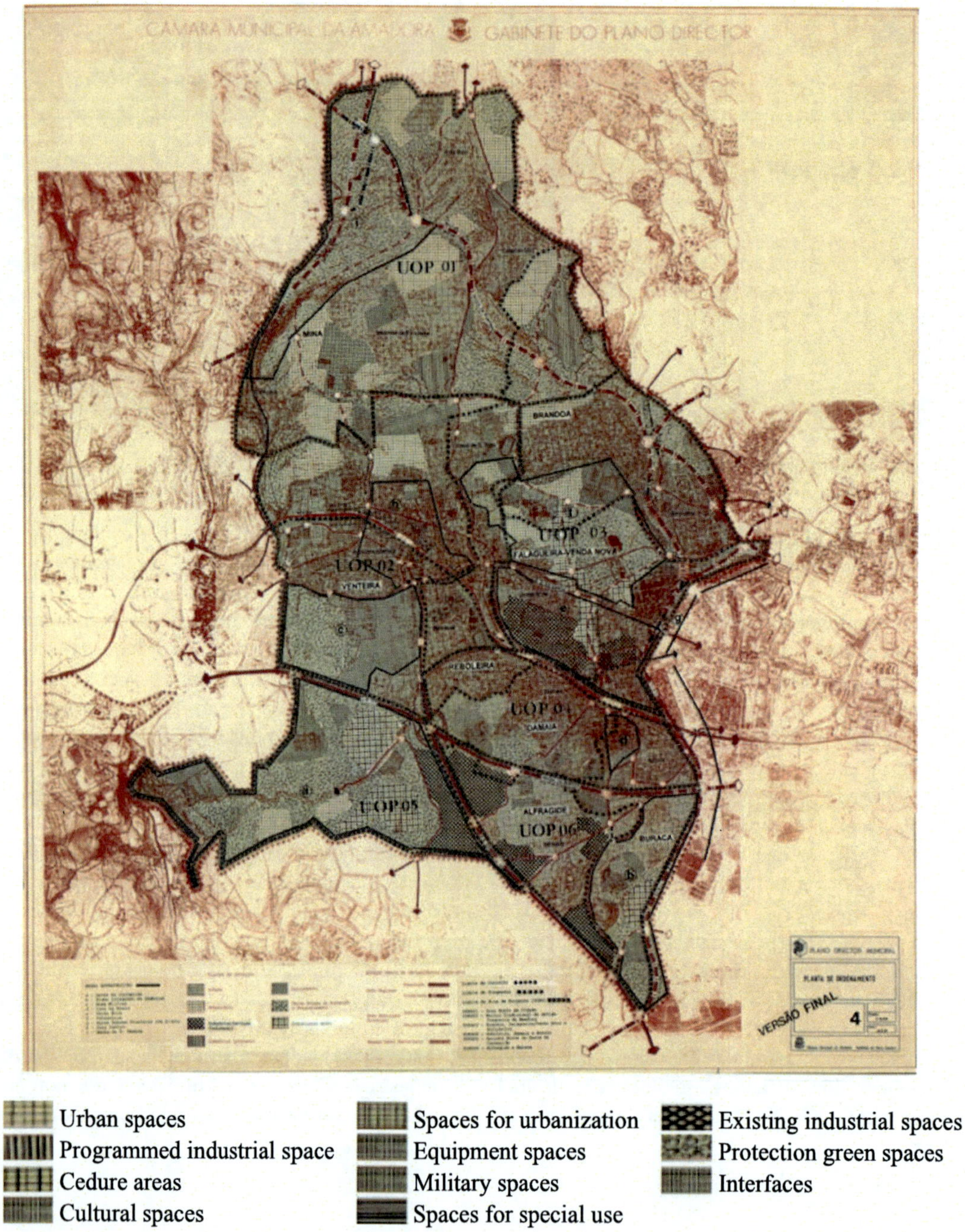

Fig. 3.3 Land-use chart of the Amadora PDM (1994). Source: CM Amadora (1994)

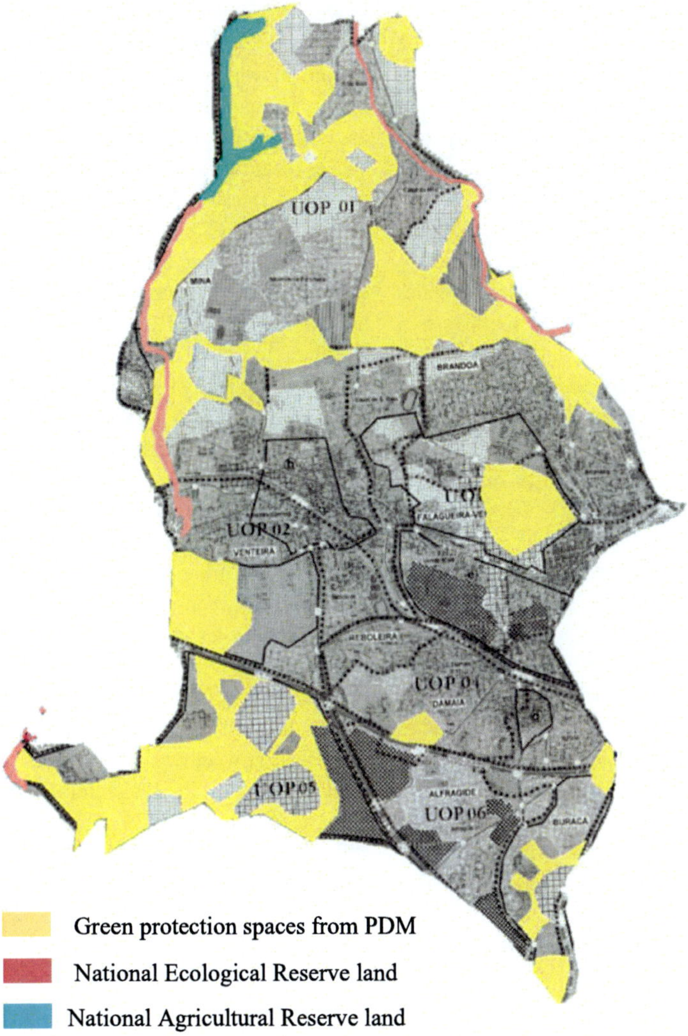

Fig. 3.4 Unbuilt areas officially established in Amadora. Source: Authors

3.2.2 The Alcochete PDM (1997)

Also approved in response to the 1982 PDM Law, in 1997, the Alcochete PDM (CMAlcochete, 1997) is another example that remains active today. The municipality was governed by the APU at the time of the publication of the PDM, and this coalition has been in power through most of the democratic period, except for three electoral cycles, including the most recent one (2017–2021). Standing at the curvature of the Tagus delta, Alcochete also mediates between the Southern Bank of Tagus on its southwestern limit and with the municipality of Benavente (outside the Lisbon Region) on its northern limit. Being a strongly rural territory, agriculture and livestock became key activities in the sixteenth century. Another key activity of Alcochete is the saltpans, which remains today with large dimensions across the riverfront, even though in other places such as Moita or Vila Franca de Xira, saltpans have declined throughout the twentieth century. Like Palmela, Alcochete also lost the municipality status, being annexed to Montijo (then called Aldeia Galega do Ribatejo) only to regain it in 1898.

The PDM includes a glossary with basic definitions, a land-use system and a set of five Operative Units. The latter include all the three main settlements—Alcochete, Samoucou and São Francisco, alongside the land assigned for their physical expansion—as well as two villages, Fonte da Senhora which is assigned for urban restructuring and Passil, assigned an extension area (Fig. 3.5).

In general, the PDM does not directly mention the food system, despite the considerable amount of rural land, including both agriculture and agroforestry, as well as a considerable extent of natural spaces—all of which constitute classes of land-use. The municipal territory is crossed by a large extension of land which simultaneously belongs to the RAN and to the national Institute of Water (*Instituto da Água*). The spine-shaped structure of this protected land nearly divides Alcochete into two halves, and while on the western sector agriculture is predominant, on the eastern sector agroforestry becomes dominant. The three towns of the municipality (Alcochete, Samouco and São Francisco) are all located on the western sector, as are most of the industrial estates. Saltpans do not constitute a land-use, but the terrain they occupy extends from the protected land to the riverside towards the west and is classified as a natural space. This classification is divided into two categories, the first of which follows the regulations of the REN and the second of which is oriented towards agricultural uses and public equipment. Indeed, natural spaces are customarily associated with publicly-owned and unproductive green spaces meant for the leisure and qualification of urban spaces (Marat-Mendes et al, 2021) (Figs. 3.6 and 3.7).

Rural spaces are divided according to two categories, and include spaces for agriculture, where allowed uses must conform to the protection of agricultural practices, livestock and forestry. As such, uses like support infrastructures and facilities, industrial facilities and storage, rural tourism and collective equipment are allowed in rural soils. As for REN soils, construction is only allowed for housing and facilities for support of agriculture that do not affect ecological balance.

Thus, the majority of the municipal territory includes activities directly related to the food system, even in the free spaces of its urban cores. Perhaps because of the spatial proximity between the main settlements, agricultural and natural spaces are directly linked with urbanized land and land to be urbanized. There is a clearly stated intention to preserve the agricultural production structure of the municipality, as well as its green land-cover, while natural and rural spaces, whether or not directly categorized as agricultural land, allow for agricultural activities as well as rural tourism. The two villages which are projected to grow are still separated by agricultural land, a strategy that has been proposed by planners in the Lisbon Region since Gröer, to avoid the merger of separate settlements.

Generally, there is a considerable balance in the complexity of regulations for both rural and urban (or to be urbanized) land. And since the Operative Units do not include any regulation or a specific programme, in the absence of detailed plans, the main driver in this PDM is the land-use system. Considering the passive role the public sector is expected to have in the management or intervention in rural space, although the Alcochete PDM certainly promotes a rich rural spatial system and a productive economy, it also refrains from suggesting a considerable public intervention.

Despite being approved in the late 1990s, this PDM demonstrates an awareness of sustainability agendas, including the preservation of ecological balance, as well as the preservation of the agricultural production structure and

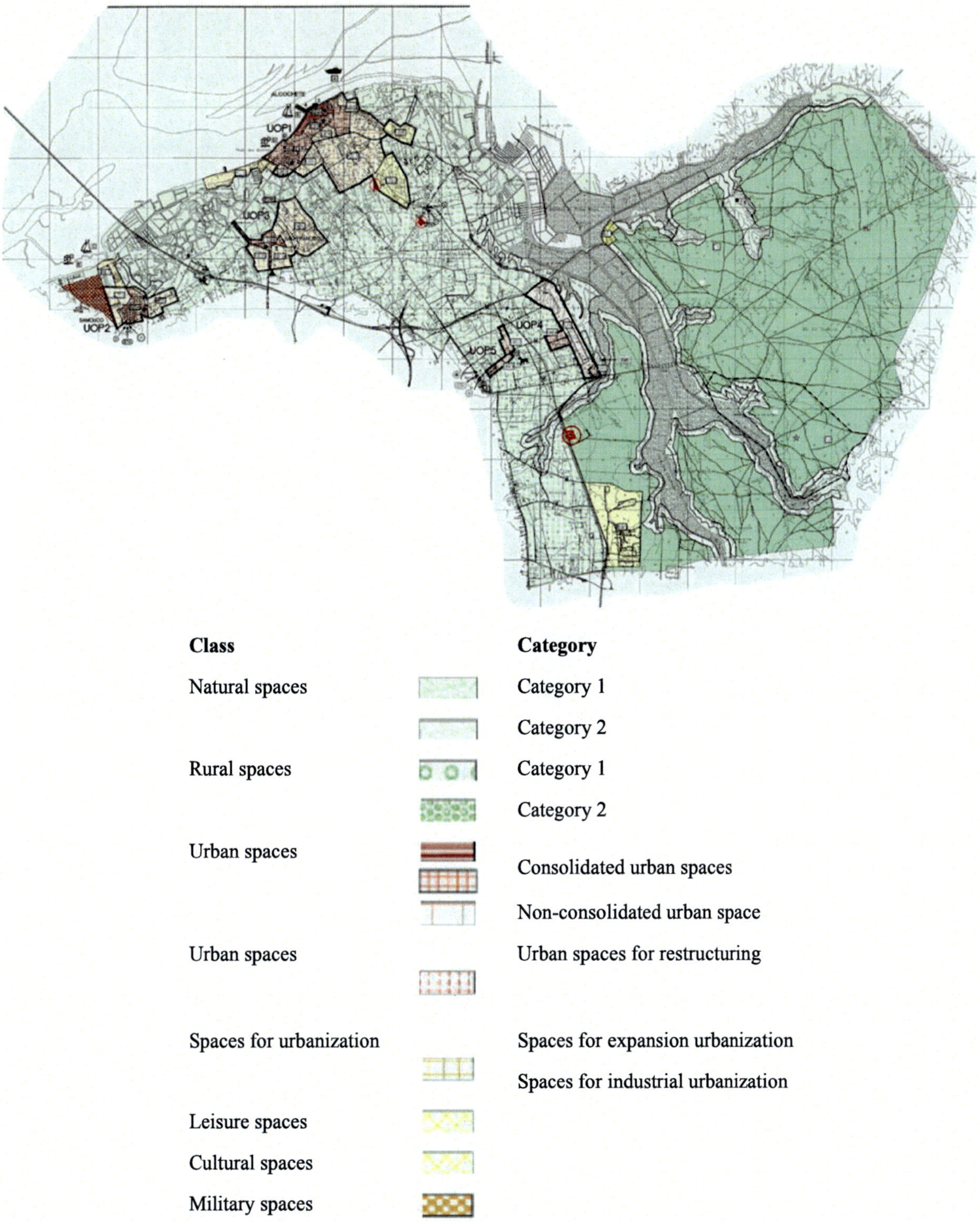

Fig. 3.5 Land-use chart of the Alcochete PDM (1997). Source: CM Alcochete (1997)

Fig. 3.6 Farm in Cercal de Baixo, between the towns of Alcochete and São Francisco. Source: Authors

of streamlines and natural drains. These protection dispositions are reiterated with respect to natural spaces, rural spaces, and industries in consolidated urban land. Dispositions for rural land must be understood in addition to the rules of REN and RAN which are both directly mentioned in the PDM and which take precedence over it.

In spaces for expansion urbanization, industrial uses are allowed, as long as they are non-polluting, which would offer an opportunity to incept in new urbanizations spaces for food industry, especially transformation or packaging, that would also benefit from proximity to Alcochete's rural land. The same is true in spaces for industrial urbanization, which is divided into three subcategories: if one of them is meant for an explosives factory, the remaining two subcategories allow for industry that is compatible with urban uses. Again, this would be a good scenario for the transformation of agricultural produce, establishing a direct link between the rural agricultural land and the urban areas, which most likely include commerce of food (supermarkets and shops).

Fig. 3.7 Street in the town of Alcochete. Source: Authors

3.3 The 1998 Spatial Planning Act

In 1998, a new Planning Act[18] was approved. A key aspect of this new Act was that it responded to a noticed absence of direction in planning practice, caused by the sparse publication of laws on regional planning or on the several municipal plans. Thus, it importantly systematized a new structure for spatial planning, ranging through different concerns and scales, from the national to the local level. The aims cut across several themes, from the social, economic and environmental realms, to concerns over physical heritage, resource management, working conditions and civil protection. As seen above, some of the key afflictions of approved PDMs in the Lisbon Region (Pereira & Silva, 2008) seem to be amended by these principles. The new Planning Act was justified by the need to ensure national-territorial cohesion and the correction of asymmetries between regions while promoting the diversities within this same territory. Environmental balance was also promoted, through the rational use of resources, directly implying biophysical aspects of space articulated with social and economic concerns. Rationalizing, refurbishing and modernizing urban systems while safeguarding and valuing rural spaces was also a goal.

In concomitance with the aims, the Planning Act provides principles, whereby responsibilities were divided between different actors. Principles like intergenerational solidarity, adequate use of natural and cultural resources, and compatibilization of territorial, social, economic and sectoral policies seem more immediately associated with public action. A fair distribution of obligations and privileges implicated in spatial planning, responsibility for territorial management including the duty to compensate, correct contract models between the

[18] Decree 48/98 (Lei de Bases da Política de Ordenamento do Território e do Urbanismo), published in Diário da República 184/1998, Série I-A, 11-8-1998.

public and the private sector and juridical safety to ensure the stability of planning processes seem more directed towards the relations between the municipality and other public or private actors. Interestingly, these principles also include the privileging of the decision-making level closer to the citizen and the encouragement of citizen participation through access to information and intervention, which seem to charge municipalities with seeking the input of their communities for planning decisions. However seductive these points may be on paper, their transposition to reality in some aspect would be difficult to attain. Balanced distribution of housing, work, culture and leisure spaces or the implementation of adequate housing policies are unlikely when municipalities have scant power to design or implement such projects and depend on the market to do so.

While the key point of this Planning Act is the indication and systematization of spatial planning of municipal, sectoral, national, interregional, regional, intermunicipal and detailed scales, the PDM remains the only mandatory instrument, thus compromising a priori the efficiency and respectability of the remainder. While proposing all these types of plans, the Planning Act also divides them into different typologies, ranging from territorial development (National Spatial Programme, Regional Plans, Intermuncipal Programmes), territorial planning (PDM, Urbanization Plans and Detailed Plans), sectoral development policies (Sectoral Plans) and planning of special nature (Special Plans).

The Planning Act also indicates who must create each of the plans. Thus, the National Spatial Programme is assigned to the Central Government, demanding approval by the National Assembly (*Assembleia da República*—AR), after inquiring about municipalities and other local institutions. Regional Plans are prepared by CCDRs with municipalities inquired, and afterwards they are approved by the regional assemblies and rectified by the Government. Intermunicipal Programmes are prepared by the involved municipalities, approved by their Municipal Assemblies (*Assembleia Municipal*—AM) and finally confirmed by the Government. PDMs are prepared by each City Council, approved by the AM, with the opinion of the CCDR and subject to rectification by the Government.

A further distinction is made between programmes and plans: the former is mandatory only for public institutions, while the latter applies to both the private and the public sectors. This means that the PDM remains the strongest of planning instruments, its elaboration being mandatory and its dispositions being binding to the private sector. The sectoral plans and policies are an important innovation, responding to the hope of promoting the integration of planning policies with wider concerns, but these include a plethora of sectors that have often not been translated into spatial plans, namely mobility, communication, energy, geological resources, education, culture, health, housing, tourism, agriculture, commerce, industry, forestry and environment.

With respect to municipal planning, it provides a basic framework from which to write a PDM. Within the land-use system, soils will be assigned classes and categories, the first referring to the basic destination—again, urban or rural—and the second referring to the use and edification regime of land according to the dominant activity. The Planning Act itself provides the key definitions for land-use classes: rural soil for land with potential for agriculture, livestock, forestry or mineral exploration as well as natural spaces of protection or leisure unoccupied by infrastructures or including non-urban infrastructures; and urban soil for land with potential for urbanization and buildings or already in that situation. It remains to the municipality to establish land use categories.

3.3.1 The Lisbon Metropolitan Area Regional Plan (2002–2009)

Already in the 1980s, the CCDR of Lisbon and Tagus Valley (CCDRLVT) started to promote the work for a plan of the Metropolitan Region, but this is a process that took nearly two decades, being finished only after the restructuring of planning policies prompted by the 1998 Planning Act. This period was marked by strong demographic fluctuations. While the Lisbon city lost about 30% of its residents, the growth of the Northern Bank of Tagus of 26.9% supersedes that of the Southern Bank, with 22.2% (Pereira & Nunes Silva, 2008). Meanwhile, there is a variation of 22% in housing, with a large proportion of second residency homes and mostly by the number of vacant houses, reaching 47% in the Metropolitan Region and 60% in Lisbon (Pereira & Nunes Silva, 2008).

In 1991, Lisbon and Porto, with their respective surroundings, areas were legally constituted as 'Metropolitan Areas', i.e. as special and mandatory associations of municipalities. However, they have proved inoperative throughout most of their existence, with difficulties affirming political validity and ensuring coordination between different municipalities and actors (Pereira & Nunes Silva, 2001).

Within the scheme proposed by the 1998 Planning Act,[19] the Regional Spatial Plan (*Plano Regional de Ordenamento do Território*—PROT) was created to implement at the regional scale the policies established by the National Spatial Plan (*Programa Nacional da Política de Ordenamento do Território*—PNPOT) and Sectoral Plans (*Planos Sectoriais*—PS) like those concerning the environment, the national motorway system, relocation programmes and the

[19] Lei de Bases da Política de Ordenamento do Território e do Urbanismo.

construction of regional and national equipment (CCDRLVT, 2002, 7), but also to mediate between these programmes and municipal spatial planning (CCDRLVT, 2002, 46). The CCDRs played an important role as consultants and supervisors for intersectoral coordination but were not obliged (or even authorized) to oversee urbanization or detailed plans or to evaluate PDMs (Pereira & Nunes Silva, 2008).

The first PROT designed for the Lisbon Region—which coincides with the AML and further includes the municipalities of Azambuja and Benavente—was finished in 2002, while the first PNPOT was presented only in 2007.[20] As such, the PROT-AML absorbed what it could from broader policies (CCDRLVT, 2002) but showed relative freedom in the territorial model that it proposed.

Its strategy was clearly based on the proposals of the Compact City as it came to be defined by architect Richard Rogers (1998), with a central area constituted by the Lisbon city and its immediate surroundings and extending, on the Southern Bank of Tagus, to the cities of Almada, Seixal and Barreiro. It furthermore sought to contain peripheral urban areas, restricting further spatial dispersion and safeguarding ecological and agricultural land (Pereira & Nunes da Silva, 2008).

Its key aims were environmental improvement and sustainable development; cohering the urban system to decrease regional contrasts, balk suburbanization through multipolar structuring, improved mobility and competitive transportation (CCDRLVT, 2002). These aims were mostly collected from Sectoral Plans. Applying these goals to the specific territory of the AML, the PROT-AML proposed:

> Environmental sustainability through natural preservation and the creation of a Metropolitan Structure for Environmental Protection with the key areas, corridors and ecological links for regional environmental improvement, through protection of water, landscape and rurality as metropolitan elements;
>
> Metropolitan qualification, by containing urban sprawl, reaffirming the centre of the AML and the protection of natural resources and protected areas, and establishing new centres with mobility networks and logistical structure;
>
> Social and territorial cohesion, by improving urban conditions, slum clearance, suburban refurbishing and equity of access to equipment, housing and services;
>
> Organization of the metropolitan transportation system, through the creation of the 'Metropolitan Authority for Transportation' to strengthen collective transportation, especially through train and river, and to complete the radial mobility structure (CCDRLVT, 2002: 15).

Written in the early 2000s, the PROT-AML aimed to give the Lisbon Region an European and Iberian capital, taking advantage of the potentials of its geographic location for Euroatlantic business (CCDRLVT, 2002). The goal may have been interrupted by shifting political and financial conditions within the country from 2007 onwards, leading to the precocious withdrawal of the Regional Plan.

The territorial strategy of the territorial model was based on reaffirming the central role of Lisbon within the metropolitan area, using its recently increased access to the Southern Bank of Tagus to revitalize its critical areas. To lay out conditions for achieving this, the PROT-AML first created a map classifying territorial dynamics.

In the Northern Bank of Tagus, the situation was more polarized, with the Amadora-Lisbon-Loures-Vila Franca de Xira belt in critical urban condition and the Lisbon-Oeiras-Cascais (former PUCS territory) in steady development. Extensive green areas were located in the more peripheral areas of Mafra, Sesimbra, Palmela and the Montijo exclave, countering the more densely urbanized areas (CCDRLVT, 2002). The PROT-AML also included an Environmental Strategy, which acknowledged the quality of the metropolitan agrarian and forestry spaces, to be protected as a feature of local character against excessive urbanization or infrastructure needs (CCDRLVT, 2002). One way to ensure this protection was the Metropolitan Structure of Environmental Protection and Recovery, which mapped and managed specific parts of the territory but also promoted adequate activities to integrate these areas into the remaining urban structure.

From this 'diagnosis', the PROT-AML took the goals of internationalization and integration between Lisbon and the Southern Bank centres (Almada, Seixal and Barreiro), by defining as a fundamental element the construction of the Chelas–Barreiro Bridge (CCDRLVT, 2002: 41). It was stated that the Chelas–Barreiro crossing should allow motorways and trainways, and although the trainway was fundamental for the development of the AML, the PROT-AML left undecided which of the two mobility types must be the priority of this public work. Remarkably, in the territorial dynamics map (Fig. 3.8) Chelas was signalled as a 'problematic space', whereas the Barreiro municipality was classified with 'problematic spaces' as well as 'critical urban areas' and was further pointed out as an 'area for recovery and refurbishing'. It is likely that the PROT-AML expected that a major public work like the Chelas–Barreiro bridge would promote social and territorial cohesion by directing investment towards mismanaged council housing estates and half-derelict industrial facilities in these areas, while also optimizing geographical conditions (existence of valleys and available vacant land). However, the bridge was never constructed (Fig. 3.9).

Besides this infrastructural element, the PROT-AML also programmed social change by developing specialized and complementary centres of economic activities and by

[20] Its revised version was approved in 2019.

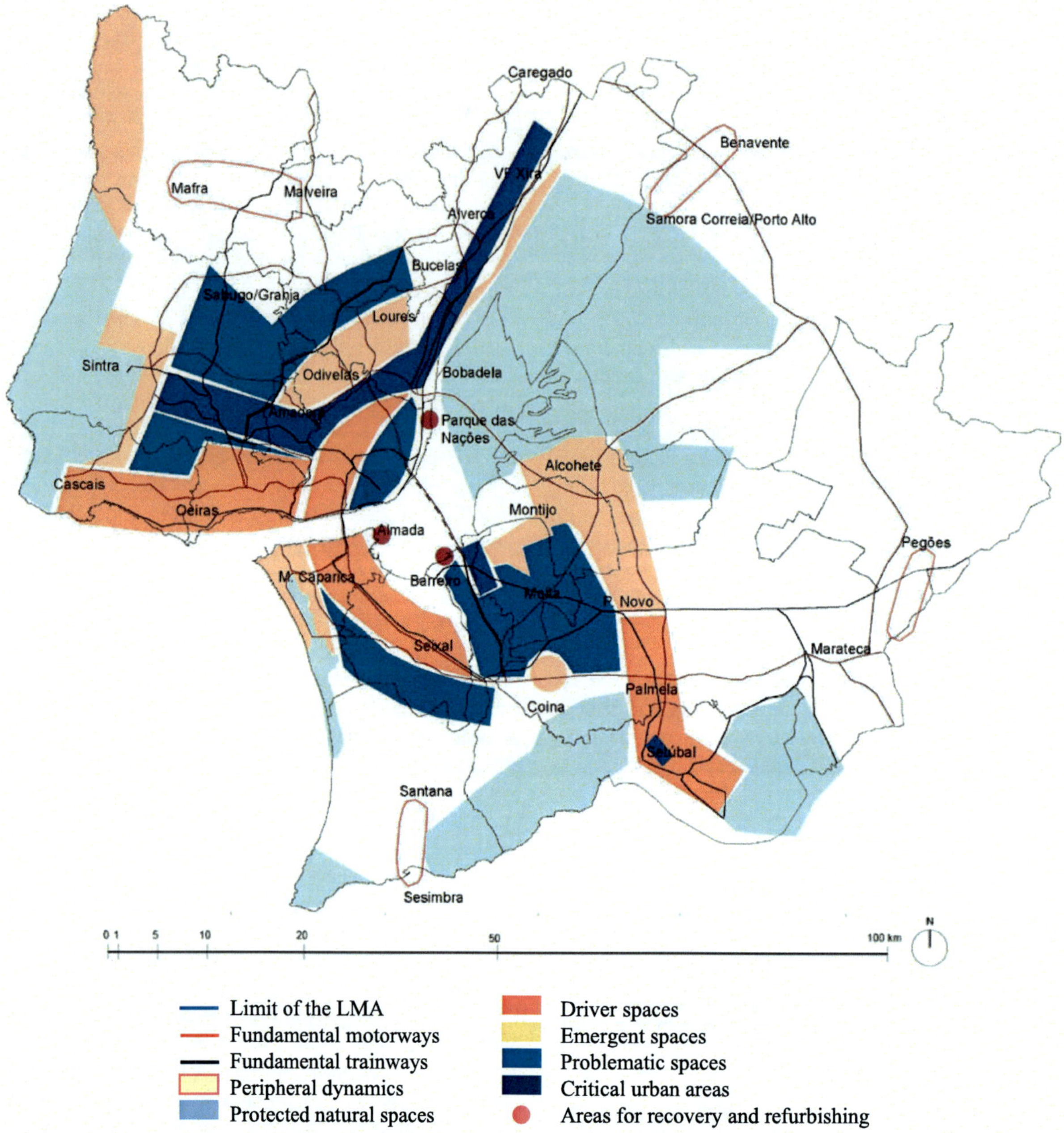

Fig. 3.8 PROT-AML, territorial dynamics. Source: Authors (Adapted from CCDRLVT, 2002, 32)

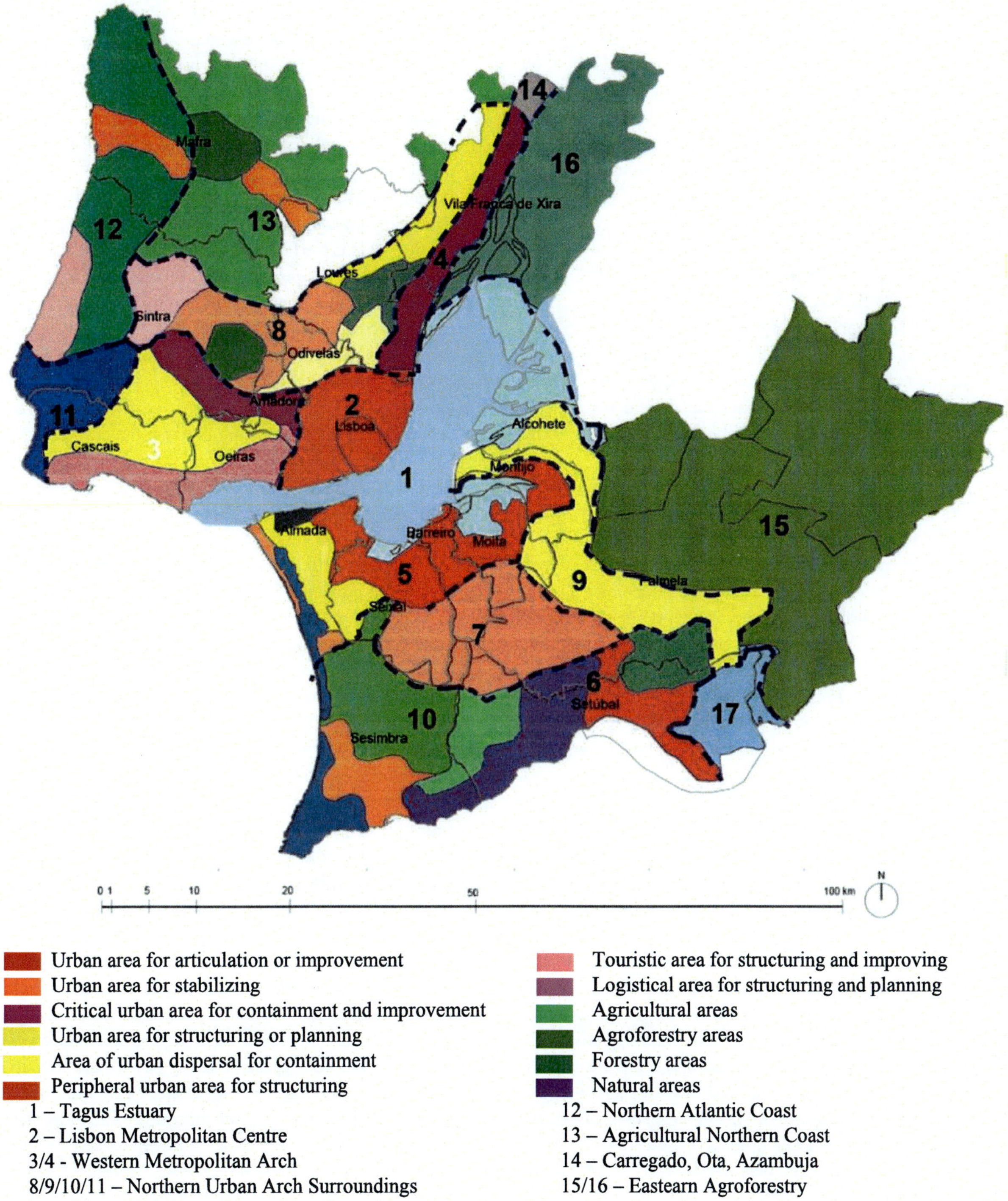

Fig. 3.9 PROT-AML, territorial units and model scheme. Source: Authors (Adapted from CCDRLVT, 2002, 44 and 48)

safeguarding the ecological metropolitan structure (CCDRLVT, 2002: 41). To do so, the regional territory was divided not in specific categories of land-use established by municipal planning instruments, but rather in territorial units for integral intervention and containing several land-uses. There were 17 territorial units within the PROT-AML.

These territorial units have specific conditions and propositions, and most include more than one land-use (CCDRLVT, 2002), and are the following:

1. Tagus estuary, to be reaffirmed as the real centre of the AML (CCDRLVT, 2002: 47);
2. Lisbon Metropolitan Centre, the Lisbon city, 85% of which is already integrally urbanized and which presented several problems, including a decaying centre, derelict storage and industrial facilities and several axis of deficient housing quality (CCDRLVT, 2002: 47–50);
3. Western Metropolitan Space, the axis between Algés and Cascais, generally filled with low-density housing[21] and important clusters of service economy (particularly in Oeiras); as well as the axis between Amadora and Sintra, with poor population and deficient housing and urban conditions, as well as illegal housing clusters and derelict industrial and storage facilities (CCDRLVT, 2002: 50–51);
4. Sacavém–Vila Franca de Xira Axis, which includes considerable amounts of housing, as well as chords of industrial facilities parallel to the riverside and the trainway, many of which were derelict or in need of replacement (CCDRLVT, 2002: 52);
5. Southern Riverfront, which includes chords of riverfront and seafront settlements in Almada, Seixal, Barreiro and Montijo. All of these areas were locus to suburbanization after the construction of the 25 de Abril Bridge,[22] development was polynuclear and fragmented, typically in high-density models and mixed with other land-uses, mostly industrial, although industrial facilities in these areas were increasingly abandoned and derelict. Moreover, mobility structures within these areas were deficient and there was a hierarchy between the municipalities with direct access to Lisbon and those which did not have such access. As a result, Almada (which receives the 25 de Abril Bridge) was already becoming more autonomous (CCDRLVT, 2002: 52–54);
6. Setúbal–Palmela, an urban centre constituted by the urban-industrial core of Setúbal and the agricultural area of Palmela in the north. This unit was already well served with transportation and as such was expected to link the AML to the southern region of Alentejo. The agricultural areas are valued for ensuring environmental and agricultural qualities to the Sado riverbank and ensure biological diversity and the use of soil for water depuration (CCDRLVT, 2002: 54);
7. Inland Southern Flatland, the central territory of the Setúbal peninsula divided between six different municipalities. This is the first territorial unit in which the main patterns of development were actively fought by the PROT-AML. It is mentioned the unplanned and highly disperse construction boom verified in this territory, some of it illegal, creating what the planners call a 'spatial promiscuity' between housing and other types of unplanned uses, mainly industry and storage, which were becoming derelict. On the other hand, the emergence of Coina as an industrial core is noticed and maintained by the PROT-AML (CCDRLVT, 2002: 54–55);
8. Northern surrounding arch; a peripheral area of transition, which includes degraded areas of extraction industry (Pero Pinheiro—Montelavar), areas for environmental protection (Serra da Carregueira) and heritage protection (Belas); as well as several urban areas in need of consolidation. It further includes the Loures floodplain, an important agrarian area as well as other agricultural fields, more or less disperse (CCDRLVT, 2002: 55–56);
9. Eastern transition space; a plain territory between Alcochete, Pinhal Novo (Palmela) and Palmela, dominated by agriculture of several types and with low-density housing and settlements. Population is mostly constituted by former urbanites who use the Vasco da Gama Bridge to commute to Lisbon. In the two key urban settlements, Alcochete and Pinhal Novo, tourism and leisure are important activities (CCDRLVT, 2002: 56);
10. Arrábida/Espichel/Sesimbra woods; a set of autonomous spaces that are fundamental for environmental protection. The Arrábida Mountain includes cliffs, beaches, marines, pastures, forestry and agricultural areas; the Sesimbra Woods include forestry areas; the Azeitão (Setúbal) agricultural area includes several farms where traditional agriculture is still practiced, with important outputs like cheese and wine. Further included is the urban chord Sesimbra–Santana–Albufeira Lagoon, which includes activities like fishing (including the Sesimbra port, one of the most important nationally) and tourism/leisure (CCDRLVT, 2002: 56–57);
11. Sintra Mountain; which includes the Sintra Mountain Natural Park and the old urban settlement of Sintra, a large-scale forestry area complemented by a low-density urban area (CCDRLVT, 2002: 57);

[21] This is a direct result of the fact that this was the territory planned by the PUCS.

[22] Originally the Salazar Bridge, opened during the dictatorship, in August 6, 1966.

12. Northern Atlantic Coastline; the Atlantic agricultural area in the Northern Bank of Tagus, from the Sintra Mountain to Mafra (and which, from Mafra, extends to the outside of the AML), mostly constituted by small-scale agriculture, which was being abandoned and replaced by unplanned industrial facilities (CCDRLVT, 2002: 57–58);
13. Northern agricultural inland; large agrarian land extending beyond the AML, with specific agricultural and forestry systems, fundamental for the production of wine, cheese and vegetables with an already established market. The Malveira urban area is instrumental to support this agricultural system. The Mafra Reservation is important for environmental safeguarding. Other unplanned uses were starting to be verified, namely industrial (CCDRLVT, 2002: 58);
14. Carregado/Ota/Azambuja; territory immediately north of the AML, but not included in it, key land-uses are industry (including extraction) and logistics (CCDRLVT, 2002: 58–59);
15. Eastern agroforestry; a territory mainly used for cork oak plantation, which was already being replaced by irrigation agriculture; partially outside of the AML, it includes the cord Samora Correia–Porto Alto–Benavente (CCDRLVT, 2002: 59–60);
16. Tagus Wetland; the most important agricultural area of the AML, particularly characterized by irrigation agriculture that must be protected (CCDRLVT, 2002: 60);
17. Sado Estuary; only the Northern Bank of which belongs to the AML, which has great environmental importance (CCDRLVT, 2002: 60).

Most territorial units comprise several sub-units, allowing for complementary land-uses to be conceptualized and for relations between different units to be articulated.

Considering agricultural and agroforestry land as rural and urban and economic activities spaces as urban, the model of the PROT-AML concentrates urban soils around the Tagus Estuary, in a divergent sprawl on the Northern Bank of Tagus (from Lisbon to Cascais, Sintra, Ericeira and Vila Franca de Xira) and a convergent sprawl on the Southern Bank (from Almada, Seixal, Barreiro, Montijo and Alcochete to Setúbal). This semi-continuous urban area is loosely surrounded by ecological corridors. All around this urban area, the rural area is kept in the Northern and Eastern zones.

In this context, the PROT-AML promoted a diffuse distribution of cores of third sector cores and they were obviously more frequent within the urban central zone of the Lisbon Region, although there were some examples within the rural crown. Cores for industry and logistics were located in the peripheries of this central urban area and as such were also close to the rural crown.

Two key infrastructures are to be noticed: the first the Chelas–Barreiro Bridge and the second, the Ota Airport, which, despite being outside the AML is defining to its infrastructure plan, of particular importance for the Lisbon–Vila Franca de Xira axis. All of the key areas for these infrastructures were classified by the PROT-AML itself as 'critical urban areas' or as 'problematic spaces'. Interestingly, neither of these two projects was followed through, and although interest in the Chelas–Barreiro Bridge has recently emerged again, the second airport of the AML is now planned for Western Montijo, reusing a pre-existing military aerial basis.

The territorial strategy advanced by the PROT-AML had a clear pattern based on a specific set of points:

- A centre in the Lisbon city with cores of services-equipment clustered in the surrounding crown, becoming more disperse the more peripheral they are;
- Industrial cores and circuits are more distant from urban centres and located in peripheral areas;
- Links to the outside of the region to Torres Vedras (through Loures), to Alenquer and Azambuja (through Vila Franca de Xira), to Samora Correia–Benavente (through Vila Franca de Xira and through Montijo);
- Links between the two banks of Tagus through three bridges (two of them pre-existing), to Almada, Barreiro and Alcochete, all converging in Lisbon;
- A rural ring, restrained to the peripheral areas of the region.

When reading the PROT-AML, nearly 20 years after its completion, and considering how much of it will never come to pass, one must understand the conditions that the regional planning team encountered. In its introductory chapters, the PROT-AML notes the following situation:

> The metropolitan model of settlement and urbanization is evolving from a compact urban structure, a radial structure organized around railways on the Northern Bank and a set of riverfront centres on the Southern Bank of Tagus, to an increasingly radial and polynuclear network, despite maintaining the predominance of its centre—the city of Lisbon—with respect to employment, services and specialized equipment (CCDRLVT, 2002: 10).

Although a plethora of competent plans have been designed throughout the years and the first attempts at an integral regional planning were attempted already at the time of Agache, Pacheco and Gröer, such efforts were continuously undercut, mostly by the Central State. Indeed, this is one aspect where there is great continuity between the dictatorship and the democracy, for while the latter has created a more complete and truly modern model for spatial planning, it proved unable to resolve the key problem associated with the lack of efficiency and strength of planning: its political weakness, unrespectability and isolation from many sectoral policies (Ferrão, 2011). Indeed, in 2009 the PROT-AML was ordered a revision, which was never approved—the legal status of the PROT-AML is thus by now unclear, for although its original version was to be implemented until 2022, the lack of approval of the revision makes it unclear whether or not it should be considered still active.

In the absence of planning, the territory ended up being transformed autochthonously, often without the involvement of the public sector in any capacity, or sustained by PDMs, many of them not properly updated. The PROT-AML diagnostic is very significant because it notes a change in an occupation pattern that runs through the whole of the twentieth century. Accordingly, the Lisbon Region would be loosening its highly centralized character, and other centres were emerging, particularly around specific types of activities developing in specific municipalities. The PROT-AML seems to identify this change as being generally positive—and there are good reasons that it would. One of them is the fact that merely 5 years prior, architect Richard Rogers had published his 'Cities for a Small Planet', an anthology of lectures in which he discussed his proposed model for a sustainable urban environment, the compact city. Conceived to contain suburban sprawl, decrease car-dependency, improve urban ecological conditions and favour a more circular metabolism (Rogers, 1998), the compact city model was promptly accepted as a planning paradigm, and it was even promoted by the EU and minutely tested and critiqued by several authors (Burton et al, 1996; Williams et al, 2000).

Read against the basic principles advanced by Rogers, the PROT-AML can be said to generally correspond to them, although the most important one is left out: namely, the consideration of the metabolic performance of the territory. As such, the key strategic goals of the PROT-AML were always treated as more or less self-contained realities, as if their stipulation through land-uses would ensure an ecologically balanced territorial system. However, as even Rogers was already noticing, the matters of urban environment, transportation, population density and land-use are closely knitted together and co-determine each other.

Rogers also noticed the importance of mixed land-uses within the context of a polynuclear urban structure, and although the system of Territorial Units classified in the PROT-AML mostly comprised mixed-use territorial units, this was not done in the urban centres and other urban areas, but rather in peripheral cores. Here, the PROT merely confirms the dynamics already at play within the territory. But by not acknowledging its metabolic aspect, the plan missed an opportunity to promote a more sustainable land-use in urban areas, particularly in their relationship with activities and land-uses traditionally understood as rural.

Within the polarized urban–rural relation inherent to the PROT-AML, the potential of productive landscapes helps to improve urban sustainability. The ecological corridors—and in general the Environmental Structure (*Estrutura Metropolitana de Protecção e Valorização Ambiental*)—seem to predict only 'natural' (i.e. unproductive) green spaces within urban areas, and even these were suggested with little complexity as if they would merely continue whatever types or ornamental gardens or woods areas are already in existence.

Thus, despite its briefly stated commitment to sustainability, the PROT-AML sought to develop the territory without calling into question the very opposition between urban and rural land. This ended up translating into a horizontal metabolism, with biomass entering urban perimeters from outside and with no use for waste in close-by agricultural activities.

When disclosing the very aims of spatial planning at a regional scale (CCDRLVT, 2002), an a priori emerges immediately: that infrastructure and equipment were the most powerful factors for structuring the territory. But this mention of infrastructure tended to be understood as 'grey' infrastructure, with 'green' infrastructure being administered by the Environmental Structure, as if it could not be, by itself, another specific element of territorial articulation and cohesion.

Indeed, this limitation is as cultural as it is legal since by limiting regional planning to locate infrastructure and equipment, the Portuguese law stripped it from any real possibility of balancing and planning a sustainable territory. Particularly in metropolitan areas, where the environmental harms caused by twentieth-century urbanization were already quite considerable, it would have been fundamental that regional planning could have greater binding power over the metropolitan hinterland, a reality that must eventually be acknowledged in a country like Portugal, where the State owns little forestry or agricultural land. Regional planning should have a greater say in not only protecting the rural metropolitan structure[23] but also in promoting and transforming it as necessary to contain environmental liability—in other words, to plan it. However, as it is, the PROT-AML cannot avoid but respond to a model of territorial development and sustainability dictated by the restraints of capitalist dynamics.

Maybe it comes down to the basic aims of the plan, the ambitions that it has. It is case-in-point to remember the key aim of the PROT-AML is to internationalize the region.

[23] This is already one of the functions of the RAN.

> To give dimension and centrality, both European and Iberian, to the AML, a favourable and qualified space of euro-atlantic relations, with advanced productive, scientific and technological resources; unique natural, historical, urban and cultural heritage; a land of exchange and solidarity, especially attractive for living, working and visiting (CCDRLVT, 2002: 15).

The vapidity of the writing is perhaps more than just a result of the highly general way in which these basic aims had to be stated. Read in detail, this sentence means nothing, it merely states that the AML is a good city-region and that the PROT-AML intends to make it even better. To not even mention just how overstated some of these descriptions are, the fundamental point is that the PROT-AML wishes to emphasize whatever advantages the region may already have. The rhetoric of 'advanced technological resources', of 'exchange and solidarity' confirms what architect Ernie Scoffham (b.1939) identified as being the dynamics of 'new city states', in which cities compete between themselves, offering themselves to international investment and visitation, by building on whatever advantages they have in terms of resources or heritage (Scoffham, 1994). But the lack of social balance and environmental sustainability in many Western cities are also direct results from this competitive dynamic.

The elaboration and approval of the PROT-AML lacked any impact at the municipal level, perhaps because its emphasis on contention and refurbishing—instead of expansion—contradicted what was, at the early 2000s, the general tendency of PDMs (Pereira & Nunes da Silva, 2008). Indeed, blatant disparities between the spatial model of the PROT-AML and PDMs active at the time did not prompt a revision of the latter.

Particularly interesting is that incompatibilities have often rose around ecological corridors and other green spaces, towards which many financial groups and real estate actors have directed their attention, approving urban parcelling as touristic developments which were never intended to function as such (Pereira & Nunes da Silva, 2008).

If the Lisbon Region was to become more sustainable, a paradigm in planning practices was necessary. In this context, acknowledging the power of PROTs in establishing self-sustaining resilient city-regions is a central challenge.

3.3.2 The Lisbon PDM (2012)

Lisbon is perhaps the municipality in the country with more plans, of several types, in its history. In a sense, this is natural since, being the capital city, it naturally attracted more people and activities and thus demands a higher degree of coordination and regulation. Aside from Improvement Plans from the late nineteenth century, the attempts at designing an integral plan for Lisbon started with Étienne de Gröer (1938–1948) and with the GEU (1959), although none of these would be approved. In 1967, Meyer-Heine finished his plan which, after revisions, was approved 10 years later under the democracy. Strategic Plans and PDMs were also approved alongside the key shifts in the urban structure of the capital.

In 2012, a municipal executive by the PS approved a new PDM, for Lisbon which remains active today, and which would become the key to the city's development. Considering that Lisbon is the capital city had recently undergone a depression phase, with demographic losses and economic decrease, pressures for intervention in the consolidated city were extremely high, which has often turned the PDM into the subject of polemics, from lack of compliance with its stipulations to abuse of its exceptions, highly publicized in the Portuguese press.

In order to understand the Lisbon PDM, it is useful to analyse the introduction prepared by its main intellectual mentor, Architect and Councilman for Urbanism Manuel Salgado (b.1944). According to this text, the PDM is a direct answer to the situation of the city until 2011, when the last Census was conducted, showing a city that was losing population, especially young people, while families increased. Moreover, the text acknowledges that the number of elderly people increased and the schism between the rich and the poor had become more contrasted.

The PDM acknowledges that the majority of the territory is consolidated and major infrastructural needs are met, so the key challenge was to respond to a crisis situation. The response is that Lisbon must become a socially, economically and environmentally sustainable city. This is meant to attract more families, competitiveness and investment, as well as a landscape reconstruction. Other problems identified are resource waste due to a diffuse growth pattern, air pollution and mobility problems. The grand goals are to rehabilitate, rejuvenate and balance Lisbon socially; to prioritize urban rehabilitation; make the city friendly, safe and inclusive; increase the environmental sustainability and efficient resource use in the city; promote innovation and creativity; affirm the identity of Lisbon in a globalized world and create an efficient, participated and financially sustained governance model. While all of these are perfectly legitimate, even laudable principles, it is also true that, in reality, they have been nothing short of banal justifications for secondary activity (Figs. 3.10, 3.11, and 3.12).

Indeed, the key programmatic view of the PDM is that Lisbon must recover from its crisis period by attracting tourism, easing—or not regulating—the conversion of a derelict centre into an area for picturesque promenade, protecting façades but promoting the demolition of interiors and remaining indifferent to the displacement and gentrification processes that were factually kickstarted with this

Fig. 3.10 Graça, Lisbon. Source: Authors

PDM and that suffered only a slow decrease with the COVID-19 pandemic and the subsequent halt of tourism.

From the publication of the PDM in 2012 to the pandemic period, the prices of property and rent in the centre of the city rose exponentially, with the State creating legislation favourable to the private sector and for the displacement of local inhabitants and traders, especially those with less resources (Pavel, 2020). While the introduction to the PDM promises a city that is sustainable from the social perspective, if the properties listed on the AirBnB platform for the centre of Lisbon were all at maximum capacity, there would be more visitants than residents (Pavel, 2020; Pereira et al, 2021) (Fig. 3.13).

Despite the promise of an environmentally sustainable city, the PDM also maintained the tendency, in practice since the GEU Masterplan, to classify the whole municipal territory as urban, although divided into two operative categories—'Consolidated spaces' and 'Spaces for Consolidation'—as well as six functional categories—'Central and Housing Spaces', 'Economic Activity Spaces', 'Green Spaces', 'Special Use for Equipment', 'Special Use for Infrastructure' and 'Special Use for the Riverfront'. The land-use system also includes a relevant set of dispositions for the Municipal Ecological Structure (*Estrutura Ecológica Municipal*—EEM), and an articulating system of Structuring Corridors that includes public and private areas to connect green and urban areas.

A key innovation of the Lisbon PDM, in the context of the Metropolitan Region, is that its land-use system is based on morphological criteria, distinguishing specifically between the most frequent types of urban fabrics of the city, with specific legislation, mostly to ensure continuity. To be sure, such a land-use system has been in place in the PDM for Porto since 2006 (Oliveira, 2006). There are also detailed lists of protected and to-be-protected estates, charters for cultural and natural heritage, transportation interfaces, motorway structure, bicycle circuits, criteria for parking spaces and for bicycle circuits.

While the totality of the Lisbon land is classified as urban, there are several official types of 'Green spaces', some of

Fig. 3.11 Estrela, Lisbon. Source: Authors

which include all the land used for productive activities. Typologies include 'Green spaces for leisure and production', 'Green spaces for protection and conservation', 'Green spaces for framing infrastructures' and 'Green spaces in the riverfront'.

The food system is never referred to directly, which is hardly surprising, considering that rural land is non-existent and industrial activities have decreased to a negligible amount. However, urban agriculture is referred in the 'strategical goals' to promote a sustainable and efficient city and provide continuity with natural systems and increase urban resilience. Gardens are also mentioned as a component of the central urban areas and former rural settlements. There are green spaces for agricultural production, meaning areas for conception, management and use of cultivated space and nature preservation, if compatible with other urban uses. The Municipal Charter of Heritage includes landscape heritage that includes gardens, gazebos, preserves, footrails, graveyards, farms, closes, parks and agricultural spaces. Special attention is paid to patios both public and private, safeguarded as landscape, environment and heritage features whose main use must be leisure. In this context, the PDM emphasizes the neighbourhoods of the old centre, where historically there have been practices of urban agriculture that the PDM wishes to maintain. Maintenance is due to landscape value, not productive value. Proving it is the fact that Bairro Alto, Lapa and Madragoa are included in Operative Units without dispositions regarding urban agriculture or its safeguarding, while they have abundant dispositions that clearly favour tourism, commerce and cultural functions or connections to the riverside, an obvious hotspot for urban redevelopment projects. A programme for incentives for urban agriculture is indeed planned for several peripheric areas, including Lumiar, Ameixoeira, Telheiras, Carnide, Olivais, Oriente and Marvila and Benfica.

'Green spaces for leisure and production' are defined as permeable and cultivated spaces on organic natural soil, public and private, including gardens, patios, as well as historical farms and monastery spaces, used for leisure urban agriculture, and which may integrate public equipment and

Fig. 3.12 Martim Moniz, Lisbon. Source: Authors

supporting infrastructure, as well as stores for food and drinks and tourism. Moreover, urban agriculture is encouraged here as important for increasing local food production and self-sufficiency, although the spatial implications of this emphasis on productive space are not addressed. The lack of further dispositions makes this extremely irrelevant in the general context of the PDM. The writing of the item seems to trivialize the productive aspect, and urban agricultural spaces are generally not protected by any of the Municipal Charters, nor are they granted any specific protection from future urban development.

Although some specific propositions of the PDM could offer opportunities for the food system in Lisbon, the most powerful opportunities are provided by its genesis and its goals. The fact that the PDM wishes to make Lisbon a sustainable territory is important, mostly given that sustainability is understood in environmental, social and economic terms. Given that the food system relates to the three components, it would be extremely wise to include it as a means to achieve sustainability. A conscious and consequent inclusion of the food system in the Lisbon planning system would go as far as demanding a different land-use system, making the territory more multifunctional and the landscape more diverse. The concept of landscape heritage offers a good entry-point for food system activities since the *non-aedificandi* status of most of them would make it easier to incept agricultural uses, which would further encourage an engagement of people with this type of heritage while maintaining it constantly used and informally surveyed.

The Lisbon PDM is innovative for the concepts it employs, providing instruments to assess the urban differences at play in the whole territory. This is laudable, however, in terms of its contents and its implications, it can hardly be described as a balanced plan. If it is a plan written for stopping residents from leaving and to increase their quality of life—including in the environment—it often

Operative categories	Functional categories
Consolidated spaces	Central and housing spaces
	Economic activity spaces
	Green spaces
	Special use spaces for equipment
	Special use space for infrastructure
	Special use space in the riverfront
Spaces for consolidation	Central and housing spaces
	Economic activity spaces
	Green spaces
	Special use spaces for equipment
	Special use spaces in the riverfront

Fig. 3.13 The lisbon PDM. Source: CM Lisboa (2012)

Fig. 3.14 Alto do Restelo, Lisbon. Source: Authors

seems to have actively contributed to the emergence of new problems. The privileging of urban rehabilitation instead of new construction is producing repetitive designs, while lack of compliance with the PDM seems to be tolerated on important works (Figs. 3.14 and 3.15).

Moreover, the goal of bringing people back to the city is never addressed in its full complexity on the PDM, since redevelopment and regeneration projects which have been promoted under it belong to a market that often can only be accessed by upper-class people, sometimes in conflict with the aspirations of living communities. Gentrification processes have been denounced all over the city, under this PDM. With respect to urban agriculture, despite the attention it receives in the PDM, it is superficially approached and, as we will see in Part III, a contradiction may arise when the PDM notices its cultural importance in the city centre but promotes it mostly at the periphery. This presupposes a hierarchization of activities that clearly places urban agriculture as an inadequate activity for the central areas of the city, the ones visited by tourists and the upper-classes. Thus, there is more innovation in the approach to planning than to the city, and while there are abundant signs of ecological and environmental concerns, the results do not go beyond a recognition, long overdue, of the true complexity of green spaces within cities. But both urban agriculture and green spaces are, all things considered, mostly a decorative element of the aspect the PDM really seeks to determine, that of construction.

Fig. 3.15 Parque das Nações, Lisbon. Source: Authors

3.3.3 The Moita PDM (2015)

Moita's original PDM was prepared in 1983, over a year after the publication of the first PDM Act, and the municipality, despite its relatively suburban position in the Lisbon Region, became an example of an efficient and timely planning initiative. In 1992, this PDM was adapted to the alterations brought forward by the 1990 Law for Municipal Spatial Planning. In 2015, a new PDM was approved, under the municipal administration of a socialist coalition between the Electoral Front of the United People (*Frente Eleitoral Povo Unido*—FEPU) and the APU.

The territory of this municipality extends for 55.26 km^2 divided between four civil parishes, and its population in 2011 was 66,029 inhabitants. The roots of inhabitation lie in the Middle Ages when the first written documents testify to the existence of a settlement near the current town of Alhos Vedros. Throughout the centuries, small settlements sprang at the fringes of forestry and pine woods, with two standing out by the late sixteenth century, Moita and Sarilhos Pequenos. The Tagus River is an important landmark of the territory, as it breaks the northwestern area into two parts through a reentrance, from which several '*esteiros*' (water channels) are carved. By the mid-twentieth century, Moita concentrated on important food-related businesses, some of which depended directly on this relation between the land and the river, like saltpans and oyster collection. Others included several types of livestock, the cultivation of potato, cauliflower, orange, cabbage and wine, as well as smoked meats and cheese.

Over time, as urbanization grew, a lot of the territory of Moita was transformed. Currently, urban cores like Moita, Alhos Vedros, Baixa da Banheira and neighbourhoods like Vale da Amoreira or Quinta da Prata share the territory with small villages like Gaio-Rosário relatively untouched by modernization, and with low-density suburbs promoted by the private sector, some of which are not completed. This urban fragmentation that afflicts many of the inland territories of the Southern Bank of Tagus is mediated by rural land, where agricultural and forestry activities remain productive.

The food system is present mostly through dispositions for rural—and specifically agricultural—space. 'Rural soil' includes soils for agriculture and livestock with significant territorial and dynamic dimensions that justify its permanence and sustainability, but also partially cultivated land, riverside areas including mallards, vivariums, saltpans and beaches, areas for hydric network protection, infrastructure areas, collective equipment, industry, storage, commerce, services and tourism not included in urban perimeters.

Each different category of rural space has specific indications, which include spaces for agriculture and livestock, periurban agricultural spaces, natural spaces, mixed-use spaces, infrastructure, equipment, rural settlement housing and motorways. These categories apply according to the dominant uses in the land, forms of settlement and structuring levels. Construction is generally limited to supporting facilities and low-density housing, as well as structures for environmental and biophysical protection (Figs. 3.16, 3.17, 3.18, and 3.19).

In natural spaces (on rural soils), facilities for environmental education and leisure are allowed, as well as protection of saltpans and aquaculture, providing some opportunity for creating structures and initiatives in which environmental education is provided through food-related projects that focus on the links between the community and nature. Furthermore, in mixed-use spaces (on rural soils again), cultural and scientific equipment is allowed, opening the possibility of research relating to forms of agriculture and food production in the municipal territory. The category of Natural Spaces for Protection of the Hydric Network (on urban soils) also provides opportunities for cultivation which could play an important role in the local food system.

The Moita PDM acknowledges the importance of agriculture not only as an activity but also as a generator of specific forms of settlement which must be categorized according to their specificity. Although more could be done to achieve a more multifunctional landscape in the municipality, several aspects of the PDM already suggest some flexibility in the type of activities allowed in urban soils. Green and protection urban spaces are key to this problem, but the fact that the PDM is so attentive to agricultural issues makes it probable that it could easily be used for important food system strategies.

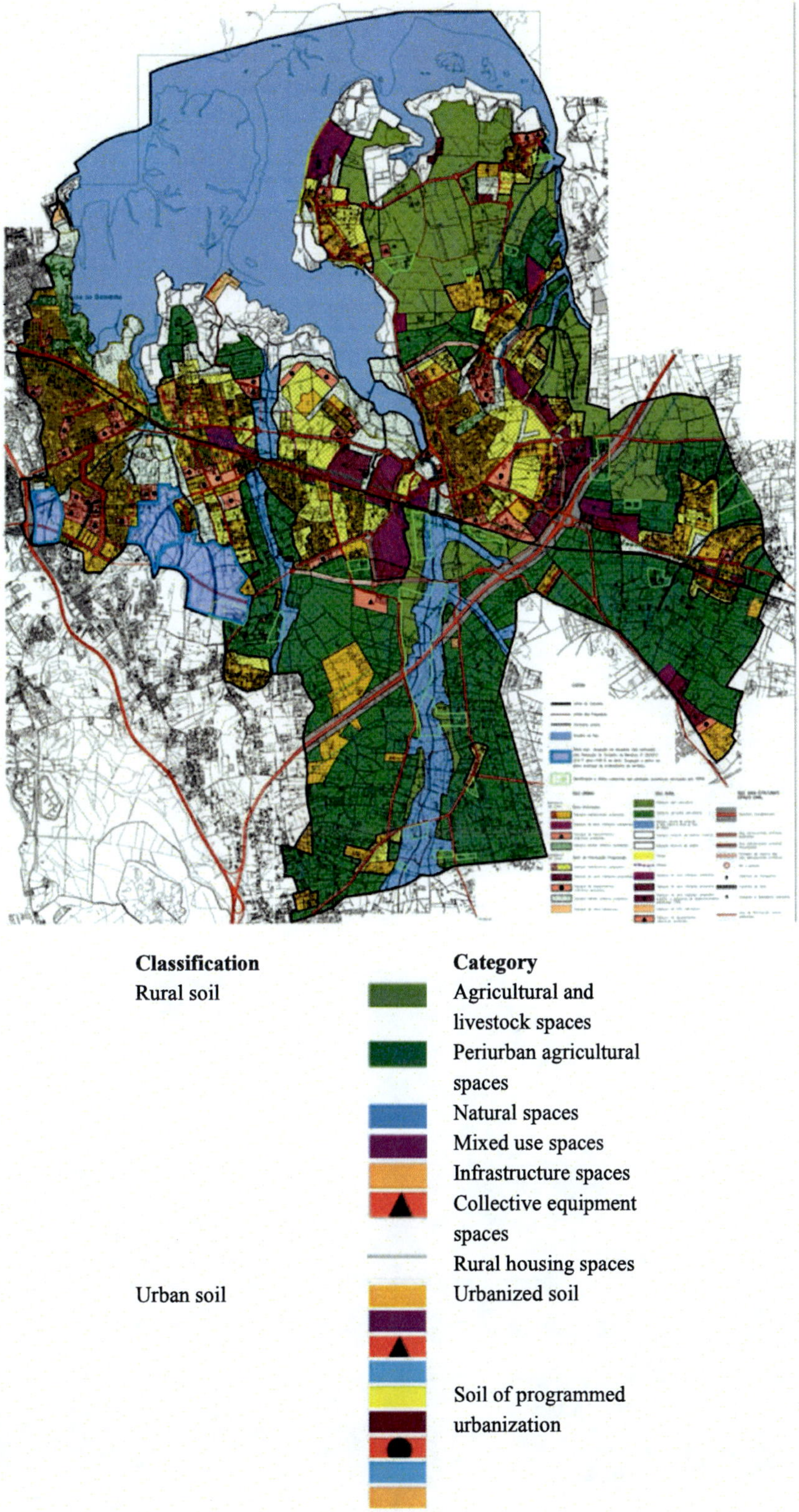

Fig. 3.16 Land-use chart of the Moita PDM (2015). Source: CM Moita (2015)

Fig. 3.17 A street and the Tagus River reentrance, Alhos Vedros, Moita. Source: Authors

Fig. 3.18 Moita centre with rural remnants and the bullfight arena surrounded by modern buildings, Moita. Source: Authors

Fig. 3.19 Barra Cheia, a village in Moita. Source: Authors

3.4 The 2014 Spatial Planning Act

In 2014, a new Planning Act[24] was published. While in many aspects it is more of a clarification of the previous Planning Act, it brought important changes, namely the establishment of the EEM, integrating green-fields to value environmental protection of natural resources, relieving the pressure to urbanize unoccupied land (Mourão & Marat-Mendes, 2015). There are five further indications with new content that were not present in the 1998 Act. This is the case with sustainable development; territorial resilience; the safeguarding coastal areas and riverbanks; territorial regeneration including the refurbishing of degraded or illegal urban areas; and the promotion of mobility for those with reduced mobility. It presents a definition of sustainable development:

> fulfilling needs of the present without compromising future generations, including preservation of natural resources and cultural heritage, capacity of ecosystems to be productive on the long run, rational and balanced management of territory to mitigate regional asymmetries, promotion of territorial cohesion, sustainable production and consumption of energy, safeguarding biodiversity, biological balance, climate and geological stability, harmonizing human life and the environment (DL31/2014, Art.2-a).

This definition encompasses environmental, social, cultural, economic and territorial aspects, understood in spatial and chronological terms. However, as it means to provide guidelines for policy and not design, it lacks specifications on how such goals are to be spatialized. Although it is significant that the 2014 Planning Act includes concerns for sustainability, these are incepted into a planning structure and a distribution of powers in existence for much longer. Considering environmental threat and social injustice, the challenge would be not to incept sustainability into the existing framework, but rather to reconceive planning under a sustainability perspective, promoting urban design in strategic areas such as access to resources, since urban sustainability is not only a matter of land-use, but of proper urban design (Codoban & Kennedy, 2008; Kennedy et al, 2011; Marat-Mendes et al, 2021).

The new Planning Act seems considerably anxious about demonstrating a democratizing spirit, emphasizing the Constitutional right to private property—perhaps a result of having been prepared and approved by a right-wing government. It charges the State, the Autonomous Regions and municipalities to promote public territorial policies and monitor their implementation while highlighting warrants of equity and transparency for land-related problems and ensuring the existence of public spaces for infrastructures, equipment and collective areas. Although it also establishes that everyone has the right to a rational territorial management so that public interest must be the major drive behind it, it specifies nothing on how to resolve the long-standing tensions between these two rights it so clearly expresses.

With respect to the planning hierarchy, this new Planning Act maintains the scheme proposed by its predecessor, while adding some new planning instruments. In the intermunicipal scale the Intermunicipal Director Plan (*Plano Director Intermunicipal*—PDIM), the Intermunicipal Urbanization Plan (*Plano de Urbanização Intermunicipal*—PUIM) and the Intermunicipal Detailed Urbanization Plan (*Plano de Pormenor Intermunicipal*—PPIM) were added. Although the practical possibilities of these plans are very seductive and may contribute towards stronger cooperation between municipalities, ultimately laying the foundation for regional-scale planning, these new planning instruments are variations of already existing instruments, namely the PDM, the Detailed Plan and the Urbanization Plan, taken into an intermunicipal context. It is also specified that Special Programmes apply to fundamental goals of public interest and of resources, including coastal plans, programmes for protected areas and programmes for bayous and firths. However, the planning instruments hierarchy was also not fundamentally changed.

One of the aspects that has been more strongly highlighted with respect to this new Planning Act and its corresponding Juridical Regime is the elimination of spaces for urbanization. These existed in nearly all PDMs and corresponded to land, usually rural, where urbanization was allowed. The apparent goal of the Government at the time is to reduce speculation and the overvaluing of properties in these spaces in comparison with others where urbanization is precluded. A five-year transition period is allowed for the urbanization of spaces yet classified as such, and if this urbanization does not happen, the land is automatically classified as rural. Yet, the outcome of these options may be just the opposite of what the Government announced. Indeed, by decreasing the areas where urbanization is allowed, those where it really is obviously become even more valuable and prone to speculation, while the owners of small and medium properties who may lack the funds to conduct urbanization in their land will see their property devalued after the transition period (Varino, 2015).

3.4.1 The Setúbal PDM (2019)

Setúbal had a first Municipal Masterplan right after the Revolution, in an attempt to resolve the continuous revisions of the Urbanization Plan for the city whose first version had been presented in 1944 by João António Aguiar, but which

[24] Decree 31/2014 (Lei de Bases Gerais da Política Pública de Solos, Ordenamento do Território e do Urbanismo) published in Diário da República 104/2014, Série I, 30–5-2014.

underwent several polemics with the local population—due to its aggressive demolition proposals—leading to constant revisions and reformulations (Faria, 1981; Lôbo, 1995; Pereira, 2007). Another key source of conflict was Aguiar's classification of the northern floodplain orange groves as land for urbanization, which threatened the rural component of the municipality (Pereira, 2007).

Being a key city of Lisbon Region and its link to the Alentejo region, southwards, Setúbal has a long history, officially becoming a city in 1860. Its dynamic urban life partly results from being what Kostof (1991) terms 'natural harbour cities', which have sweeping backdrops with offset street-sweeps. Setúbal's urban structure was defined, until the nineteenth century, by east–west main streets, and key changes were induced by the construction, in the same century, of the port and of Avenida Luísa Todi parallel to the coast, which replaced the defensive line and created a modern "boulevard" prepared for urban life (Faria, 1981; Pereira, 2007). Until the 1950s, the landscape of Setúbal was defined by dispersed residential neighbourhoods housing a precarious and dispossessed workforce, separated by farmsteads and other rural land belonging to the local middle-class (Faria, 1981).

Modernization in Setúbal depended largely on industrial development, with a strong presence of the preserve industry, particularly canned fish. Several neighbourhoods or shanty towns emerged in the periphery to house the working-class whose income was uniformly low (Faria, 1981). Indeed, low wages and deprivation were constant in the first half of the twentieth century in Setúbal, causing a continuous "chronic housing crisis" (Faria, 1981: 110), a key motivation for the city council to seek a planning solution. However, the contrasts between Aguiar's vision and the aspirations of the population, as well as the tendency of the New State to ignore popular needs to ensure the interests of the industrial bourgeoisie in accessing low-paid workers rendered nearly two decades of planning efforts completely fruitless (Faria, 1981).

This may have prompted the agility of the municipality, in the post-revolutionary scenario, to seek a planning policy applied to the overall municipality, even at a time when no legal framework was available to guide such plan (Fig. 3.20).

Fig. 3.20 Setúbal centre. Source: Authors

While in 1994 Setúbal approved its PDM, in accordance with the 1990 Law on Municipal Spatial Planning, this instrument remained active until very recently. Considering the importance of Setúbal as an articulation between the Lisbon Region and that of Alentejo, as well as the importance and dimension of its industrial and portuary structures, such an outdated planning situation was worrisome.

In 2019, a new version was finally advanced, being currently in the period for public consultation (CMSetúbal, 2020). It establishes a strategic approach, seeking to promote Setúbal as a competitive centre with superior urban functions, as a platform for portuary, logistic and business activities, as a place for cultural and natural tourism, strongly compromised with environmental quality. The Municipal Ecological Structure has among its functions leisure, production and environmental protection, as well as the promotion of a sustainable use of the territory and the preservation of biodiversity, and the articulation of urban and rural land. Ecosystem services are to be promoted in these spaces, with a specific mention of food production, including vegetables and livestock. The reducing of urban metabolism is also mentioned directly, with the PDM promoting regeneration, renovation and refurbishing in both urban and rural soil. Urban agriculture and harvesting of rainwater are both references in the mitigation of climatic change (Fig. 3.21).

It importantly acknowledges the diversified rural landscape within the municipality, predicting three types of agricultural land as well as a land-use category for rural settlements and areas of disperse building. A general description notices the western area of the municipal territory, filled with agricultural zones mediating with urban zones—predominantly residential—and with forestry zones; then a nearly continuous chord of agricultural land extends towards the city centre, surrounded by forestry. The centre itself is surrounded by agricultural land and, to a lesser degree, by forestry areas. The eastern area is dominated by forestry, but there is an important presence of agriculture surrounding areas of disperse building, stemming almost continuously from the eastern end of the Setúbal city. Some of these areas are included in the RAN and thus subject to its restrictions.

As often happens, the majority of food-related aspects are contained in the regulations for rural land, which is directed for agriculture, forestry and livestock, as well as natural spaces, thus precluding urbanization and restricting constructions beyond those for individual residence and support for agricultural activities. The PDM specifically classifies as agricultural all soils which currently have such use—in order to preserve it—seeking to counter fragmentation of cultures and the destruction of agricultural land, ensuring also the protection of associated built structures such as wells, threshing floors, waterwheels, windmills and stone walls.

There are three subcategories for agricultural land. The first, Agricultural Spaces for Production, are those with the highest capacity, integrated with the RAN. Construction is fairly restricted, and compatible activities include livestock, forestry, agroindustry, rural tourism or agricultural commerce. The second consists of Other Agricultural Spaces, where agriculture is the dominant activity, but there is no inclusion in the RAN. Restrictions on construction do not change, but compatible activities also include exploration of energetic resources, transformation and commerce of agricultural produce, camping and trailing, hotels and leisure spaces. The third subcategory, Other Agricultural Spaces—Farms, applies to a specific set of contiguous properties, west of the Setúbal city and west of the Azeitão town, whose morphological, typological, landscape and heritage aspects all must be integrally preserved. Compatible uses include also winehouses, mills, cheese dairies, artisanal industry, social and cultural equipment, camping and trailing, commerce and services.

The land-use of Rural Settlements applies to specific compact areas in territories where agriculture, forestry or nature conservation is undertaken, including Aldeia da Portela, Aldeia de São Pedro, Aldeia Grande, Grelhal and Portinho da Arrábida. In these areas, consolidation, urban structuring, functional upgrading are priority actions, eventually justifying detailed plans. Compatible uses are housing, services, industry (if compatible with housing), tourism, leisure equipment and agricultural supporting structures.

Another important land-use when considering food-related activities is that of Urban Low-Density Spaces, which are located on periurban territories, presenting degrees of urban–rural hybridity, with the coexistence of agriculture, urban buildings, equipment and infrastructure. The PDM proposes the safeguard the pattern of lowest space occupation, privileging individual detached houses and the completion of urban fabrics, while specifically safeguarding existing practices of periurban agriculture (Fig. 3.22).

While the Setúbal PDM embodies many of the aspirations that seem implicit in the 2014 Planning Act, emphasizing environmental preservation and sustainability strategies, the PDM is also very attentive to the existing reality of its territory. Although there is no clear food legislation that the PDM comes to spatialize, its minute assessment of the territory leads it to a nuanced account of the diversity of rural realities, encompassing different scales and types of agriculture in their relation with different scales and types of settlement. The promotion of urban agriculture and water harvesting in urban areas is also an important step towards better use of resources—including land—although the PDM could go further in establishing or promoting links between urban agricultural spaces and their connection with the rural agricultural soils, which are already significantly continuous, being interrupted only in the Setúbal city itself (Table 3.1).

Classification	Category
Rural soil	Agricultural spaces
	Forestry spaces
	Natural spaces
	Rural settlements
	Areas of dispersed building
	Touristic spaces
	Infrastructure and equipment spaces
	Exploration areas for geological and energetic resources
	Industrial activities spaces
Urban Soil	Central spaces
	Housing spaces
	Urban spaces with low density
	Economical activities spaces
	Special use spaces

Fig. 3.21 Land-use chart of the Setúbal PDM (2019). Source: CM Setúbal (2020)

Table 3.1 Plans analysed in Part I from the Lisbon Region, 1937-2019

Plan/Foreplan	Date	Condition	Authors	Scale
Agrarian Colony of Santo Isidro de Pegões	1937–52	Approved and fully implemented	Junta de Colonização Interna (JCI) Plan: Mário Pereira and Henrique de Barros Architectural design: Henrique Brando Albino and José de Oliveira Trigo Public buildings: Eugénio Correia Agrological study: J. Botelho da Costa and António Sardinha de Oliveira Hydrological study: Engineer Teixeira Duarte	Local
Regional Foreplan from Moscavide to Vila Franca de Xira	1946–55	Approved as a Foreplan, partly implemented	Direcção-Geral dos Serviços de Urbanização (DGSU) Étienne de Gröer, Nikita de Gröer and Fernando Lorenzini Borges Campos	Intermunicipal/Sub-regional
Vila Franca de Xira Urbanization (Fore) Plan[25]	1946	Not approved	Direcção-Geral dos Serviços de Urbanização (DGSU) Miguel Jacobetty Rosa and João Guilherme Faria da Costa	Local
Palmela Urbanization (Fore)Plan	1948	Approved for revision	Direcção-Geral dos Serviços de Urbanização (DGSU) João António Aguiar	Local
Palmela Urbanization (Fore)Plan—Remodelling	1952	Approved as a foreplan, partly implemented (street-system)	Direcção-Geral dos Serviços de Urbanização (DGSU) João António Aguiar	Local
Lisbon Region Masterplan	1959–64	Not approved, informally followed at lower planning scales	Ministry of Public Works—Directorate-General of Urbanization Services	Regional
Almada Masterplan and Partial Plans	1950–65	Approved, not implemented	Gabinete de Urbanização da Câmara Municipal de Almada (GU-CMA) Chief-architect: José Rafael Botelho	Local
Santo António dos Cavaleiros Urbanization	1964–1989	Approved and implemented	ICESA Fernando Reaes Pinto	Local
Amadora PDM	1994	Approved, still active	Câmara Municipal da Amadora (Amadora City Council)	Municipal
Alcochete PDM	1997	Approved, still active	Câmara Municipal de Alcochete (Alcochete City Council)	Municipal
PROT-AML	2002	Suspended in 2007	CCDRLVT	Regional
Lisbon PDM	2012	Approved, still active	Câmara Municipal de Lisboa (Lisbon City Council)	Municipal
Moita PDM	2015	Approved, still active	Coalition: FEPU—Frente Eleitoral Povo Unido (Electoral Front of the United People) + APU —Aliança Povo Unido (United People Alliance) + PCP/PEV—Partido Comunista Português/Partido Ecologista os Verdes (Portuguese Communist Party/Ecologist Green Party)	Municipal
Setúbal PDM	2019	Approved	CDU—Partido Comunista Português/Partido Ecologista Os Verdes (Portuguese Communist Party/Green Party)	Municipal

[25] Plan is analysed in Chap. 9.

Fig. 3.22 Street in the town of Setúbal. Source: Authors

References

Abrantes P, Fontes I, Gomes E, Rocha J. (2016) Compliance of Land Cover Changes with Municipal Land Use Planning: Evidence from the Lisbon Metropolitan Region (1990–2007). Land-Use Policy 51:120–134. http://doi:https://doi.org/10.1016/j.landusepol.2015.10.023

Baptista A, Henriques J (1985) Os custos sociais de urbanização e o financiamento dos municípios da Área Metropolitana de Lisboa. Sociedade e Território 3:31–37

Bordalo A (2007) Bairro do Alto dos Moinhos: estudo de caso. MSc Arquitectura: Cultura Arquitectónica Portuguesa, ISCTE, Lisboa

Burton E, Jenks M, Williams K (1996) The Compact City. A Sustainable Urban Form? Routledge, London

Cabral J, Portas N (2011) As transformações da regulação: processos e actores. In: Portas N, Domingues A, Cabral J (eds) Políticas Urbanas II. Fundação Calouste Gulbenkian, Lisbon, pp 232–267

CCDRLVT (Comissão de Coordenação e Desenvolvimento Regional de Lisboa e Vale do Tejo) (2002) Plano Regional de Ordenamento do Território – Área Metropolitana de Lisboa PPROT-AML. Published in Resolução do Conselho de Ministros n.o 68/2002. Diário da República, Série I, Nº 82, 8 Abril 2002, 3287–3328. http://www.ccdr-lvt.pt/pt/plano-regional-de-ordenamento-do-territorio-da-area-metropolitana-de-lisboa/54.htm

CCDRLVT (Comissão de Coordenação e Desevolvimento Regional de Lisboa e Vale do Tejo) (2008) Conceitos e Objectivos. In: Reserva Ecológica Nacional (REN). Law 239/2012, published in Diário da República, 1.ª série — N.º 212, 2 de Novembro de 2012, 6308–6345. http://www.ccdr-lvt.pt/content/index.php?action=detailfo&rec=1347&t=Conceito-e-Objectivos

CMAlcochete—Câmara Municipal de Alcochete (1997) Plano Director Municipal de Alcochete. Published in Diário da República n.º 193/1997, Série I-B, 22–8–1997

CMAmadora—Câmara Municipal da Amadora (1994). Plano Director Municipal da Amadora. Published in Diário da República 142/1994, Série I-B 22–6–1994

CMLisboa—Câmara Municipal de Lisboa (2012) Plano Director Municipal De Lisboa. Published in Diário da República 168/2012, Série II, 30–8–2012

CMMoita—Câmara Municipal da Moita (2015) Plano Director Municipal da Moita. Published in Diário da República, Série II—N.o 128–3–7–2015

CMVFX—Câmara Municipal de Vila Franca de Xira (2009) Plano director municipal De Vila Franca De Xira. Warning 20905/2009. Diário da República, Series 2–N.224–November 18th

CMSetúbal (2020). Revisão do Plano Director Municipal de Setúbal (in process). See https://www.mun-setubal.pt/pdm-2020/#1581510268663-c4cebb2e-3631

Codoban N, Kennedy C (2008) The Metabolism of Neighbourhoods. Journal of Urban Planning and Development 134(1):21–31

d'Almeida P, Marat-Mendes T, Toussaint M (2020) Portugal's Rising Research in Architecture and Urbanism: The Influence of International Research Centers and Authors. Journal of Urban History 1–28. https://doi:https://doi.org/10.1177/0096144220968078

d'Almeida PB, Marat-Mendes T (2021) A atualidade da investigação científica em arquitetura e urbanismo desenvolvida no LNEC (1961–1979) face ao desafio da sustentabilidade. Cidades Comunidades & Territórios 15 – 40

DGADR (Direção-Geral de Agricultura e Desenvolvimento Rural) (2015) Enquadramento. In: Rede Agrícola Nacional. ttps://www.dgadr.gov.pt/component/content/category/index.php?option=com_content&view=article&id=199:ran&catid=10:ambiente-e-ordenamento&Itemid=309

Faria CV (1981) Novo fenómeno urbano – aglomeração de Setúbal. Assírio e Alvim, Lisboa

Ferreira, BM (2010) (In)Formar a cidade contemporânea. MSc dissertation, Iscte Instituto Universitário de Lisboa, Lisbon

Ferrão J (2011) O ordenamento do território como política pública. Fundação Calouste Gulbenkian, Lisbon

Fiúza F (2013) Torres de habitação em Alfragide [Atelier Conceição Silva]. Passagens 1:172–177

Gonçalves F (1974) Legislação Urbanística Portuguesa 1926–1974. LNEC, Lisboa

Gonçalves F (1978a) A mitologia da habitação social – o caso português. Cidade Campo 1:21–83

Grande N (2012) O Ser Urbano nos Caminhos de Nuno Portas/The Urban Being: on the Trails of Nuno Portas. Imprensa Nacional Casa da Moeda, Guimarães

Kennedy C, Pinctel S, Bunje P (2011) The study of urban metabolism and its applications to urban planning and design. Environmental Polution 159:1965–1973

Kostof S (1991) The City Sahaped. Urban Patterns and Meanings Through History. Thames & Hudson, London

Leite I (2007) Francisco da Conceição Silva: para uma compreensão da obra e do grande atelier/empresa – 1947/1975. MSc dissertation, Faculdade de Letras, Universidade de lisboa, Lisboa

Lopes S, Borges JC, Fernandes R; Marta-Mendes T (2021) Rediscovering modern 'civil architecture': The case of Bela Vista neighbourhood in Setúbal. Paper presentes at the International Conference Optimistic Suburbia II, Iscte Instituto Universitário de Lisboa, 16–18 June 2021.

Lôbo MS (1995) Planos de urbanização – a época de Duarte Pacheco. FAUP Edições, Porto

Lopes RG (1990) Planeamento Municipal e Intervenção Autárquica no Desenvolvimento Local. Escher, Lisboa

Marat-Mendes T, Borges JC, Lopes SS, Pereira MM (2020) Where the fields have no name: Urban-rural transitions in the Lisbon Region planning history. Cidades, Comunidades & Territórios 41:105–132

Marat-Mendes T, Borges J, Dias A, Lopes R (2021) Planning for a Sustainable Food System. The potential role of urban agriculture in Lisbon Metropolitan Area. Journal of Urbanism: International Research on Placemaking and Urban Sustainability. ID: 1880960 DOI:https://doi.org/10.1080/17549175.2021.1880960

Mourão J, Marat-Mendes T (2015) Urban Planning and Territorial Management in Portugal. Antecedents and Impacts of the 2008 Financial and Economic Crisis. In: by Knieling J, Othengraffen F (eds) Cities in Crisis – Socio-Spatial Impacts of the Economic Crisis in Southern European Countries. Routledge, London, p 157–171. https://doi.org/10.4324/9781315725048

Oliveira V (2006) The morphological dimension of municipal plans. Urban Morphology 10(2):101–113

Pavel F (2020) Em Que Casa Fico? Reflexões Acerca do Direito à Cidade e à Habitação em Tempo de Covid-19. Finisterra 55 (114):203–206

Pereira M (ed, 2007) Setúbal – a cidade e o rio. eGeo, Setúbal.

Pereira M, Nunes da Silva F (2008) Modelos de ordenamento em confronto na área metropolitana de Lisboa: cidade alargada ou recentragem metropolitana? Cadernos Metrópole 20(2):107–123

Pereira M, Silva CN (2001) As Grandes Áreas Urbanas – contributos para a definição de alternativas ao modelo institucional vigente. Proceedings do Seminário Território e Administração – Gestão de Grandes Áreas Urbanas. CEGPR/CEG/CESUR/LNEC, Lisbon, pp 73–89

Pereira SM, Matos M, Carreira M (2021) Alojamento local e COVID-19: Inquérito aos titulares/gestores (report). DINÂMIA'-CET, Lisboa

Portas N (2005a) Arquitectura(s). História e Crítica, Ensino e Profissão. FAUP, Porto

Portas N (2005b) Arquitectura(s). Teoria e Desenho, Investigação e Projecto. FAUP, Porto

Portas N, Domingues A, Cabral J (2007) Políticas urbanas – tendências, estratégias e oportunidades. Fundação Calouste Gulbenkian, Lisboa

Pothukuchi K, Kaufman JL (2000) The Food System. J Am Plann Assoc 66(2):113–124. https://doi.org/10.1080/01944360008976093

Rafacho A (2011) Do plano a ação. O envolvimento dos atores no planeamento do espaço urbano. Vale da Amoreira na Moita. MSc dissertaion, FAUL, Lisbon

Rogers R (1998) Cities for a small planet. Basic Books, London

Scoffham E (1994) An agenda for urban design in the new City State. In: Scoffham E (ed) The New City State. Centre for Urban Design – University of Nottingham, Nottingham

Silva SKV (2018) O plano de urbanização da Amadora de Faria da Costa: o projecto e a sua implementação. MSc dissertation. Universidade Lusíada, Lisboa

Tulumello S, Ferreira A, Colombo A, Di Giovanni C, Allegra M (2018) Comparative Planning and Housing Studies Beyond Taxonomy: A Genealogy of the Special Programme for Rehousing (Portugal). Transactions of the Association of European Schools of Planning 2:32–46. https://doi.org/10.24306/TrAESOP.2018.01.003

Varela R (2019) A people's history of the Portuguese Revolution. Pluto, London

Varino P (2015) A especulação imobiliária no novo RJIG. Cadernos Poder Local – Instrumentos de Gestão Territorial em Análise. Seixal, Câmara Municipal do Seixal, pp 33–35

Williams K, Burton E, Jencks M (eds) (2000) Achieving Sustainable Urban Form. E & FN Spon, London

Part II
Feeding the Conurbation

Today, the Lisbon Region is defined by two conurbations sprawling around the estuary of the Tagus River, extending inwards to other significant—albeit smaller and more isolate—towns and villages on a more rural landscape of forestry and agriculture. In its 18 municipalities live nearly a third of the continental Portuguese population. The rural crown lying beyond the conurbations no longer functions as a proper hinterland, its productive potential being insufficient for feeding 3 million people year round.

But this has not always been so. Changes in demography, eating habits and the physical structure of the metropolitan region have taken place, mostly in the twentieth century, with the collapse of traditional economy and society. In this chapter, we put this change into perspective, depicting the characteristic of the past physical structure and its inherent food system organized in a local foodshed around a large consumption centre (the Lisbon city). For such exercise, we start in Chap. 4 by analyzing from this perspective, two surveys conducted by architects on the national territory, highlighting the depiction of the Lisbon Region, first in the late 1950s and afterwards at the early 2000s.

In Chap. 5, we move to contemporary times, highlighting some of the results from a survey conducted in the SPLACH Project focusing on the spaces of food system phases, be them built or unbuilt spaces, and which covers all phases from production to transformation, distribution, commerce, consumption and waste disposal.

Finally, in Chap. 6, we briefly overview the traditional eating habits of the Portuguese and highlight the traditional cooking of the Lisbon Region, noticing the key changes in dietary habits brought forward on the 1960s and 1970s. Moreover, we observe the key findings of recent agricultural statistics made available on Portugal, which allow an observation of the current state of food-related activities in the land-uses and the economy of the Lisbon Region.

4 The Rural Habitat and Architectural Imagination

4.1 Introduction

As seen previously, there has been a recognition in the general culture of the importance of food systems. This phenomenon runs alongside the increased ethical and cultural awareness of the limits of natural resources and the general concern for non-materialistic forms of pursuit of happiness and the subsequent recognition of sustainability as a necessary political and social goal (Tibbs 2011; Marat-Mendes and Borges 2019).

To ask whether architecture and urban design can contribute to the changes preconized in this shift of values, even within the narrower scope of the food system, is in a sense futile. For across time and space, key dietary needs, namely food and water with their spatial demands, have shaped the human habitat. Prior to the rise of agrifood economies of scale, local-based foodsheds were configured by economic, political and transportation structures, and in turn they were a key force in organizing the territory: in cities, trails for food distribution defined axes of circulation and often named after the food activity or product they ensured (Steel 2008; Salvador 2019). Nowadays, the traces of the determining role of the food system can be difficult—if not impossible—to identify, especially in metropolitan cities. The polarization between urban and rural territories and activities, intensified by official land-use schemes, has redeveloped many premodern settlements and erased many spaces typically connected with food production. Moreover, from the perspective of labour, there has been a steep decrease in agriculture in the West since the 1970s (Fischer-Kowalski and Haas 2014), which leaves the rural economy at natural disadvantage in the metropolitan context, where cities concentrate jobs more strongly on the third and fourth sectors.

An observation of the urban forms of the past—or even of the recent past, in a case such as Portugal, which remained in a backward state well into the twentieth century—demonstrates the wide array of built structures and spatial arrangements necessary for the food system as well as the ways in which such structures have evolved and changed throughout the centuries.

Architect Carolyn Steel (b?), who has developed extensive research on food and cities, recalls that while contemporary civilization is markedly urban, the majority of our ancestors led predominantly rural lives as hunters, farmers, serfs, peasants or otherwise (Steel 2008). It is their history that remains on the built structures from the surroundings of contemporary metropolitan cities, in the rural architecture of ancient villages and their farms, some of which no longer serves its original purpose, and might even be abandoned. As we saw in Chaps. 1 and 2, in Portugal, these spaces were for long excluded from planning policies, as keeping them from urban development would freeze their form and functioning in time.

Economies of scale, including in the realm of the food system have undeniably produced important changes in the human habitat (Steel 2008). Indeed, they have fulfilled the modern aspiration to organize the territory according to the functioning of the production system (Weaver 1984) and have long been ensured through systems of exploitation of countries in the Global South (Castro 1946; Moorcroft 1972). Considering these two aspects, a reorganization of our food systems is not only fundamental to halt environmental decline but also for achieving social justice and responding to social pressures for sustainable transitions (UN 1987; Tibbs 2011; Marat-Mendes and Borges 2019). Local food and short supply chains have been promoted in articulation with international supply flows in official agendas (FAO 2019), but more radical approaches have emphasized that local supply, within the city-region, ought to become self-sufficient (Viljoen and Bohn 2014) and must contribute to sustainable transitions by promoting a more balanced use of the territory (Marat-Mendes et al. 2021).

Surely, it is nearly impossible to imagine a fully self-sustained regional food system, precisely because urban populations have been growing steadily and are predicted to continue to do so, with the UN predicting that by 2050, 80%

T. Marat-Mendes et al., *Atlas of the Food System*,
https://doi.org/10.1007/978-3-030-94833-7_4

of the world population will be living in cities. Yet, the recent outbreak of COVID-19 has shaken many taken-for-granted assumptions about our way of life. The virtues of compact and dense cities suddenly seem less certain and the tide of unemployment and poverty resulting from lockdown policies threaten to intensify the abandonment of the countryside, towards the typically higher employment opportunities of urban centres. In Portugal, there have been accounts of migrations *to* the countryside, but this phenomenon appears to be circumscribed to an affluent middle-class whose jobs can be easily done remotely (Ruela 2020). Such movements of return to the countryside run parallel to the recent (and politically ambiguous) emergence in popular culture of 'cottagecore' taste, which focuses on pastoral and agrarian aesthetics, with motifs such as homemade food, dried and fresh flowers, houseplants and a general preference for natural colour palettes with neutrals and warm tones (Reggev 2020). Good examples are portrayed or represented in films such as Luca Guadagnino's *Call me by your name* (2017), Mike Newell's *The Guernsey Literary and Potato Peel Pie Society* (2018), Ari Aster's *Midsommar* (2019), Pixar's *Burrow* (2020) or Taylor Swift's album *Folklore* (2020). It is possible that such reappraisal—however superficial—of agrarian life may be influenced by the increasing cost of urban life, from stratospheric rents in main cities to public services whose prices keep rising as they keep being privatized. Still, it is too soon to tell if the pandemic crisis will accelerate the urbanization pressure over already dense cities, or if it will affect the culture so much that a retreat to the more relaxed—and spacious—reality of the countryside to become a true alternative.

The political ambiguities of a non-environmentally conscious retreat to the countryside, with its classist overtones, is not new to cottagecore. In Moorcroft's (1972) anthology, Malcom B. Wells highlights how the cost of environmental mistakes is paid by those left behind—the working class, the urban poor, the ethnical minorities, doomed to "breathe the worst of air, drink the worst of the water, eat the worst of food, and live in the meanest of houses" (quoted in Moorcroft 1972: 433).

Authors who have researched the role of nature and ecology in contemporary society seem to agree that a reorientation of our worldview is necessary, and that part of this reorientation has included a shift from nature-preservation to food production (Purdy 2016; Napawan 2016). However, to avoid furthering social inequality, such a shift must be considered structurally and must encompass prospects for territorial change to clarify its ideological overtones and its practical end goals and implications.

Thus, the plethora of built structures and spatial arrangements necessary for the different phases of the food system demand, in order to adapt to current physical and social conditions, diversified and creative design solutions, which can make an effective improvement of current realities. Such solutions must take heed of not only the demands of food production, transformation or consumption but also of water storage and irrigation, waste recycling and the promotion of public space and conviviality (Parham 2015). Infrastructure systems for transportation of produce or waste are also determinants for sustainability since a lot of current food transportation is ensured through heavy motorized vehicles. To these CO2 emissions, one may add those resulting from 'food miles', i.e. the distance ran by someone to access a food-selling point, often ensured by car (Moudon et al. 2013).

Architecture, urban planning and urban design, all can play an important part in imagining how to materialize a reorientation of our attitudes towards nature and food production. Indeed, this has happened before in history: just as the rise of industrialization and conurbation was running out of control in the most industrialized nations, urban planners and designers of several orientations, from the City Beautiful to the Garden City movements, from the 'Chartre d'Athénes' to urbanism of dictatorial regimes, all agreed upon the need to include urban parks as reservoirs of green space amidst the dense city (Kostof 1991). Today, the tendencies in need of counterbalancing have shifted, and thus it is necessary for architects to remain open to new concerns, new problems and new solutions, be it in the design of buildings or in their intervention on larger scopes, as those of urban design and even spatial planning where, as seen in Part I, much remains to be done in order to promote sustainability.

In 1956, in an essay that has since become emblematic, Alison and Peter Smithson discuss the important relation between mass production, advertisement and architecture which, they argued, most of their peers had failed to correctly assess:

> Already the mass production industries have revolutionized half the house—kitchen, bathroom, utility room, and garage—without the intervention of the architect, and the curtain wall and the modular prefabricated buildings are causing us to revise our attitude to the relationship between architect and industrial production (Smithson and Smithson 1974: 44).

The message remains important today, for when architects are unable to realize the social aspirations and values, they risk seeing their work become obsolete and their tasks fulfilled by others. If such an agenda is to favour food production as a form of environmental quality (Purdy 2016) and to promote a reorganization of the food system at the scale of the city-region (FAO 2019), retrieving an operative notion of a local or regional foodshed (Getz 1989; Steel 2008; Salvador 2019), there is a vast and diversified tradition from which to learn from. In the case of the Lisbon Region, such tradition does not belong only to the rural past, but also to the present, with its greater emphasis on city life.

In the past, Portuguese architects have often looked to the territory itself and its occupations and uses in order to make sense of their physical and cultural landscape, as we shall discuss here. Such exercises also proved important as sources of inspiration and orientation with regards to designing for the present and the future (Marat-Mendes et al. 2021b). In the following sections, we overview two important surveys conducted by architects in twentieth-century Portugal, both of which present paradigmatic portraits of the Portuguese territory. In each of them, we highlight surveyed examples that depict food-related structures and activities which do exist in both surveys, although this was not their theme. Afterwards, we observe them in their specific context —whereas in Part I we observed the history of planning, here, we will take the surveys as an opportunity to observe the changes that did take place according to or in spite of planning instruments. From their comparison, key changes in the Portuguese social, environmental and spatial reality can be identified. From this identification, we present a brief account of the key socio-metabolic changes that took place in the territory of the Lisbon Region, emphasizing particularly the relationships where the food system has a more determinant role.

4.2 The Survey on Regional (or Popular) Architecture

The contemporary Lisbon Region coincides with what was the Roman Ager Olisiponense, from the centuries V to II bC, and much of it remained rural throughout the centuries, eventually fostering a specific cultural identity—the 'saloia' culture (Cabrita 2008). 'Saloia' (feminine) or 'saloio' (masculine) is a regionalism which lacks a specific translation to English but is akin to 'rustic' or 'rural', directly expressing the link between this culture and the agrarian surroundings of the Lisbon city. Indeed, the 'saloios', i.e. the people from the 'saloia' region, have been historically linked with the urban economies of the region, to which they supplied produce and workforce, thence guaranteeing their own sustenance (Cabrita 2008). These regional small economies were quite dependent upon agriculture and thus, settlements always presuppose a link with cultivation spaces —the relations of these elements within a settlement indeed constitute a popular and organized morphology where food is a key defining element.

A key element for any discussion about the material aspect of this rural culture is the elaboration of architectural surveys, which have played a fundamental role in Portuguese architecture of the twentieth century. Although most studies tended to focus on the more remote countryside, in the inland, some attention was eventually given to the agrarian surroundings of urban settlements, which until the 1974 Revolution remained underdeveloped in comparison with other Western cities.

Portuguese rural architecture spiked the curiosity of an important Brazilian architect, Lúcio Costa (1902–1998) who by the 1930s was already suggesting, in his article 'Documentação necessária' (*Necessary documentation*), that autochthonous rural architecture in Portugal was a high-quality cultural repository yet to be acknowledged and properly studied, suggesting its inventory (Costa 1937). For nearly a decade, this proposal was not given any specific attention by architects, although it stood close to the emerging interests of other knowledge areas.

Particularly representative is the pivotal works on rural research published by Geographer Orlando Ribeiro, an unsurmountable influence on the Portuguese survey practice, who takes a particular interest in the rural settlement as the landscape element that revealed the historical evolution of countryside life (Marat-Mendes and Cabrita 2015). Ribeiro got started on this survey in the 1930s, but it eventually remained incomplete. An outline of the envisioned study was published as a contribution to Human Geography, seeking to explain the relation between different types of rural settlement, natural and social conditions, property regimes, forms of land labour and lifestyles (Ribeiro 1939). This methodology builds upon the contribution of his former professor, Geographer Albert Demangeon (1872–1940), who had proposed in the early 1900s methodologies and survey techniques for geographical inquiries into rural communities (Marat-Mendes and Cabrita 2015).

In Ribeiro's (1991) rural habitat survey proposal, beyond personal trips he had taken in the country, the key elements for analysis were selected cartography, at the scale of 1:100,000 in extensions of 100 km^2 and in loco photography. The author notably emphasized the need for an interdisciplinary approach to the rural settlement, fusing morphological interpretation of the habitat with the study of its formation and origin (Ribeiro 1991; Anastácio and Marat-Mendes 2016; Marat-Mendes and Cabrita 2018).

Studies conducted at the time (and even earlier) by Ethnologists also contributed much to the construction of knowledge on the Portuguese rural habitat. Notably, Portuguese ethnologist Jorge Dias (1907–1973) coordinated several inquiries into agro-pastoral communitarianism in Vilarinho das Furnas (Dias 1948) and Rio de Onor (Dias 1953), leaving many portraits of rural housing and agricultural structures. With several collaborators, Dias also researched more specific issues, including the functioning of granaries and dry farming (Oliveira et al. 1994) and irrigation systems (Galhano and Dias 1986). Similar studies were also conducted in Portuguese villages by foreign anthropologists, as is the case with Wateau's (2000) study on water-access and irrigation systems on the Minho region, or O'Neill's (2011) study on social inequalities inherent to communitarian agriculture in the Trás-os-Montes region.

In the same decade, a 'Survey on rural habitat' was promoted by the Lisbon Superior Institute for Agronomy, conducted by agronomist engineers Eduardo Alberto Lima Basto (1875–1942), António de Faria e Silva (b?) and Carlos Silva (b?). The first volume was published in 1943 and the second one in 1947. A third volume, which included the Lisbon Region, was censored by the dictatorship to prevent the disclosure of the evidently poor life-conditions, and remained unpublished until 2002, when the whole survey was republished (Basto et al. 2002). The key influences on this survey were German, Italian and English studies of the rural economy, and thus it included information on local life-conditions, construction techniques and basic notions of housing architectural form (Diamantsou-Kremesi and Marat-Mendes 2012; Marat-Mendes and Cabrita 2015).

At the time Ribeiro (1939) published his guidelines for the rural habitat survey and Lúcio Costa his article on Portuguese rural architecture, architects Philip Goodwin (1881–1935) and Kidder Smith (1913–1997) curate the 1943 MoMA exhibition *Brazil Builds—Architecture Old and New 1652–1942,* which included examples of autochthonous, colonial and modern architecture but decidedly treated them at the same level. The exhibition catalogue was known to Portuguese architects through Nuno Teotónio Pereira, who acquired it in 1945 (Milheiro 2012). Its approach surprised the Portuguese, who saw in it a search for authenticity and timely feeling (Tostões 1997). Conditions were ripe for the upcoming efforts to study the rural habitat. The interest, evident among geographers inspired by Demangeon's work, and also in architecture, was largely exemplified by the eclectic selection in the Brazil exhibition, and all of these had an impact upon the sensibility of a new generation of architects (Tostões 1997; d'Almeida 2006; Milheiro 2012; Cabrita and Marat-Mendes 2019).

In 1947, Francisco Keil do Amaral, who was also the head of the Portuguese Architect's Union (*Sindicato dos Arquitectos*), published an article in direct response to the one Lucio Costa had published nearly a decade ago. The title, 'A necessary initiative' (*Uma iniciativa necessária*), makes this relation explicit and also instates a programmatic aspect. In 1947—the same year of Keil's article—architect Fernando Távora published his 'The problem of the Portuguese house' (*O problema da casa portuguesa*), where he stressed the functional and pragmatic aspects of traditional architecture, a repertoire he would incorporate in his own architectural practice, to counterpart an unmistakably modern sensibility (Tostões 1997) and Raul Lino had been publishing his theoretical views and design proposals on 'the Portuguese houses' for long (Lino 1918, 1933). Yet the most energetic promoter of an extensive survey on the countryside and its architecture would be Keil who campaigned from 1947 onwards to gather from the State the necessary financial support to conduct a survey on rural architecture, a goal which was met with failure after failure for nearly a decade.

Only in 1955 did the Architect's Union manage to access financial resources, and the work for the Survey on Portuguese Regional Architecture (*Inquérito sobre a Arquitectura Regional Portuguesa*) was officially launched. The country was divided into six zones—Minho, Trás-os-Montes, Beiras, Estremadura (which included the Lisbon Region), Alentejo and Algarve—assigning a team of architects and interns for each zone. In 1958, all the survey was completed, with over 10 000 photographs collected. The results, alongside a mock-up of the future publication, were presented to Oliveira Salazar, the head of the New State (Tostões 1997). In 1961, the survey was published in book form (AAVV 1961), but its title was altered to 'Popular Architecture in Portugal' (*Arquitectura Popular em Portugal*), without Keil's knowledge or approval (Esteves and Mestre 1987).

Keil do Amaral was notorious for his diplomatic personality, which allowed him to often mediate between the institutionalized power of the dictatorship and the architects who opposed it and saw it as stifling for the modernization of society and architecture (Tostões 1997). Thus, he was able to promote an "honest, vivid and healthy regionalism" (Amaral 1999, 126), by apparently agreeing with the strongly anti-urban stance of the New State. Predictably, the survey concluded without reservation that there was no 'Portuguese architecture', not only because there were considerable regional variations, but also because several zones already witnessed an uprooting of their autochthonous building characteristics (Tostões 1997).

In 1948, the UIA Congress, held in Lisbon, affirmed the incorporation of the values of modern architecture in Portugal—despite resistance from the regime. However, this happened already in the second postwar, a time when the key concepts of modernism itself were already under critique and fierce debate.

In the mid-twentieth century, many European countries started to promote surveys on their vernacular architectural traditions, including Germany, France, Italy, England, Spain and Greece (Marat-Mendes et al. 2021b). In Greece, for instance, in the 1950s were founded the Archives for Vernacular Architecture and in the following decade, the architect Panayotis Michelis (1903–1969) coordinated a monograph on 'The Greek Traditional House', launching a tendency for this research subject (Diamantsou-Kremesi and Marat-Mendes 2012). In 1972–1974, this finally resulted in an extensive 'Survey of the Greek Traditional Architecture', conducted by architects and historians. More methodologically defined, this survey encompassed the identification of factors that determined the structures and forms of the settlement configuration and image, as well as the identification of the individual character of each settlement (Diamantsou-Kremesi and Marat-Mendes 2012).

Although there existed previous researches even in Portugal—namely from geographers and anthropologists—on rural habitats, the survey teams followed no pre-defined housing types, did not focus solely on housing structures and sought to adapt their survey to the existing architectural diversity (Marat-Mendes and Cabrita 2015). The regional architecture survey teams achieved a magnificent inventory, also attempting to classify existing building types in each zone, without systematizing the overall results (Marat-Mendes and Cabrita 2015).

While there was a common set of goals guiding the teams in their fieldwork, the final result was heterogeneous in specific subjects, structure and presentation (Marat-Mendes and Cabrita 2015). This may hamper particular comparisons between the materials of different regions; it also justifies (and even denounces) the inspirational aspect of the survey. Despite that, some methodological procedures may be inferred, namely the use of cartography, photography as the key resource, elaboration of drawings of architectural and urban settings, description of materials and construction techniques and deductive reasoning (Marat-Mendes and Cabrita 2015).

In this context, the influence of Orlando Ribeiro in the regional architecture survey can hardly be overstressed. Through informal contacts and indirectly through his published works (the book of the Regional Survey contains no systematized bibliography) he will inspire architects towards an interdisciplinary approach to visited settlements (Marat-Mendes and Cabrita 2015). Architect Nuno Teotónio Pereira, who integrated the team for Zone 4—Estremadura, wrote a memorial to Ribeiro after his death, in which he calls him "the great master of out Geography" (Pereira 1995: 98), highlighting his understated contribution to modern architectural practice, informing architects on how to become "agents of territorial intervention" (Pereira 1995: 98).

The 1961 Survey indeed marked a turning point in the understanding of modern architecture, from the international vision represented by CIAM and by the 1948 Congress, to a more localized and context-sensitive approach. While the survey is primarily focused on the popular, autochthonous and vernacular architecture of the countryside, searching for a national building tradition, in practice, would not be achieved. Examples depicted clear regional contrasts, and emphasis was laid on examples which could anticipate modern architectural form (Tostões 1997; Milheiro 2012).

At this time, as we saw in Chaps. 1 and 2, spatial planning was still focused surely on urban cores, assuming, perhaps abusively, that rural settlements were undergoing no significative changes. Architects involved in the 1961 survey indeed pioneered a direct relationship between the rural realities of the country (and their forms) with the proposal for new, modern and often urban spaces.

In 1956, the Portuguese representation at the CIAM X, in Dubrovnik, strongly testified to the influence of the survey, which had started in the year prior, but was still unpublished (Marat-Mendes and Cabrita 2015; Marat-Mendes and Borges 2017).

The Portuguese team submitted a project for a collective rural housing estate for Rio de Onor, a village located in the Trás-os-Montes region, near the borderline with Spain. This project was signed by architects Viana de Lima (1913–1991), Fernando Távora and Octávio Lixa Filgueiras and it drew from survey materials, specifically from Zone 2 (Trás-os-Montes), whose survey was conducted by Filgueiras, Carlos Carvalho Dias (b.1929) and Arnaldo Araújo (1925–1982). The architects of the project adapt a rural house design for a collective architectural form capable of (modern) mass reproduction. Underlying the strategy is not only a wish to promote modernization of architecture and of habitat forms but also knowledge about the structural and symbolical aspects of rural housing, encompassing the everyday relationships between housing spaces and inhabitants, as well as between the house, the settlement and circulation routes (Marat-Mendes and Borges 2017; Marat-Mendes and Cabrita 2018).

The Dubrovnik Congress was to be the last CIAM gathering. It was organized by Team 10,[1] a group of younger architects mostly originary from countries that had remained peripheral to the Modern Movement, like England, Italy and Holland. Team 10 was formed within CIAM, but in 1959, the same group would dissolve this institution, promoting different approaches to architecture and urban planning. Specifically, Team 10 prompted architecture to be more attentive to the pre-existing conditions of the territory, and to the needs of regional organization, considering several settlement scales and functions, from urban to rural (Mumford 2000; Van Den Heuvel and Risselada 2005; Borges 2017; Baía 2020). A significant influence on the intellectual approach of Team 10 is to be found in the work of Patrick Geddes, whose Valley Sections these architects redesigned in the 1954 'Doorn Manifesto' (Van Den Heuvel and Risselada 2005). While most Portuguese architects remained distant from Team 10—only Amâncio Guedes (1925–1995) would officially join. The 1956 project for Rio de Onor presented in Dubrovnik resonates strongly with the key

[1] Team 10 is founded within CIAM. Its original members are Aldo Van Eyck, Alison and Peter Smithson, Jaap Bakema, Giancarlo Di Carlo, Georges Candilis and Shadrach Woods. Over the years, they would be joined in their meetings by several other architects, including Ralph Erskine, Amâncio Guedes, Herman Hertzberger, Josep Antoni Coderch, Rolf Gutmann, Grief Grung, Oskar Hansen, Chales Polonyi, Jerzy Soltan, Oswald Mathias Ungers, John Voelcker, Stefan Wewerka, among many others.

criticisms of Team 10 to pre-WW2 CIAM urbanism (Borges 2017; Marat-Mendes and Borges 2017).

Another example of the influence of the 1961 Survey, usually less noticed in Portuguese architectural historiography, is the Chelas Urbanization Plan, coordinated between 1962 and 1972 by architect Francisco Silva Dias, who integrated the team for Zone 4. Most notably in the first neighbourhood to be built in Chelas, called Zone I, planners and architects converged in designing a plethora of loosely defined public spaces, reinventing at a high-density urban scale the streets, squares and closes encountered in rural settlements (Borges and Marat-Mendes 2019, 2020).

The Portuguese 1961 Survey and its Greek counterpart share a similar spirit, but they are separated by important practical differences. While in Greece the survey inspired a strong tradition of academic research into the urban morphology of vernacular settlements, in Portugal this did not come to pass (Diamantsou-Kremesi and Marat-Mendes 2012), possibly due to the lack of a clearly stated methodology. Moreover, in Greece, the survey was developed in a scholarly context, while in Portugal it was promoted and ensured by the Architect's Union. Yet, perhaps the most striking contrast between these two examples is that while the Greek survey resulted in legislation for protected settlements and for special rules concerning future construction (Diamantsou-Kremesi and Marat-Mendes 2012), in Portugal the 1961 Survey was mostly an inspiration for modern design, a repository of references which, in themselves, would not become particularly cherished or protected.

In Porto, the Survey gained central importance in the imagination of young architects, in part because several architects who taught at the Fine Arts School at the time or had recently graduated from it, did participate in the fieldwork. This led to the development of further studies, including on the areas of Barredo (Távora 1969) or Rio de Onor (Fernandez 1964, 2011), which also happened in the aftermath of the Greek Survey. However, in Lisbon, the key figures of the regional architecture Survey were not at the time linked to the University, which was more closely controlled by the New State.

Indeed, even at the time, the world portrayed by the Survey on Regional Architecture was seen to be vanishing, and years later, Teotónio Pereira (2000) goes as far as to suggest that such vanishing did start to take place in the years immediately following the publication of the survey.

4.2.1 The Food System Seen Through the Survey on Regional Architecture (1961)

The Survey on Portuguese Regional Architecture allows for the identification of several building types and land-uses that were directly related to the food system of the Lisbon Region, as well as to water provision, including in the 'saloia' area of the Lisbon Region, covered in Zone 4, by architects Nuno Teotónio Pereira, António Pinto Freitas and Francisco Silva Dias.

Among the most frequent structures photographed within the region are examples of water-mills, windmills, ports, fountains, public wells, fields and their limiting walls, farmyards, porches, granaries, salt pans, haystacks, taverns and shipyards. The key constructive techniques, from white plaster and stone window-frames to dry stone walls and even hay cover, are all depicted in the survey. Although most photographs are taken from the exterior of buildings, there are occasional photographs of interiors, most notably of kitchens. Not only do these photos relate to food preparation activities, but they testify to a particular feature of the rural house, namely, that its central space is indeed the kitchen. This had been highlighted by the Portuguese team that participated in CIAM X.

Another very important element, and the most frequently photographed, was the typical 'saloia' house, which recurs in several variations, with varying sizes, and often surrounded by ploughed fields. Although most photographs clearly depict an artisanal agriculture, it was also one that clearly still existed as an economic activity and displayed signs of ongoing labour. This is also very clear in many of the photographs that depict settlements, instead of individual buildings. Often, agricultural fields surround building clusters, and even intermingle in the built structure: in such situations, a threshing floor becomes similar to a square or a plaza, a yard for chickens becomes a porch. The spaces where people spend their everyday life, and where they encounter their neighbourhoods, often function for food production as well.

The images depict an environment where agrarian activities ran in an artisanal way, using the sun as the key power source. Untouched by industrialization, buildings used similar materials, those available in the area, and the landscape was shaped by continuous agrarian labour. Despite the picturesque aspect of most images, the territory was abundantly marked by its productive capacities. In many cases, the ploughed fields and the cultivations were true extensions of the housing unit. The agrarian aspect was possible to observe very clearly in the broader landscape photographs, where housing was only one of several types of buildings—others include facilities for agricultural support, wells for accessing water, ovens and granaries.

The S. Mamede Chapel in Janas, Sintra, is heavily featured, with its porch meant for cattle blessing in the worship of Saint Mames and its surrounding precinct, which harboured an agricultural street fair and was home to a specific ritual of blessing of the cattle (Marat-Mendes et al. 2020). Another notable example of the Lisbon Region in the Survey is the Ilhéus housing complex in Picanceira, Mafra, a slab of

23 serial small houses in 'saloio' style, built in the nineteenth century for agricultural workers brought in from the Azores archipelago. Given its proto-industrial character, and the impressive composition resulting from the repetition of the simplest elements (see Fig. 5.7), it became one of the strongest images from the 1961 Survey and a particular source of interest and influence for architects (Pereira 2000) who sought modern solutions for urban housing. The hemispherical exterior oven is particularly unique to this structure, intended for baking bread, while an interior division within the house was meant for stocking edibles—a detail the surveyors took notice of in the plans they drew (Figs. 4.1, 4.2, 4.3, 4.4, 4.5, 4.6, 4.7, 4.8, 4.9, 4.10, 4.11, 4.12, 4.13, 4.14, 4.15, 4.16, 4.17 and 4.18).

Fig. 4.1 Covered well, Sintra (Janas). Source: Inquérito sobre a Arquitectura Regional Portuguesa (1955–1960). Zona 4 (Arquitectos Nuno Teotónio Pereira, António Pinto de Freitas e Francisco da Silva Dias) © Ordem dos Arquitectos

Fig. 4.2 House, Sintra (Assafora). Source: Inquérito sobre a Arquitectura Regional Portuguesa (1955–1960). Zona 4 (Arquitectos Nuno Teotónio Pereira, António Pinto de Freitas e Francisco da Silva Dias) © Ordem dos Arquitectos

Fig. 4.3 Salt pans and salt houses, Alcochete. Source: Inquérito sobre a Arquitectura Regional Portuguesa (1955–1960). Zona 4 (Arquitectos Nuno Teotónio Pereira, António Pinto de Freitas e Francisco da Silva Dias) © Ordem dos Arquitectos

Fig. 4.4 House appearance of the region, Seixal (Fernão Ferro). Source: Inquérito sobre a Arquitectura Regional Portuguesa (1955–1960). Zona 4 (Arquitectos Nuno Teotónio Pereira, António Pinto de Freitas e Francisco da Silva Dias) © Ordem dos Arquitectos

Fig. 4.5 Windmill, Mafra (surroundings Ericeira). Source: Inquérito sobre a Arquitectura Regional Portuguesa (1955–1960). Zona 4 (Arquitectos Nuno Teotónio Pereira, António Pinto de Freitas e Francisco da Silva Dias) © Ordem dos Arquitectos

Fig. 4.6 Tide mill, Seixal (Paio Pires). Source: Inquérito sobre a Arquitectura Regional Portuguesa (1955–1960). Zona 4 (Arquitectos Nuno Teotónio Pereira, António Pinto de Freitas e Francisco da Silva Dias) © Ordem dos Arquitectos

Fig. 4.7 Houses, Sintra (Ral). Source: Inquérito sobre a Arquitectura Regional Portuguesa (1955–1960). Zona 4 (Arquitectos Nuno Teotónio Pereira, António Pinto de Freitas e Francisco da Silva Dias) © Ordem dos Arquitectos

Fig. 4.8 Houses, Sesimbra (Alfarim). Source: Inquérito sobre a Arquitectura Regional Portuguesa (1955–1960). Zona 4 (Arquitectos Nuno Teotónio Pereira, António Pinto de Freitas e Francisco da Silva Dias) © Ordem dos Arquitectos

Fig. 4.9 Sanctuary of Our Lady of the Cape, Sesimbra (Cabo Espichel). Source: Inquérito sobre a Arquitectura Regional Portuguesa (1955–1960). Zona 4 (Arquitectos Nuno Teotónio Pereira, António Pinto de Freitas e Francisco da Silva Dias) © Ordem dos Arquitectos

Fig. 4.10 Stores, Mafra (Ericeira). Source: Inquérito sobre a Arquitectura Regional Portuguesa (1955–1960). Zona 4 (Arquitectos Nuno Teotónio Pereira, António Pinto de Freitas e Francisco da Silva Dias) © Ordem dos Arquitectos

Fig. 4.11 Interior of a store, Montijo (Pegões). Source: Inquérito sobre a Arquitectura Regional Portuguesa (1955–1960). Zona 4 (Arquitectos Nuno Teotónio Pereira, António Pinto de Freitas e Francisco da Silva Dias) © Ordem dos Arquitectos

Fig. 4.12 House, Montijo (Pegões). Source: Inquérito sobre a Arquitectura Regional Portuguesa (1955–1960). Zona 4 (Arquitectos Nuno Teotónio Pereira, António Pinto de Freitas e Francisco da Silva Dias) © Ordem dos Arquitectos

Fig. 4.13 Oven inside a kitchen, Sintra (Assafora). Source: Inquérito sobre a Arquitectura Regional Portuguesa (1955–1960). Zona 4 (Arquitectos Nuno Teotónio Pereira, António Pinto de Freitas e Francisco da Silva Dias) © Ordem dos Arquitectos

Fig. 4.14 Interior of a kitchen, Sintra (Assafora). Source: Inquérito sobre a Arquitectura Regional Portuguesa (1955–1960). Zona 4 (Arquitectos Nuno Teotónio Pereira, António Pinto de Freitas e Francisco da Silva Dias) © Ordem dos Arquitectos

Fig. 4.15 Inside a tide mill, Seixal (Rouxinol). Source: Inquérito sobre a Arquitectura Regional Portuguesa (1955–1960). Zona 4 (Arquitectos Nuno Teotónio Pereira, António Pinto de Freitas e Francisco da Silva Dias) © Ordem dos Arquitectos

Fig. 4.16 Well, Mafra (Pobral). Source: Inquérito sobre a Arquitectura Regional Portuguesa (1955–1960). Zona 4 (Arquitectos Nuno Teotónio Pereira, António Pinto de Freitas e Francisco da Silva Dias) © Ordem dos Arquitectos

Fig. 4.17 Covered well, Sintra (Janas). Source: Inquérito sobre a Arquitectura Regional Portuguesa (1955–1960). Zona 4 (Arquitectos Nuno Teotónio Pereira, António Pinto de Freitas e Francisco da Silva Dias) © Ordem dos Arquitectos

Fig. 4.18 Covered well, Sintra (Fontanelas). Source: Inquérito sobre a Arquitectura Regional Portuguesa (1955–1960). Zona 4 (Arquitectos Nuno Teotónio Pereira, António Pinto de Freitas e Francisco da Silva Dias) © Ordem dos Arquitectos

4.3 The Urban Transition and the Rise of the Lisbon Conurbation

The marks of rural architecture have faced tremendous pressures, under the guise of the new building practices of modernity, with demolitions to ensure new construction at a larger scale, and the conception of modern urbanism is inherently dependent upon tabula rasa redevelopment (Cabrita 2008). A cultural crisis followed this situation, where the formal imitation of the past existed alongside ideals of innovation, novelty, technology and short-term goals (Marat-Mendes and Cabrita 2015; Anastácio and Marat-Mendes 2016).

The 'saloia' vernacular architecture presents a wide but very definite set of key housing types, often resulting from variations and transformations of a basic unit, augmented to respond to local economic or labour changes (Cabrita 2008). Part of this 'saloio' territory was covered by the PUCS, which proposed the preservation of the existing rural structure, including old farmsteads, villages and thorps (Pereira 2009). However, these territories were soon to be perceived for their speculative potential—and while they were sometimes occupied with approved projects (diverging from PUCS stipulations), most of the land-use conversions in these areas, as in others within the Lisbon Region, was through illegal instalments, mostly of housing, but episodically also of industrial facilities (Cabrita 2008; Marat-Mendes 2009; Anastácio and Marat-Mendes 2016). This shows that even under a plan formally approved by the State (in 1947), the hinterland of Lisbon proved almost impossible to preserve, following a pattern verifiable in the West (Parham 2015) even despite its specific political conditions and limitations.

Within the Lisbon city itself, large rural estates remained in existence until the 1950s, and only with the formation of the GTH were these urbanized, originating the new urban areas of Olivais Norte, Olivais Sul and Chelas, the notable cases of the Lisbon city, where municipal planners worked under a significative housing programme and through modern and large-scale urban and architectural design (Portas and Mendes 1992; Borges and Marat-Mendes 2019).

While many of the key towns and cities of the region received urban plans, designed by competent architects, as we saw in Part I, these were often undercut for being based on outdated information or for lack of technical and financial means for municipalities to implement them. Thus, the regional landscape transformed rapidly under the combined interests of landowners as well as real estate and construction companies, gathering the support of the financial system and dominant economic groups, operating both in the legal and informal markets (Pinto and Guerra 2019). Although housing supply increased to face the demands of migrations from the continental countryside as well as from people coming or returning from the former colonies, liberated during the Revolutionary period, the real housing needs were not met, allowing prices to rise and excluding the working classes. Soon, overcrowding, unsanitary housing, construction in courtyards and block interiors, as well as the sprawl of entire neighbourhoods filled the severe housing deficit in the two metropolitan regions, leading to massive conversions of rural land and the expansion of urban cores towards their periphery (Pereira and Buarque 1995; Guerra 2011; Pinto 2017; Pinto and Guerra 2019).

Over time, the urban model of the Lisbon Region started to change, sometimes under planned transformation, other times in spite of planning goals. The radial metropolitan space of the first regional visions—from Gröer's to the 1964 PDRL—where a large periphery depended upon a central city, started to be replaced by an expansive dynamic, urbanizing the buffers and voids among radial development axes, a process led by road infrastructure at several scales, and by the availability of land for urbanization in municipal planning (Portas et al. 2007). Given the differences in infrastructure quality, the new metropolitan space was prone to the coexistence of contradicting trends, combining dispersal and agglomeration, monofunctional fabrics with mixed land-use areas, new values of proximity and new dependencies and the frequent reorganization of flows across the overall space, as new services, equipment or job-creation centre sprout in different spaces (Portas et al. 2007).

With the adherence of Portugal to the European Economic Community (EEC) in 1986, one decade after the country underwent its democratic transition, the architectural profession was also greatly transformed. There may be many reasons for this, but some do stand out. First, that an affluent middle-class is effectively created by the new political regime, a process in which the acquisition of housing by purchase had a structuring role (Drago 2017; Pinto and Guerra 2019). The other is that while spatial planning policy sometimes stemmed from the Central State, its intervention tended to be sectoral, while the minute transformations of space were determined at the municipal level, promoting the creation of not only new housing areas but also of services, equipment and commercial spaces (Portas et al. 2007). With respect to the latter, the shopping mall emerged as a new form of public space for this new consumer society, often redesigning in its interior the traditional public spaces of the city (plazas, streets) but reframing them so that not only leisure and consumption become indistinguishable, but public space was also inherently privatized (Cachinho 2007). Architecture played a key role in this transformation. Throughout this process, Portuguese architects became permeable to international trends, significantly diversifying the programmatic and aesthetic scope of national architecture, and resulting in a rising internationalization of several key figures. On the other hand, the idea that architects contributed to the programming of activities in space, taking a critical approach to existing spaces and actively participating in urban planning vanished: architects became the providers of consumption good, and focus increasingly on the building and on executing a request (Figueira 2016).

Nuno Portas was, again, a key figure in the establishment of the Portuguese modern historiography, and he even wrote the introduction of the Portuguese edition of Bruno Zevi's (1918–2000) 'Storia dell'Architecture Moderna' (History of Modern Architecture), particularly focusing on modern Portuguese examples (Zevi 1973; Portas 2006; Toussaint 2009). Besides Portas' many publications, local guides (Ferreira et al. 1987; Berger et al. 1994; Fernandes and Cannatá 2002; Toussaint et al. 2013, 2018) and monographs on specific architects helped to solidify a vision of twentieth-century architecture in Portugal, which had swerved significantly from the type of architecture and settlement that had been portrayed by the 1961 Survey. With increasing velocity, Portugal ceased to be an overwhelmingly rural country with only two significant cities and became home to a complex network of territorial situations, where villages in the countryside were no longer the reality of the majority of the population, as they had been during the first half the century—indeed, the territory has gained larger villages, larger towns and new urban realities, including two metropolitan conurbations, a set of non-metropolitan conurbations caused by diffuse and disorganized urbanization mostly on the Atlantic coast, and medium cities with all kinds of development processes (Portas et al. 2007). The PDM, as the legal document that binds spatial decisions and which, as we saw in Chap. 3, covered all the territory, often presided over the formation of these new habitats, although it often tended to focus strongly upon urbanization processes, more than the remaining activities with territorial impact. This, as we saw, could be construed as a control from the State upon the private sector, a conflict which remains without resolution until today.

For nearly four decades, architectural surveys were not considered particularly relevant. However, in September 2003, the Portuguese OA launched a survey about the Portuguese twentieth century, in partnership with the Mies van der Rohe Foundation and the Instituto das Artes (Arts Institute). The scientific coordination was ensured by architect and historian Ana Tostões.

Entitled 'IAPXX—Inquérito à arquitectura do século XX em Portugal' (Survey of twentieth-century architecture in Portugal), it intended to represent twentieth-century architecture, directly aiming at defining strategies for safeguarding built heritage. This line of action is closer to that of the 1970s vernacular survey in Greece, however without the goal of creating detailed morphological analysis compared to that conducted in Greece. The work was conducted until April 2006, and its methodology had the 1961 Survey as the central reference. The territory was divided into six regions, but these differed from those defined by the 1961 Survey—

North, Centre, Lisbon and Tagus Valley, South, Azores, Madeira. Another preponderant difference is that while the IAPXX includes the continent and the two Portuguese archipelagoes, the 1961 Survey had focused solely on the continental territory, with surveys on the popular architecture of Madeira (Mestre 2002) and Azores (Caldas et al. 2000) conducted much later and with different purposes and circumstances.

For IAPXX, each zone was assigned a team of architects and fieldwork was stipulated for 10 months. Cars, computers, digital cameras, a pre-designed database and a software for GPS identification were made available to the architects. Since the frame of reference was not geographical situation, but a time period, examples included both vernacular and popular architecture, and also author works and urban buildings across all the Portuguese territory, regardless of its context being cities, towns or villages. In total, 180,000 km were covered, and from the 306 Portuguese municipalities, 290 were included in IAPXX. The key idea of IAPXX was not to build a History of Portuguese architecture of the twentieth century, but rather to bring architecture closer to civil society, emphasizing its quality and diversity. Furthermore, the end result escapes the traditional scope and methodologies of architectural history, by underplaying issues regarding authorship, aesthetic taste or building types, thus more accurately encompassing the reality of Portuguese architecture in this century, which are not limited to the examples usually highlighted by historians or critics. Indeed, the IAPXX confirmed the particular ability of the survey study to move beyond the boundaries of dominant critical discourse. In the end, over 5000 works were identified and 100,000 photographs were collected in the database, which was later posted online.[2]

A brief summary was published in a book in two languages, Portuguese and English (Ordem dos Arquitectos et al. 2006). Some of the collected examples were signed by renowned Portuguese and foreign architects, but the survey also acknowledged on the same grounds unknown works and designers, as well as representative buildings in an advanced state of degradation. Just like the 1961 Survey had shown great variety within autochthonous construction, so the IAPXX accounts for a century of architecture far more eclectic than historiography would usually concede. Furthermore, it is only possible to promote the safeguarding of paradigmatic architectural works when the totality of the built patrimony is correctly understood. Thus, the IAPXX must be understood as a tool for the "Direção-Geral do Património Cultural" (Directorate General of Cultural Heritage), the Portuguese institution for recognition and classification of real estate, in this aspect really coming closer to the approach taken by the Greek surveys of the past.

4.3.1 The Food System Seen Through IAPXX

Like in 1961, the scope of this new survey was not the food system, although it is also possible to identify built structures directly related to food and to water access. The Lisbon Region was covered in the IAPXX in two different teams: the Northern Bank of Tagus was covered by the Lisbon and Tagus Valley team, coordinated by the architects João Vieira Caldas (b.1953) and José Silva Carvalho (b.1948), while the Southern Bank of Tagus was covered by the South team, coordinated by Michel Toussaint (b.1946) and Ricardo Carvalho (b.1971). Several building examples can be highlighted for their relation with food activities, including markets, industrial factories, cooperatives for production of wine and oil, water springs, water tanks, canteens, farmstead houses with areas for sustenance agriculture among others. The identified examples belong not only to urban centres but also to the countryside.

More in tune with an industrialized socio-metabolism, the images show several scales of buildings. Some equipment justified a collection of photographs and drawings; nevertheless, some images do not do justice to the quality of architectural works. Many food-related urban forms may have escaped this inventory. For instance, sustenance gardens in family urban houses or farms, in some cases, may not have been even photographed, although they were there. This situation is here more evident than in the 1961 Survey, which focused on buildings, settlements and open spaces, while the IAPXX lays emphasis directly on buildings. In the case of the manufacturing structures linked to the food system, many were located in peripheral areas of the cities, near the riverbed or railway stations; however, many examples are nowadays abandoned and some are explored for tourism. The water infrastructures serve the population clusters or these manufacturing structures. Now, agricultural land is no longer an extension of buildings, especially since it is much more valuable as urban property.

Another important aspect is the variety of architectural aesthetics encountered, ranging from traditional designs to modern ones, to the characteristic style of the New State, 'Português Suave' (Soft Portuguese), a mesh of neoclassic and Art Deco aesthetics (Almeida and Fernandes 1986) (Figs. 4.19, 4.20, 4.21, 4.22, 4.23, 4.24, 4.25, 4.26, 4.27, 4.28, 4.29, 4.30, 4.31, 4.32 and 4.33).

[2] http://www.iap20.pt/Site/FrontOffice/default.aspx

Fig. 4.19 Regional Winery, Sintra (Colares). Source: IAPXX - Inquérito à Arquitectura do Século XX em Portugal. Zona Lisboa e Vale do Tejo (Coordenação João Vieira Caldas) © Ordem dos Arquitectos

Fig. 4.20 Vila Amélia, Odivelas. Source: IAPXX - Inquérito à Arquitectura do Século XX em Portugal. Zona Lisboa e Vale do Tejo (Coordenação João Vieira Caldas) © Ordem dos Arquitectos

Fig. 4.21 Oyster Depuration Center, Moita. Source: Inquérito à Arquitectura do Século XX em Portugal. Zona Sul (Coordenação Michel Toussaint) © Ordem dos Arquitectos

Fig. 4.22 Quimigal Industrial Complex, Barreiro. Source: Inquérito à Arquitectura do Século XX em Portugal. Zona Sul (Coordenação Michel Toussaint) © Ordem dos Arquitectos

Fig. 4.23 Official Slaughterhouse, Vila Franca de Xira. Source: IAPXX - Inquérito à Arquitectura do Século XX em Portugal. Zona Lisboa e Vale do Tejo (Coordenação João Vieira Caldas) © Ordem dos Arquitectos

Fig. 4.24 Silos, Almada (Trafaria). Source: Inquérito à Arquitectura do Século XX em Portugal. Zona Sul (Coordenação Michel Toussaint) © Ordem dos Arquitectos

Fig. 4.25 Duarte Pacheco Viaduct, Lisboa. Source: IAPXX - Inquérito à Arquitectura do Século XX em Portugal. Zona Lisboa e Vale do Tejo (Coordenação João Vieira Caldas) © Ordem dos Arquitectos

Fig. 4.26 Alcântara Maritime Station, Lisboa. Source: IAPXX - Inquérito à Arquitectura do Século XX em Portugal. Zona Lisboa e Vale do Tejo (Coordenação João Vieira Caldas) © Ordem dos Arquitectos

Fig. 4.27 Fish market, Setúbal. Source: Inquérito à Arquitectura do Século XX em Portugal. Zona Sul (Coordenação Michel Toussaint) © Ordem dos Arquitectos

Fig. 4.28 'A Camponesa', Lisboa. Source: IAPXX - Inquérito à Arquitectura do Século XX em Portugal. Zona Lisboa e Vale do Tejo (Coordenação João Vieira Caldas) © Ordem dos Arquitectos

Fig. 4.29 Campo de Ourique market, Lisboa. Source: IAPXX - Inquérito à Arquitectura do Século XX em Portugal. Zona Lisboa e Vale do Tejo (Coordenação João Vieira Caldas) © Ordem dos Arquitectos

Fig. 4.30 Snack-Bar Galeto, Lisboa. Source: IAPXX - Inquérito à Arquitectura do Século XX em Portugal. Zona Lisboa e Vale do Tejo (Coordenação João Vieira Caldas) © Ordem dos Arquitectos

Fig. 4.31 University Cafeteria, Lisboa. Source: IAPXX - Inquérito à Arquitectura do Século XX em Portugal. Zona Lisboa e Vale do Tejo (Coordenação João Vieira Caldas) © Ordem dos Arquitectos

Fig. 4.32 Economic Kitchen, Lisboa. Source: IAPXX - Inquérito à Arquitectura do Século XX em Portugal. Zona Lisboa e Vale do Tejo (Coordenação João Vieira Caldas) © Ordem dos Arquitectos

Fig. 4.33 Water tank, Barreiro. Source: Inquérito à Arquitectura do Século XX em Portugal. Zona Sul (Coordenação Michel Toussaint) © Ordem dos Arquitectos

4.4 A Perspective on the Transformation of the Lisbon Region

There are several specificities which make a comparison between the surveys of 1961 and 2006 difficult to sustain. As explained, they were conducted with different scopes. The first clearly privileged rural architecture, most of an autochthonous character, while the scope of the second was instead limited by a defined timeframe, and included urban and rural examples, anonymous and by architects, from little-known to well-recognized.

Yet, the context of both surveys allows a certain degree of comparison. Namely, because it is known that by the late 1950s, when the Survey on regional architecture was conducted, the types of settlements depicted testified to the general conditions of the large majority of the population, although this situation was about to change, precisely in the time period after the Survey publication in 1961 (Pereira 2000). The diversity of situations in the IAPXX, over half a decade later, expresses a more complex society, with different lifestyles and different habitats. But under what circumstances did such transformation occur? What is the history behind the urban, social and environmental transition that we may observe in these surveys?

Within the particular history of Portuguese modernization, with its delays and political fluctuations, industrialization had two key moments: the late nineteenth century and the 1950s–1960s period (Barreto 2017).

In the Lisbon Region, as seen in Chap. 1, most of the territory was used for rural activities by 1900s (Marat-Mendes et al. 2015), a situation which persisted well into the 1960s. Although agriculture was a key source of income for most of the population, the country's food system had never been self-sufficient. Throughout the nineteenth century, conflicts over the property of communal land, inequalities between large landowners and small proprietors ran alongside shy attempts at agrarian modernization and protectionist measures to promote the growing of wheat (Barreto 2017). Even thus, a key failure of the First Republic (1910–1926) was its inability to deal with several 'sustenance crisis' all over the country, including the capital city. People starved due to decreased external trade of foodstuffs and agricultural utensils, culminating with the wave of uprisings and lootings in Lisbon, Porto and Matosinhos, with Lisbon and its surroundings declared under siege in 1917 (Abreu 2017; Mónica 2021). In the early twentieth century, the Lisbon Region—alongside Alentejo and Ribatejo—become attractive for capitalist agrarian exploration and a commercial organization, but in the remainder of the country sustenance agriculture still predominated, and even in the capital's hinterland, technological delays could be verified until the 1960s (Barreto 2017). Throughout this time, salaries of rural workers remained exemplarily low and seasonal work was common (Barreto 2017).

During the New State period, there were important works developed in rural territories, including the improvement of irrigation systems and the infrastructure for the distribution of water, key examples of which can be found in the North of the country, in the Alentejo region on the south and on the island of Madeira's *levadas* (water ditches) (Watteu 2000). Industrialization of agriculture, belated and slow, aimed mostly at replacing imports with nationally produced products, albeit of inferior quality—which expresses not only the political conditions of the time but also the frail and divided outlook of the rentier bourgeoise, who totally lacked industrial culture or tradition (Lopes 2020).

In the late 1950s, when the fieldwork for the regional architecture Survey was conducted, the territory of the Lisbon Region was still a foodshed in the proper sense of the term, a hinterland from which the city gathered an important part of its food, and flows of food produce was organized and maintained to a significant amount within the region, travelling a relatively short distance by carriage, train or boat (Marat-Mendes et al. 2014, 2015; Salvador 2019). The importance of productive activities in peripheral economies and the proximity to a large consumer market in the Lisbon city were mentioned frequently in the descriptive documents of many planning instruments, even if such activities receive no planning stipulations. In Chap. 9, we will see this matter directly stated in a planning document from the 1950s for the extension are north of the Lisbon city.

The slow pace of development, as well as the lack of participation in WW2, which protected Portugal from being blitzed, created conditions for premodern and rural settlements to linger in time, even in the metropolitan regions, which were always exposed to higher pressures for suburbanization. Because of that (or in spite of that), the history of architectural ideas and practices in the country was unmistakably influenced by perceptions and collective representations of the rural world, allowing architects to find a specifically Portuguese approach to a modernist attitude, at a time when it was resisted by the official power of the dictatorship. It is interesting to notice that when Portugal is mentioned in the second edition of architectural critic Kenneth Frampton's (b.1930) 'Critical History of Modern Architecture', it is in precisely the chapter dedicated to Critical Regionalism, a banner proposed by this author to designate architectural cultures where modernity was strongly negotiated with local tastes, aesthetics and practices (Frampton 1985).

By 1960, 70% of the urban population of Portugal already lived either in Lisbon or Porto, leaving a meagre 30% to the remainder of cities and towns of the country (Barreto 2017). While the 1961 Survey was largely focused on the rural space of the country, this corresponded to the majority of the territory and to the life experience of an overwhelming segment of the Portuguese population of the time. Such contrast was broken with the proletarianization from the 1950s and 1960s onwards. In 1961, the dictatorship launched a Colonial War, which between combat and desertion, led countless men away from rural labour. This means an additional, and quite strong, wave of countryside abandonment, to which the New State often responded by converting agricultural land to forestry (Filipe 2014; Barreto 2017).

Architects, called to provide solutions for growing cities, and galvanized by the experience of the 1961 Survey, sought rural architecture as a source of formal and functional precedent for modern architectural form. Paradoxically, the inspiration upon a local tradition led Portuguese architects to concerns similar to those of the international architectural debates of the time. The rural housing project presented in CIAM X echoes with the Team 10 'Doorn Manifesto', when it calls for the design of settlements for territorial situations within a regional unit, from urban to rural (Marat-Mendes and Borges 2017). The key challenge of Team 10 for designing 'habitat' instead of thinking in terms of city, town or village, or of ready-made distinctions between urban and rural, remains a suggestive and relevant vision for the role of architecture and urbanism and an important challenge for contemporary planning structures. The definition of 'habitat' was important enough in the intellectual history of Team 10 for the group to promote minute discussions on its definition within CIAM and afterwards (Van Den Heuvel and Risselada 2005; Marat-Mendes and Borges 2017; Borges 2017; Baía 2020). The synchrony of the Portuguese with these concerns, even if indirect, testifies to a creative ability of those architects to convert the anti-urban prejudices that still thrived in the rhetoric of the New State into a stepstone from which to envision a different world—or at least a different habitat. Such a concern is even more important considering that the country was about to undergo important changes in the coming years.

Beyond scale and significance, there is another distinguishing aspect between the dynamics of industrialization in Portugal in the late nineteenth century and in the 1950s–1960s period. Namely that it was only the latter that significantly impacted the landscape, specifically that of the largest industrial belts, Lisbon, Porto and Setúbal (Barreto 2017). This economic acceleration prompted demographic transformations and an unprecedented rural exodus, the basis of the 'macrocephalic' concentration of development in large cities, countered by the lack of a network of alternatives in mid-sized cities (Lopes 2020: 208). The structure of Portuguese labour went from agriculture to services, and this transformation deepened the inequalities existing in the territory. From the 1960s onwards, the urban middle-class and the petty bourgeoisie had their social ascension, and service economy with technical services, liberal professions, public servants, merchants, teachers, among others, multiplied, giving a feeling of modernization, although such important changes, in reality, occurred mostly in urban settings and largely benefited Lisbon alone (Barreto 2017).

The shift in agricultural work is abundantly clear (Table 4.1) and explains how entire villages lost population—sometimes to complete abandonment—either for the industrial belts or for emigration (Filipe 2014; Barreto 2017).

The impact of these changes in the rural territories within the Lisbon Region was to be felt almost immediately. In the words of Nuno Teotónio Pereira (1990, 54), who would play an important role in surveying the physical conditions of rural Lisbon, "this was a world in lockstep, based on a local culture settled through generations and which started to collapse [...] with the generalization of the motorcar, agricultural machinery, diffuse industrialization and with the development of services, the uniformization of taste and mentality through the television and with the massive emigration for Europe".

Indeed, throughout the twentieth century, there was an escalating dissolution of rural life, first with the decline of agriculture as the main activity even in the countryside and afterwards by the growing importance of urban and suburban life without significant links with rurality (Ferrão 1996). The decline of agricultural work in Portugal was not, however, countered by an increase of industrial jobs, but rather by the growth of the service sector—by 1992, service jobs represent 55.2% of existing labour, while agriculture represents 11.6% (Lopes 2020). This shift, whose fundamental last steps were verified already in the democratic era, is perfectly reflected in the IAPXX, whose results were only possible in a country that has modernized and urbanized.

From the 1960s onwards, in both Banks of the Tagus, there was a great diversity of neighbourhoods of both legal and illegal origin, ranging from extensions of detached houses to apartment buildings, to a chaotic mixture of both, to shanty towns. As seen in Chap. 2, the emergence of these neighbourhoods was already clear when the Lisbon Region Masterplan was designed, between 1959 and 1964, but their presence was blatantly ignored in its territorial model (Pereira and Nunes da Silva 2008). This process helped to solidify the importance of private property for owners and developers since these were the key group of the national middle-class during the New State, its basis of support (Pinto and Guerra 2019; Agarez 2020).

With the 1974 Portuguese Revolution, municipalities became empowered to control territorial change, by defining a land-use scheme and by controlling and authorizing or prohibiting land-use conversion (Lopes 1991). But not only did this often have no impact in halting the sprawl of illegal construction in the metropolitan regions (Mourão and Marat-Mendes 2015), it also failed to account for the aspirations of the inhabitants of the rural landscapes being converted into new urbanizations (Cabrita 2008). Furthermore, it discontinued the extensive productive landscape that many spatial planning instruments had envisioned. This is particularly impressive with the PUCS, since it was indeed approved and remained active until 1992, despite

Table 4.1 Portuguese population and agrarian labour

Year	Total population in millions	Percentage of agrarian population in the active population
1900	5.0	65
1910	5.5	59
1920	5.6	no data
1930	6.4	55
1940	7.2	52
1950	7.9	51
1960	8.3	47
1970	8.1	32

Source Barreto (2017: 37)

undergoing several revisions. The rural wedges imagined by Gröer under Howard's influence were aggressively disrespected and urbanized, breaking over agricultural fields and rural villages which suddenly became engulfed in tall buildings or illegally constructed neighbourhoods (Cabrita 2008; Marat-Mendes 2009; Anastácio and Marat-Mendes 2016).

Architecture and urban design testify to these social tendencies. Social housing programmes, which in the 1960s had become a key source of architectural work (Borges and Marat-Mendes 2020), were downsized and made available only for those in conditions of poverty (Borges et al., 2022). However, the policy of incentive to acquire a house in the private market through subsidized credit and fiscal benefits, which lasted until 2002 (Pinto and Guerra 2019), had the potential to generate more work for architects. With this shift of urbanization towards the private sector, architecture ceased to be understood as a public service and instead became a consumer good, another element that contributed to form the value of private property. At the same time, there were frequent misconceptions regarding the role of spatial planning instruments. As seen in Chap. 3, while PDMs were binding for both the public and the private sector, the idea that they could not override rights of private property led to massive land-use conversions and to the construction of housing for speculative purposes, resulting in a housing surplus on the Lisbon Region (Mourão and Marat-Mendes 2015).

Already during the New State, but even more clearly in the democratic era, the construction sector played a key role in the Portuguese Gross Domestic Product, as happened in nations with late and incipient industrialization, an economic context in which home ownership reproduces stable conditions for expanding private wealth within the economy (Pinto and Guerra 2019).

Food critic Jane Grigson (1928–1990) criticized in 1978 the urbanization of rural peripheries of cities, suggesting that the alienation of local food production meant that local resources and the rights to them were taken away from urbanites (Parham 2015). A process of this kind did in fact take place in the Lisbon Region, and urbanizations spang not only in face of agricultural land—as happened in Santo António dos Cavaleiros, in Alverca or in Palmela—but also in their place. The regional foodshed extended, it increasingly became international, and the increase of metropolitan residents naturally meant an increase in dependency on food produced outside.

The democratic era favoured municipal development and sought to make municipalities responsible for the management of the totality of the territory, but failed to halt the spree of illegal urbanizations (Pereira and Nunes Silva 2008; Pinto and Guerra 2019; Agarez 2020) and also failed at starting a tradition of strategically intervening in the usage of rural space. Thus, the maintenance or erasure of food production and of the circular metabolism it entailed within the Lisbon Region, was left in the hands of owners, developers and the negotiation between interests, with the State and the community taking little—if any—real part in the process.

The alliance between powerful social groups of the bygone regime with new financial capital boosted the construction sector, already vivified by lack of housing and equipment which, having lagged in the New State, were becoming a key economic driver, mobilizing jobs and investment (Pinto and Guerra 2019).

After Portugal joined the European Economic Community (EEC), accessing housing by purchase became possible because of the liberalization of credit and subsidized credit, which intensified the importance of the real estate sector soon to become one of the most prosperous Portuguese industries (Pinto and Guerra 2019). These policies effectively constructed a large middle-class who favoured home ownership but also promoted the wealth of private banks, who counted on long-term debt payments (Drago 2017). In this new social paradigm, rural society and rural life lost their importance, and local-base networks of services or resources also become less solid, since conditions for linear metabolisms become more accessible, including with respect to food (Steel 2008; Spaargaren et al. 2014). As will be seen in Chap. 6, the effects of this erasure of the hinterland are closely associated with dietary habits as well.

There is a continued increase in the importance of urban centres in the rapidly reconfiguring Portuguese society (Filipe 2014). In the Lisbon Region, this meant that the traditional knowledge on the built environment accumulated over the centuries in the 'saloia' region was overpowered, and its inhabitants remained as mere spectators of the rise of new, markedly urban, landscapes (Cabrita 2008). The Portuguese society, its labour market and its everyday life transformed, the rural world ceased to be associated with the productive realm and became increasingly symbolical (Costa 1999). In this complex process, the village is no longer understood as an inhabited place but becomes associated with memory, where old practices, meaning and identities still have a place, often turning it into the happy counterpart of an urban life that is existentially empty and driven by economic pursuit (Filipe 2014). Such ambivalence can also be observed in the somewhat romantic status the Portuguese are often said to dedicate to the idea of the 'house' or the 'household'—an idea that Oliveira Salazar exploited in his attack of collective housing, stating the people's true wish was for a private house to leave to their offsprings (Salazar 1933; d'Almeida 2015; Borges and Marat-Mendes 2020) and which may itself be another remanent of rural life. Indeed, family and home play a very important role in Portuguese culture, as does the transfer of wealth and property between generations (Pinto and Guerra 2019).

However, this phenomenon is also underlined by economic rationality, with property, and specifically house property, being seen as an economic asset to counter the weak support available from the welfare state, a phenomenon recurrent in Southern Europe, where precarity, low wages and a strong informal economy exist alongside weak welfare policies (Pinto and Guerra 2019).

The abandonment of rural places is not particular to Portuguese modernity, being indeed found all over Europe, although there are some fundamental problems that rise in the national context. Spain and Greece also had a strong problem with rural abandonment (Steel 2008; Filipe 2014), which is interesting considering the 1970s survey on vernacular architecture and the careful regulations created to protect rural settlements (Diamantsou-Kremesi and Marat-Mendes 2012). In Portugal, there are cases of villages abandoned after the construction of dams in their vicinities, because of the collapse of specific industries, particularly mines, to which one may add the ageing of the population and lack of its renovation, considering that most descendants will live their lives elsewhere (Filipe 2014). Other factors include unemployment or lack of employment opportunities, the choice of cities for their greater quality of life and the concentration of the Portuguese economy and of the Universities on coastal areas, all of which contribute to attract, and often fixate, the new generations away from the countryside (Sousa 2011).

The abrupt transition from a rural to an urban world was strongly perceived in Portugal, and it left a profound mark on the country's culture, perceivable still today, when weekends are filled with travelling TV shows which go from town to town in the countryside, always highlighting the nostalgic beauty of seemingly deserted villages and the savory desserts, cheeses or smoked meats typical of each region, and nearly always reducing the contemporary countryside to a somewhat patronizing touristic postcard of folklore chants, embroidered garments and pastoral delicacies.[3]

A more enlightened and thought-provoking approach to this transformation can be found in many literary and even cinematic works of the Portuguese democratic era. One case that may be highlighted is the second novel by Portuguese writer Lídia Jorge's (b.1946) 'O cais das merendas' (The wharf of luncheons), set in a small rural village on the Algarve coast, which is filled with tourists during the summer but becomes empty during the winter months (Jorge 1982). The local community, a limited group of friends and acquaintances, meet and take walks and luncheons on the village, reminiscing about autochthonous histories and traditions, but these always seem disturbed by habits, language and ideas from outside. A fundamental aspect of this novel was its language, full of artifice and foreignisms, which often spill into behaviour itself. Although there is a synopsis, namely the death of a young girl, the novel ends by enigmatically affirming that among the witnesses to this crime, some "could not remember how a bird flies" (Jorge 1982: 251). While there is no moral lesson inherent to the novel, this final passage seems to signify the collapse of this nature-bound rural world, whose memory drowns under the tide of a larger change to which no community, no matter how remote, can escape.

In a sense, it can be affirmed that a key social transformation brought forward by the Revolution was that Portugal utterly accepted and promoted a metabolic transition. The traditional rural world vanished and society became predominantly urban, modern, affluent, European. Salazar's constant longings for the purity of the countryside contributed to the rural world being seen as an outdated and restrictive reality, but such change was international and nearly impossible to counter.

The complicated processes of modernization in the twentieth century introduced great changes in the 'saloia' culture and the rural territories of the Lisbon Region. 'Saloia' culture can still be identified in the Lisbon Region through several aspects, namely traditions, religious customs, festivals, pilgrimages, language, lifestyles, anachronistic forms of agriculture, vernacular architecture and settlement morphology (Cabrita 2008). But is its decadence a sign of the inevitable decay of everything rurality stands for? Does it demonstrate an inescapable surrender to the global capitalism that controls everything from outside, from the rent prices we pay to the food we eat? Or is it possible that, as historiography allows us to know better the ideology of the dictatorship and the motivations of its rhetoric, we may learn to reimagine rurality as an integral part of metropolitan spaces?

The surveys of 1961 (Regional Architecture) and 2006 (IAPXX) testify to a powerful metabolic transition, and allow us to visualize, sometimes with impressive detail, the meanings and implications of such transition. Indeed, socio-metabolisms express fundamental aspects of social life and of the landscape. The comparison between visual sources of different eras indeed allows us to conceptualize the impacts of change and to 'step into' worlds we can no longer inhabit, but may still learn to interpret. Circular metabolism and local economies, as well as local availability of resources, were replaced by linear metabolisms which pollute in order to bring from afar what could be produced nearby and produce tremendous amounts of waste without any proper (re)use, adding to the already exacerbated environmental problems faced by cities. Yet, if the two surveys

[3] Examples include 'Aqui Portugal' (Here Portugal) in RTP1; 'Domingão' (Big Sunday) in Sic; and 'Somos Portugal' (We are Portugal) in TVI. The first is the National Broadcaster, while the remaining two are privately-owned.

allow us to observe the direction or orientation of change, they may also allow us to envision the direction towards which further change may be directed.

References

AAVV (1961) [2004]) Arquitectura Popular em Portugal. Ordem dos Arquitectos, Lisbon

Agarez R (ed) (2020) A habitação apoiada em Portugal. Fundação Francisco Manuel dos Santos, Lisboa

Abreu L (2017) O Serviço de Aviação do Corpo Expedicionário Português na Grande Guerra: Da Vontade à Realidade. MSc dissertation, Faculdade de Letras da Universidade de Lisboa, Lisboa

Almeida PV, Fernandes JM (Dir.) (1986) História da Arte em Portugal: a Arquitectura Moderna, vol 14. Alfa, Lisboa

Amaral K (1999 [1947]) Uma iniciativa necessária. In: CML (ed.) Keil do Amaral, o Arquitecto e o Humanista. CML, Lisboa, pp 125–126

Anastácio M, Marat-Mendes T (2016) Inquéritos à Arquitectura Popular em Portugal: uma aproximação metodológica. In: Actas do 1° Colóquio Internacional Arquitectura Popular, Município de Arcos de Valdevez, 3–6 April 2013, 225–243

dos Arquitectos O, Afonso J, Roseta H, Tostões A, Wilfried W, Portas N, Carapinha A, Rodeia JB (2006) IAP XX: Inquérito à Arquitectura do Século XX em Portugal, AO – CDN, Lisboa

Baía P (2020) A Recepção do Team 10 em Portugal. Circo de Ideias, Porto

Barreto A (2017) Anatomia de uma revolução – a Reforma Agrária em Portugal 1974–1976. D. Quixote, Lisbon

Basto EAL, Silva AF, Silva C (2002) Inquérito à habitação rural–volume 3. Imprensa Nacional–Casa da Moeda, Lisbon

Berger FG, Toussaint M, Bissau L (1994) Guia de Arquitectura Lisboa 94. Associação dos Arquitectos Portugueses, Lisboa

Borges JC, Marat-Mendes T (2020) Conversa com Francisco Silva Dias sobre o Plano de Urbanização de Chelas. Cidades Comunidades & Territórios 41:1–23

Borges JC, Marat-Mendes T, Lopes SS (2020) Chelas Zone J revisited: urban morphology and change in a recovering neighbourhood. In: Strappa G, Carolitti P, Leva M (eds) Urban Substrata and City regeneration. Proceedings of the 5th ISUFitaly International Conference, Rome, 19–22 Februray 2020, pp 751–760

Borges JC, Marat-Mendes T (2019) When Lisbon Met the Team 10 Cluster City. Joelho 10:86–99.

Borges JC (2017) The dissolution of the modern complex. MSc Dissertation, Iscte Instituto Universitário de Lisboa, Lisbon

Cabrita MA (2008) Território e identidade. MSc Dissertation. Iscte Instituto Universitário de Lisboa, Lisbon

Cabrita MA, Marat-Mendes T (2019) A identidade cultural saloia e os novos usos para o território. In: d'Encarnação J (Coord) Dos patrimónios de Cascais. Associação Cultural de Cascais, Cascais, pp 163–178

Cachinho H (2007) L'espace public dans la banlieue de Lisbonne: de la rue au centre commercial. Sud-Ouest Européen 24:51–62

Caldas JV, Tostões A, Filipe JS, Fernandes JM, Janeiro MJ, Barcelos N, Mestre V (eds) (2000) Arquitectura popular dos Açores. Ordem dos Arquitectos, Lisbon

Castro JD (1946[1984]) Geografia da fome - O dilema brasileiro: pão ou aço. 10th edition, Antares, Rio de Janeiro

Costa L (1937) Documentação necessária. Revista Do Serviço Do Património Histórico e Artístico Nacional 1(2):31–40

Costa PR (1999) Agricultores e risco ambiental: atitudes, valores e práticas num contexto de revalorização social dos espaços rurais. MSc dissertation. FEUC, Coimbra

d'Almeida PB (2006) Victor Palla e Bento d'Almeida. Obras e Projectos de um Atelier de Arquitectura, 1946–1973. MSci dissertation. Faculdade de Ciências Sociais e Humanas da Universidade Nova de Lisboa, Lisboa

d'Almeida PB (2015) Bairro(s) do Restelo. Panorama Urbanístico e Arquitectónico. Caleidoscópio, Casal de Cambra

Diamantsou-Kremesi C, Marat-Mendes T (2012) Issues on architectural surveys – The Inquérito à Arquitectura Regional Portuguesa. In: Proceedings from International Conference Surveys on Vernacular Architecture – their significance in 20th century architectural culture. ESAP, Porto, May 2012, pp 202–218

Dias J (1953) Rio de Onor, comunitarismo agro-pastoril. Instituto para a Alta Cultura, Lisbon

Dias J (1948 [1981]) Vilarinho da Furna, uma aldeia comunitária. Imprensa Nacional – Casa da Moeda, Lisbon

Drago A (2017) Is this what the democratic city looks like? local democracy, housing rights and homeownership in the Portuguese context. Int J Urban Reg Res 41(3):426–442

Esteves J, Mestre V (1987) A partir de uma conversa com o arquitecto Silva Dias a propósito do inquérito à arquitectura regional portuguesa. In: Figueira J, Nunes J, Milheiro AV, Graça Dias M (eds) J-A Jornal Arquitectos, Antologia 1981–2004. Ordem dos Arquitectos, Lisboa, pp 96–99

FAO (Food and agriculture organization) (2019) FAO framework for the Urban Food Agenda. FAO, Rome. https://doi.org/10.4060/ca3151en

Fernandes F, Cannatà M (2002) Guia da Arquitectura Moderna – Porto. Asa, Porto

Fernandez S (2011) Rio de Onor 1963–1965. Joelho 2:38–49

Fernandez SL (1964) Recuperação de aldeias em Rio de Onor. Work for the obtention of the Architect Diploma (CODA). Escola Superior de Belas Artes do Porto, Porto

Ferrão J (1996) Três décadas de consolidação do Portugal demográfico 'moderno.' In: Barreto A (ed) A situação social em Portugal. ICS, Lisbon, pp 1960–1995

Ferreira FC, Carvalho JS, Ponte TN, Silva FJ (1987) Guia Urbanístico e Arquitectónico de Lisboa. Associação dos Arquitectos Portugueses Secção Regional Sul, Lisboa

Figueira J (2016) Arquitectanic – os dias da troika. Note, Lisbon

Filipe M (2014) Aldeias Abandonadas, Património Imaterial e Desenvolvimento Local: estudos de caso na AML. Novas Edições Académicas, Chinisao

Fischer-Kowalkski M, Haas W (2014) Exploring the transformation of human labour in relation to socio-ecological transitions. In: Beblavý M, Maselli I, Veselková M (eds) The future of labour in Europe. Centre for european policy studies, Brussels, pp 56–84

Frampton K (1985) Critical History of Modern Architecture. Thames & Hudson, London

Galhano F, Dias J (1986) Aparelhos de elevar a água da rega – contribuição para o estudo do regadio. D. Quixote, Lisbon

Getz A (1989) Urban foodsheds. Letter to Peter Bird Martin. Institute of Current World Affairs. In: http://www.icwa.org/wp-content/uploads/2015/08/AAG-2.pdf

Guerra I (2011) As Políticas de habitação em Portugal: à procura de novos caminhos. Cidades Comunidades Territórios 22:41–68

Jorge L (1982) O cais das merendas. Europa-América, Sintra

Kostof S (1991) The city sahaped. Urban patterns and meanings through history. Thames & Hudson, London

Lino R (1918 [2018]) A Nossa Casa: apontamentos sobre o bom gosto na construção das casas simples. Colares editora, Sintra

Lino R (1933 [2007]) Casas Portuguesas: alguns apontamentos sobre o arquitectar das casas simples. Cotovia, Lisboa

Lopes JT (2020) Sociedade: um século desigual e de contratempos. In: Rosas F, Louçã F, Lopes JT, Peniche A, Trindade L, Cardina M (eds) O século XX Português. Tinta da China, Lisbon, pp 205–244

Marat-Mendes T (2009) O PUCS e os vazios planeados. Novas oportunidades para o ordenamento sustentado da Costa do Sol. In: Pereira M (ed) O Plano de Urbanização da Costa do Sol - Uma visão inovadora para o território. Câmara Municipal de Oeiras, Oeiras, pp 92–122

Marat-Mendes T (2020) Designing for sustainability: retrieving a systemic role for urban form. Urban Morphology 24(2):235–238

Marat-Mendes T, Borges JC (2017) A integração da sustentabilidade no ensino da arquitetura: a sua contribuição na consolidação do papel social do arquitecto. Paper presented at the international conference 'Projetar a cidade com a comunidade: reflexões sobre processos participados', FAUTL, Lisbon, 8–9 June 2017

Marat-Mendes T, Borges J (2019) Mapping sustainability transitions in contemporary culture. Mátira Digital 7 (Novembro 2019 – Outubro 2020): 1017–1051

Marat-Mendes T, Cabrita MA (2015) Morfologia Urbana e Arquitectura em Portugal. Notas sobre uma Abordagem Tipo-Morfológica. In: Oliveira V, Marat-Mendes T, Pinho P (eds) O Estudo da Morfologia Urbana em Portugal. FEUP, Porto, p. 65–94

Marat-Mendes T, Cabrita MA (2018) Recovering the habitat concept within urban morphology. In: Colomer V (ed) Proceedings 24th ISUF 2017: city and territory in the globalization age. Proceedings 24th ISUF 2017: city and territory in the globalization age. Editorial Universitat Politècnica de València, València, pp 1303–1311

Marat-Mendes T (coord), Mourão J, d'Almeida PB, Niza S (2015) Water and Agriculture Atlas: Lisbon Region in 1900–1940. Atlas da Água e da Agricultura. Região de Lisboa em 1900–1940. Iscte - Instituto Universitário de Lisboa, DINÂMIA'CET-IUL, Lisboa

Marat-Mendes T, Mourão J, Bento d'Almeida P, Niza S, Ferreira D, (2014) Água dá, água leva. Cidades, Comunidades e Territórios 28 (Jun/2014):56–87. https://doi.org/10.7749/citiescommunitiesterritories.jun2014.028.art04

Marat-Mendes T, Borges J, Dias A, Lopes R (2021a) Planning for a sustainable food system. The potential role of urban agriculture in Lisbon Metropolitan area. J Urban Int Res Placemaking Urban Sustain. ID: 1880960 https://doi.org/10.1080/17549175.2021.1880960

Marat-Mendes T, d'Almeida PB, Borges JC (2021b) Food system photographic surveys: a necessary urban design agenda. In: Bonacho R, Pires MJ, Lamy ECCS (eds) eFood – Experiencing food: designing sustainable and social practices. Routledge, London

Mestre V (2002) Arquitectura popular da Madeira. Argumentum, Lisboa

Milheiro AV (2012) Nos trópicos sem Le Corbusier. Lisbon, Relógio d'Água, Lisboa

Mónica MF (2021) O meu país – notas sobre o nacionalismo. Relógio d'Água, Lisboa

Moorcroft C (ed.) (1972) Designing for survival. Archit Des 413–433

Moudon AV, Drewnowski A, Duncan GE, Hurvitz PM, Saelens BE, Scharnhorst E (2013) Characterizing the food environment: pitfalls and future directions. Publ Health Nutr 16(7):1238–1243. https://doi.org/10.1017/S1368980013000773

Mourão J, Marat-Mendes T (2015) Urban planning and territorial management in Portugal: antecedents and impacts of the 2008 financial and economic crisis. In: Othengrafen F, Knieling J (eds.) Cities in crisis. Socio-spatial impacts of the economic crisis in Southern European cities, Routledge, London, pp 157–171

Mumford E (2000) The CIAM discourse on urbanism, 1928–1960. The MIT Press, Cambridge-Massachussetts

Napawan NC (2016) Complexity in urban agriculture: the role of landscape typologies in promoting urban agriculture's growth. J Urban: Int Res Place Urban Sustain 9(1):19–38. https://doi.org/10.1080/17549175.2014.950317

Oliveira EV, Galhano F, Dias J (1994) Espigueiros portugueses: sistemas primitivos de secagem e armazenamento. D. Quixote, Lisboa

O'Neill BJ (1984 [2011]) Proprietários, lavradores e jornaleiros – desigualdade social numa aldeia em Trás-os-Montes. Afrontamento, Porto

Parham S (2015) Food and urbanism—the convivial city and a sustainable future. Bloomsbury, London

Pereira NT (1990) As casas da Picanceira. Pedra e Cal 5:54

Pereira NT (1995 [1996]) Orlando Ribeiro, mestre também de arquitectos. In: Tempos, lugares, pessoas. Público, Lisbon, pp 98–100

Pereira NT (2000) As casas da Picanceira. Pedra e Cal 2(5):54

Pereira NT, Buarque I (1995) Prédios e vilas de Lisboa. Livros Horizonte, Lisboa

Pereira M (2009) O Plano de Urbanização da Costa do Sol – o pioneirismo de um plano sub-regional. In Pereira M (ed) O Plano de Urbanização da Costa do Sol - Uma visão inovadora para o território. Câmara Municipal de Oeiras, Oeiras, pp 24–41

Pereira M, Nunes da Silva F (2008) Modelos de ordenamento em confronto na área metropolitana de Lisboa: cidade alargada ou recentragem metropolitana? Cadernos Metrópole 20(2):107–123

Pinto TC (2017) Moving to a new housing pattern? New trends in housing supply and demand in times of changing. The Portuguese Case. Critical Housing Analysis 4(1):231–141

Pinto TC, Guerra I (2019) Housing policies, market and home ownership in Portugal: beyond the cultural model. Cidades Comunidades & Territórios 39:101–114

Portas N, Mendes M (1992) Portugal architecture 1965/1990. Editions du Moniteur, Paris

Portas N, Domingues A, Cabral J (2007) Políticas urbanas – tendências, estratégias e oportunidades. Fundação Calouste Gulbenkian, Lisboa

Portas N (2006) A oportunidade do IAPXX e uma interpretação dos Anos 4. I AAVV. IAP XX Inquérito à Arquitectura do Século XX em Portugal. Ordem dos Arquitectos – CDN, Lisboa, pp 51–60

Purdy J (2016) After nature: a politics for the anthropocene. Harvard University Press, Cambridge, Massachusetts

Ribeiro O (1939) Inquérito ao habitat rural. IAC, CoimbraRibeiro O (1987) A formação de Portugal. Ministério da Educação, Lisboa

Ribeiro O (1991) O mundo rural. In Opúsculos Geográficos, volume IV. Fundação Calouste Gulbenkian, Lisbon

Ruela R (2020) Regresso ao campo. Visão, 1433, August, pp 28–47

Reggev K (2020) What exactly is cottagecore and how did it get so popular? architectural digest. https://www.architecturaldigest.com/story/what-exactly-is-cottagecore

Salazar AO (1933) Conceitos económicos da nova Constituição — Discurso radiodifundido da U. N., em 16 de Março. In: Discursos, vol I, Coimbra Editora, Coimbra, pp 199–204

Salvador MS (2019) Shaping the city through food: the historic foodscape of Lisbon as case study. Urban Des Int 24:80–93

Smithson A, Smithson P (1956 [1974]) But today we collect ads. Architecture d'aujourd'hui 344:44–45

Sousa PRB (2011) As encruzilhadas do Despovoamento. Interior, Jovens e Emprego, O caso do Concelho de Castro de Aire. MSc dissertation, FEUC, Coimbra

Spaargaren G, Oosterveer P, Loeber A (eds) (2014) Food Practices in Transition. Routledge, London

Steel C (2008 [2013]) Hungry city—how food shapes our lives. Vintage, London

Távora F (1969) Estudo de renovação urbana do Barredo. Câmara Municipal do Porto, Porto

Tibbs H (2011) Changing cultural values and the transition to sustainability. J Future Stud 15(3):13–32

Tostões A (1997) Os verdes anos da arquitectura portuguesa dos anos 1950. FAUP, Porto

Toussaint M, Melo M, d'Almeida PB, Alcântara MD (2013) Guia de Arquitectura de Lisboa/Lisbon architectural guide 1948–2013. A + A Books, Lisboa

Toussaint M, Melo M, Rapagão JP, d'Almeida PB, Nunes J (2018) Guia de Arquitectura do Porto (1942–2017)/Porto architectural guide (1942–2017). A + A Books, Lisboa

Toussaint M (2009) Da Arquitectura à Teoria e o Universo da Teoria da Arquitectura em Portugal na primeira metade do Século XX. PhD Thesis. Faculdade de Arquitectura ad Universidade Técnica de Lisboa, Lisboa

UN (United Nations) (1987) Our Common Future. http://www.un-documents.net/our-common-future.pdf

Van Den Heuvel D, Risselada M (eds) (2005) Team 10 – in search of a utopia of the present. nai010 publishers, Amsterdam

Viljoen A, Bohn K (2014) Second nature urban agriculture: designing productive cities. Routledge, London

Wateau F (2002) Conflitos de água e rega—ensaio sobre a organização social no vale de Melgaço. D. Quixote, Lisboa

Watteu F (2000) Conflitos e água de rega: Ensaio sobre a organização social no Vale de Melgaço. Etnográfica Press, Lisbon

Weaver C (1984) Regional development and the local community. Wiley, New York

Zevi B (1973) História da arquitectura moderna – prefácio de Nuno Portas. Arcádia, Lisboa

5 On the Road

5.1 Introduction

In this chapter we present a survey on food system physical structures in the 18 municipalities of the Lisbon Region, conducted between 2019 and 2021 for the SPLACH Project. Our key materials are visual, using photographic evidence and cartographical elements as evidence of the social activities, environmental conditions and physical configurations that shape metropolitan life. We thus follow up on the visual approach to socio-metabolism used in the projects MEMO and SPLACH (Marat-Mendes et al. 2015; Niza et al. 2016; Marat-Mendes et al. 2021).

Architectural surveys of the past have proved to be effective tools not only for putting the spatial and territorial conditions into perspective, but also to envision directions into which these might be reimagined, planned and changed. Its role as historical document as an exercise in architectural and design culture is not yet exhausted.

Even if that was not their goal, the elements collected in those surveys allow a visual characterization of the culture, the society and the economy to which they belong. The surveys discussed in Chap. 4 suggest vastly different forms of socio-metabolism, as discussed. Selective as they are, the photographs in architectural surveys denounce in the nearly 50 years that separate them in time, a transition from a fundamentally agrarian socio-metabolism to an industrialized and modern one. The territorial situations of both surveys depict assigning contrasting roles to food-related activities, thus exposing the process of territorial and social transformation whereby Portugal became closer to the economies and social mores of other Western democratic countries. The idea that Portugal was a fundamentally rural country popularized by dictatorial propaganda promoted what was perceived to be a set of features from rural housing (Anastácio and Marat-Mendes 2016), namely the low densities, the single-family houses, and the emphasis on the privacy of the nuclear family. Such a vision was obviously not sustained during the democracy, and the process of urbanization accompanied a process of social modernization of an unmistakably cosmopolitan nature.

From these processes resulted the current landscape of the Lisbon Region, where two conurbations sprawl around the Tagus Estuary, surrounded by a complex mix of rural estates, dispersed settlements and towns and industrial facilities whose future is uncertain.

Inspired by the willingness of surveyors in the past, we went on the road to find the built and unbuilt structures which, in the Lisbon Region, this time focuses specifically on the spatial conditions where food-related activities take place, considering all stages of the food system.

Considering the existing metropolitan territory, a survey on the food system is in a sense impossible to conduct at this scale. First, because the food system, having been globalized, is not wholly contained within the regional territory, it would also be impossible to feed its current population with its current available space. However, all the phases of the food system exist in the Lisbon Region, and while reliance on outside food sources has been a historical constant, the degree to which this dependence is maintained or reduced can and must always be reconsidered. In a world where environmental degradation is advancing and poverty—including food poverty—still remains, the importance of balanced use of resources, including those of the territory, must be seriously considered.

5.2 Framework

The materials presented in this survey are divided into the key phases of the food system.

The first phase is production, and it includes all the land classified as agricultural PDMs active until 2015.[1] It

[1] This was the date of the last survey on municipal land-use by the Portuguese Directory General of the Territory (Direcção Geral do Território—DGT). The methodology for interpretation of this data is presented in Chap. 6.

T. Marat-Mendes et al., *Atlas of the Food System*,
https://doi.org/10.1007/978-3-030-94833-7_5

furthermore includes urban and periurban agricultural spaces, which were surveyed in 2019–2021 by the SPLACH Project.[2] Images are selected to represent the diversity of situations encountered, as well as the different activities of food production that take place in the Lisbon Region.

The second phase is transformation. Here, the central subject includes factories and industrial complexes, but we also included slaughterhouses, storage structures as well deactivated spaces like windmills and a fish drying yard, examples of a more artisanal—and occasional—economy.

For the third phase, distribution, we present the heavy road transportation that ensures supply to urban supermarkets and hypermarkets and a fish market. However, emphasis is placed on the Lisbon Region Supply Market (Mercado Abastecedor da Região de Lisboa—MARL), in São Julião do Tojal, Loures. Located in the northern periphery of Lisbon, the MARL is the nodal point from which a gargantuan amount of food enters the region, to be distributed afterwards at closer scales. With its two gigantic platforms and its separate road system, the MARL provides an impressive view of the functioning of metropolitan life.

The fourth and fifth phases, respectively trade and consumption, are the most characteristically urban ones. We did not include private home kitchens in the survey, and most examples are related to the spaces of public or collective use as well as to economic activities. It must be highlighted that often these two phases intertwine. Municipal markets, hypermarkets and street fairs usually include both activities, as selling points are grouped together with restaurants, bars or cafés. With the normalization of online delivery platforms, strongly intensified by the COVID-19 pandemic (O'Brien 2020), a part of restaurants' business implies selling food that is consumed elsewhere, further mixing the borderlines between consumption and trade spaces. In consumption spaces such as cafés and restaurants it is also common for the use of outside space for meals (*esplanadas*). The amount of space which can be occupied with this use has been enlarged during the pandemic, to allow a more generous space between sits and tables, but such occupation of public space, sometimes rather abusive, is prior to the pandemic.

For the waste disposal phase, facilities for the treatment of solid residues and water have been selected.

Indeed, in many respects, the surrounding context of the first phases of the food system is that of a "rural country that no longer exists, with abandoned villages and farms, and which no incentive programme will ever bring back" (Anastácio and Marat-Mendes 2016). But is going back the only way to make these territories inhabited again? Or can rurality—at least in its food-production dimension—acquire new meanings when we consider that the current rate of urban growth cannot be sustained much longer?

The marks of rurality can still be encountered in many urban cores which until the twentieth century were not more than villages or farmsteads. While there is a sense of 'historical areas' that clearly distinguishes these settlements, they are surrounded, often in quite chaotic ways, by urban buildings, or interwoven with them. In such cases, the character of the 'saloia' architecture can still be clearly perceived, but it no longer holds a real sense of rurality (Figs. 5.1, 5.3 and 5.4), instead having what art historian Alois Reigl (1858–1905) called age-value (Riegl 1903), characterized by imperfection, lack of integrity, formal dissolution under natural or mechanical destruction.

There are also examples where the rural character has been maintained, with the structures demonstrating few adaptations (Fig. 5.1). In the large vineyard latifundia of the Southern Bank of Tagus, notably in Palmela and the Montijo exclave, some small-scale settlements can still be found that maintain a strong relationship with cultivation spaces. In some cases, collective buildings can still be encountered within larger farms (Fig. 5.2).

While the 'saloia' architecture of the past still remains a visible part of towns and villages all over the Lisbon Region, there are examples of more recent—even contemporary—forms of rural housing. In many cases, despite the more decidedly modern design of the housing structures, there can still be verified the presence of food production, even if at a domestic scale, probably destined for self-consumption (Fig. 5.5).

In other cases, decay is perfectly evident, even despite the lack of pressure for urbanization. It is not rare for land, both used and unused, to display the ruins of buildings, some of which will probably not last much longer. While in some cases, urbanization meant a pressure to erase or deform existing settlements (Fig. 5.1), in other places, the lack of urban development seems to have prompted a sort of abandonment of the location, resulting in the fall—sometimes quite literally—of built structures (Figs. 5.6 and 5.7).

While in many villages, there is a clear concern for protecting the heritage of rural architecture, stronger efforts are needed in most municipalities towards proper maintenance. However important it is, architectural conservation or refurbishing is only part of the problem. As discussed, in Portugal like in many other Western countries, agricultural jobs have drastically disappeared. Thus, it is not surprising that in our trips throughout the Lisbon Region, we often encountered examples of rural architecture that while physically safeguarded, no longer hold any other sort of symbolic or functional value. The intrinsic link of the rural villages and architecture, 'saloia' or not, is not to be found only in its formal characteristics, but also on its links with the land, be it for cultivations, cattle breeding or otherwise. Without

[2] See Chap. 8 in this Atlas.

Fig. 5.1 Abandoned rural house in Várzea de Sintra. Source: Authors

acknowledging these links, buildings may be saved and their architectural features maintained, but the larger process of decay continues to intensify, by reducing former productive systems to mere decors for leisure background (Fig. 5.8).

Closer attention must be given to the elements and the landscapes that comprise the food system—fields, industrial estates and other elements—which are frequently abandoned, derelict or uncultivated, a problem which, as discussed in Chap. 1, has always been verified in the national territory, with uncultivated land sometimes even representing a larger amount than cultivated land. Simultaneously, current rural architecture has departed significantly from the 'saloia' identity (Cabrita 2008) while much of what remains of the rural roots of the territory is abandoned and risks irreversible degradation. If changes are to be made in the way the urban centres of the Lisbon Region feed themselves, it would make sense to start by reappropriating what remains of the rural past, thus saving it from destruction and putting it to use anew.

On the other hand, there is a certain persistence of certain practices and certain cultivations. While Palmela and the Montijo Exclave maintain immense quantities of vineyard and Mafra remains strongly marked by either agriculture or forestry across most of its territory, in many other instances, continuity is less obvious. In the interior and the surroundings of many urban neighbourhoods, small-scale agriculture or horticulture have not been erased by urbanization, but rather resurfaced, in reshaped forms, to adapt to the new urban conditions. The persistence of these activities and the adaptability of their physical spaces demonstrate the variety and complexity of forms and practices associated with food, as well as of the creative appropriation of space by residents since in many cases, there is a degree of informality in new practices. Space for 'scaling up' these activities and forms of territorial use do exist, and steps are necessary to promote this 'scaling up' of local food within the context of local consumption, making it available to a wider section of the population.

Fig. 5.2 Well-maintained rural housing in Pinhal Novo, Palmela. Source: Authors

Many food industries have disappeared from the industrial fabric of the region, although it is clear that agriculture has been modernized. However, most existing spaces are quite conventional, seldom existing innovative practices comparable with solutions already tested and implemented in other countries (Viljoen et al. 2014; Manning 2016).

With the disappearance of the rural and regional food system and the entrance of many food-related activities into a national or international agroindustrial business model, transformation spaces which depended directly on the resources and conditions of its immediate surroundings decreased. On the other hand, only in large estates, especially the latifundia in the Southern Bank of Tagus, there is a close articulation of the early phases of the food system, with wineries surrounded by hectares of vineyard. In most examples, the production phase is segregated from the remainder.

If production and transformation are somewhat segregated, in the later phases of the food system, where the consumer has a more decisive role, different phases seem to integrate or mingle. The fact that Portuguese habits of conviviality often include food, coffee or beverages, selling and consuming are naturally associated in many spaces. In cases such as street fairs or municipal markets, this relation is easiest to see. At Feira do Relógio, one of the largest street fairs in Lisbon, happening all Sundays in Chelas, the central area is occupied with food trucks—coming to the fair is not only a moment for shopping for fruits and vegetables, clothing or knick-knacks but also a moment for socializing with friends and neighbours, which often happens over food trucks and tents. Throughout the Lisbon city and many of its surrounding towns and cities, consumption has been marked by the emergence of international food venues, from

Fig. 5.3 Rural housing in the Moita town. Source: Authors

restaurants of Vietnamese Phở, Japanese Rāmen and Sushi, Taiwanese Bubble Tea, Middle Eastern Kebab, Brazilian Açaí and 'Pastel de Vento', as well as Indian, Nepalese, Chinese cuisine, most of which are associated with local migrant communities. Some of the produce needed for these businesses is sold in specific stores—an example is given in Fig. 5.80. Furthermore, and linked with strong migrant communities, supermarkets of Russian, Chinese, Cape Verdian, Indian or Brazilian products have become more frequent, with a few stores also specializing in world foods, not so much with ethnic items, but rather with specific brands that are not regularly commercialized in Portugal.

While our survey is not focused on architecture and we sought to remain open to the unbuilt elements of the food system, in territories with larger urban fabrics, it is very evident that the food system has important and ramified physical dimensions. The plethora of urban buildings where food is made available is very considerable and integral to the images we associate with urban life—cafés, restaurants. The challenge is to remember that they are also part of a larger system without which their importance cannot be fully understood, and whose boundaries are open to change.

Fig. 5.4 Abandoned rural house in Palmela. Source: Authors

Fig. 5.5 Recent rural housing in Carvoeira, Mafra. Source: Authors

Fig. 5.6 Derelict building on rural land, Pinheiro da Cruz, Alcochete. Source: Authors

Fig. 5.7 Rural workers' housing estate in Picanceira, Mafra, partly derelict and partly refurbished. Source: Authors

Fig. 5.8 A dry stone wall in Arneiro dos Marinheiros, Sintra, typical of the 'Saloia' region. Source: Authors

5.3 Production

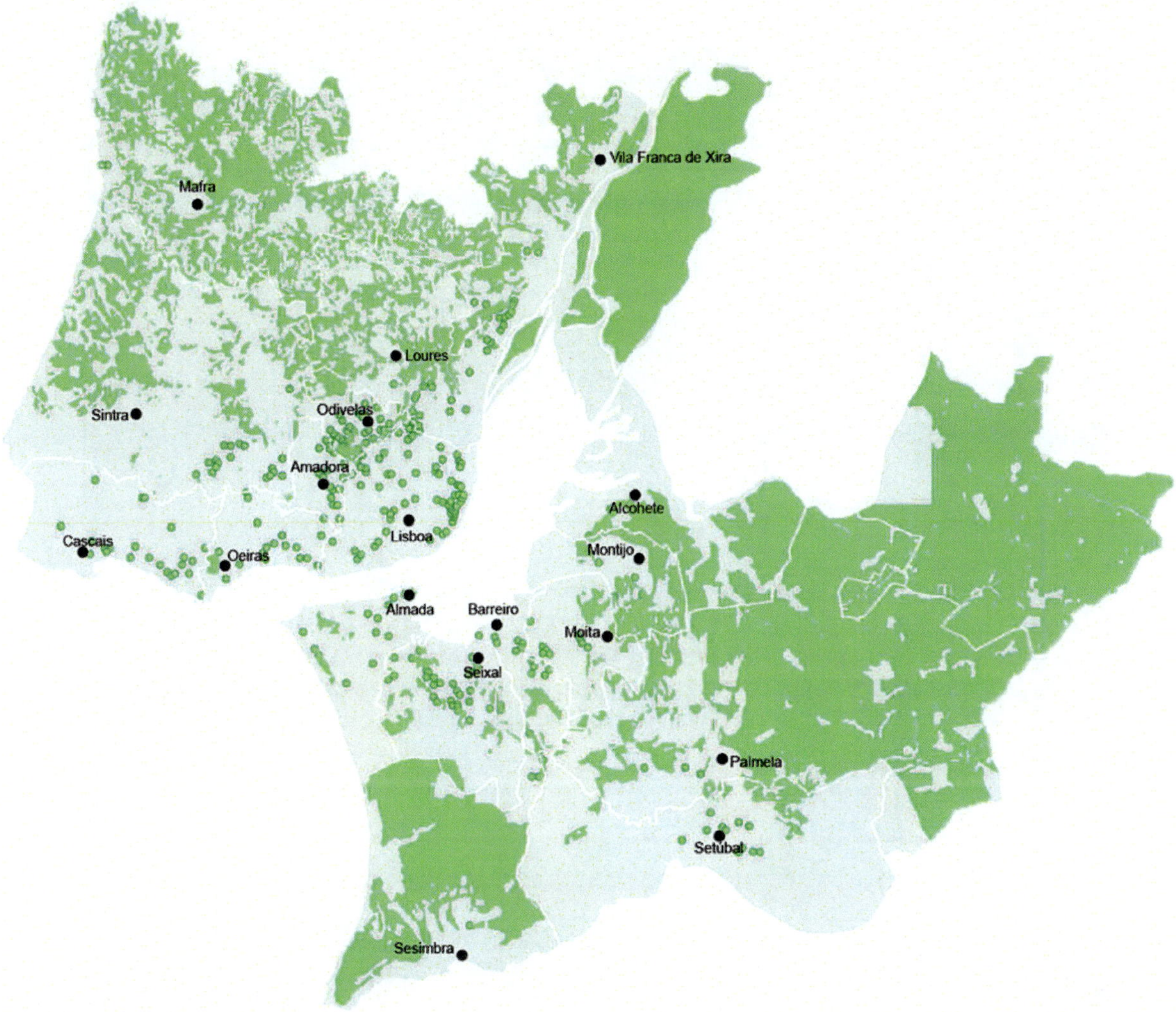

Fig. 5.9 Production: Productive spaces. Source: Authors

Fig. 5.10 Large-scale rural agriculture, Poceirão and Marateca, Palmela. Source: GOOGLE INC. Google Maps

Fig. 5.11 Small-scale rural agriculture (family-based) and agroforestry, São João das Lampas, Sintra. Source: GOOGLE INC. Google Maps

Fig. 5.12 Municipal urban agriculture, Chelas, Lisbon. Source: GOOGLE INC. Google Maps

Fig. 5.13 Informal urban agriculture, Vale da Amoreira, Moita. Source: GOOGLE INC. Google Maps

Fig. 5.14 Monoculture agricultural field in Assafora, Sintra. Source: Authors

Fig. 5.15 Livestock production in urban agriculture gardens in Arrentela, Seixal. Source: Authors

Fig. 5.16 Rural agricultural field in floodplain in Bobadela, Loures. Source: Authors

Fig. 5.17 Polyculture agricultural field in São João das Lampas, Sintra. Source: Authors

Fig. 5.18 Raising beds for agriculture, Montijo. Source: Authors

Fig. 5.19 Livestock production, Moita. Source: Authors

Fig. 5.20 Agricultural field fertilization, Moita. Source: Authors

Fig. 5.21 Rural agriculture fields with livestock production near the city centre, Oeiras. Source: Authors

Fig. 5.22 Fishing harbour, Ericeira. Source: Authors

Fig. 5.23 Urban agriculture garden, Lisbon. Source: Authors

Fig. 5.24 Polyculture agricultural fields in the Lizandro floodplain, Mafra. Source: Authors

Fig. 5.25 Livestock production, Aldeia da Lagoa, Mafra. Source: Authors

Fig. 5.26 Urban agriculture garden, Barreiro . Source: Authors

Fig. 5.27 Cooperative agricultural field in the Great Wetland of Tagus, Vila Franca de Xira. Source: Authors

Fig. 5.28 Fishing boats in Paço de Arcos, Oeiras. Source: Authors

Fig. 5.29 Vineyard in Poceirão, Palmela

Fig. 5.30 Almond trees field in Picanceira, Mafra

Fig. 5.31 Salt pans in Samouco, Alcochete. Source: Authors

Fig. 5.32 Large-scale rural agriculture fields in the Montijo Exclave. Source: Authors

Fig. 5.33 Kitchen gardens in urban backyard in Pirescoxe, Loures. Source: Authors

Fig. 5.34 Ploughed fields, São João das Lampas, Sintra. Source: Authors

Fig. 5.35 Fishing port in Samouco, Alcochete. Source: Authors

Fig. 5.36 Greenhouse production in Arneiro dos Marinheiros, Sintra. Source: Authors

Fig. 5.37 Rural agricultural field in floodplain in Santo António dos Cavaleiros, Loures. Source: Authors

Fig. 5.38 Urban agriculture garden in the Chelas valley, Lisbon. Source: Authors

5.4 Transformation

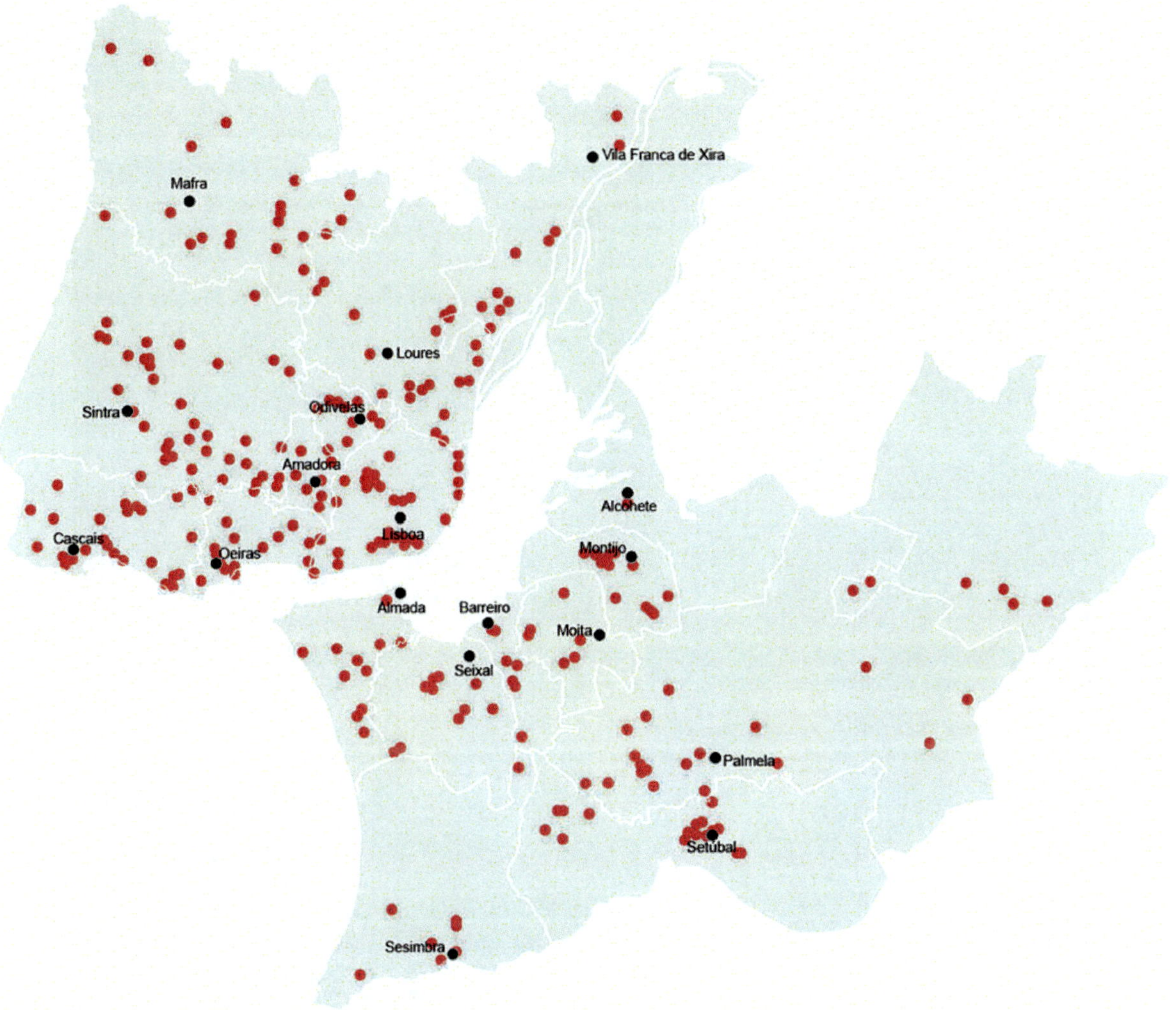

Fig. 5.39 Transformation: Industrial spaces. Source: Authors

Fig. 5.40 Cooperative winery, Santo Isidro de Pegões, Montijo Exclave Source: GOOGLE INC. Google Maps

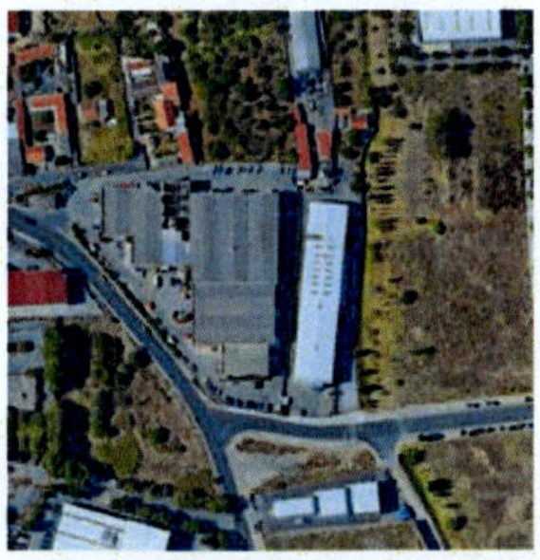

Fig. 5.41 Food processing industry, Alcabideche, Cascais. Source: GOOGLE INC. Google Maps

Fig. 5.42 Winery of Rio Frio, Palmela. Source: GOOGLE INC. Google Maps

Fig. 5.43 Food processing industry, Castanheira do Ribatejo, Vila Franca de Xira. Source: GOOGLE INC. Google Maps

Fig. 5.44 Dairy farm, Barra Cheia, Moita. Source: Authors

Fig. 5.45 Slaughterhouse, Atalaia, Montijo. Source: Authors

Fig. 5.46 Drying yard, Samouco, Alcochete. Source: Authors

Fig. 5.47 Livestock feed factory, Sarilhos Grandes, Montijo. Source: Authors

Fig. 5.48 Windmill, Palmela. Source: Authors

Fig. 5.49 Malt factory, Poceirão, Palmela. Source: Authors

Fig. 5.50 Grain storage, Lezíria Grande do Tejo, Vila Franca de Xira. Source: Authors

Fig. 5.51 Brewery, Boca da Lapa, Vila Franca de Xira. Source: Authors

Fig. 5.52 Tomato pulp factory, Pegões, Montijo Exclave. Source: Authors

Fig. 5.53 Sugar factory, Santa Iria de Azóia, Loures. Source: Authors

Fig. 5.54 Dried cod, Ericeira, Mafra. Source: Authors

Fig. 5.55 Cod factory, Rosário, Moita. Source: Authors

Fig. 5.56 Windmill, Alburrica, Barreiro. Source: Authors

5.5 Distribution

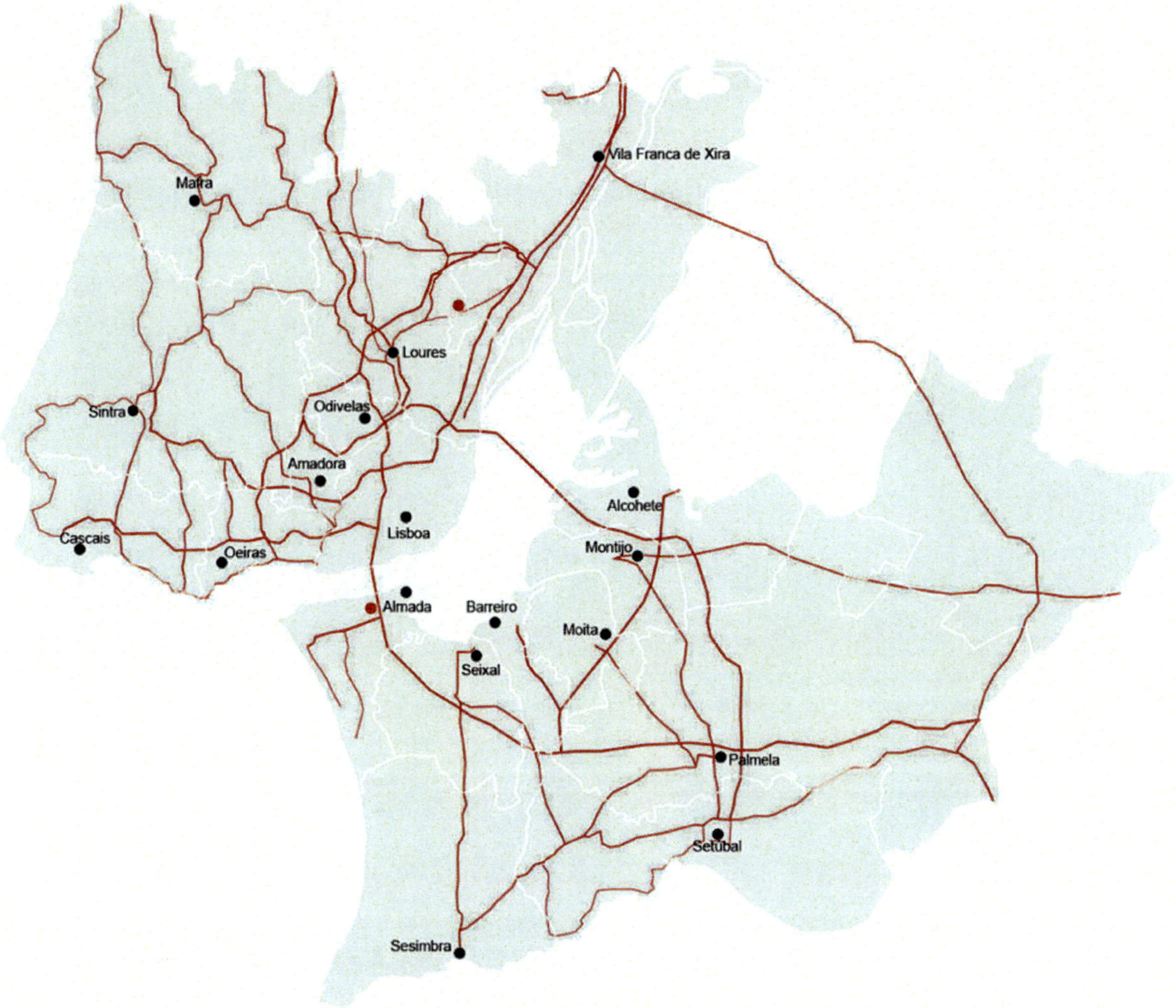

Fig. 5.57 Distribution: Supply markets and road network. Source: Authors

Fig. 5.58 Supply market, Almada. Source: GOOGLE INC. Google Maps

Fig. 5.59 Supply market of the Lisbon Region (MARL), Loures. Source: GOOGLE INC. Google Maps

Fig. 5.60 Road junction, Santa Iria de Azóia, Loures. Source: GOOGLE INC. Google Maps

Fig. 5.61 Road junction, Praias do Sado, Setúbal. Source: GOOGLE INC. Google Maps

Fig. 5.62 Interior of the supply market of the Lisbon Region (MARL), Loures. Source: Joana Benedito

Fig. 5.63 Interior of the supply market of the Lisbon Region (MARL), Loures. Source: Joana Benedito

Fig. 5.64 Supply market of the Lisbon Region (MARL) complex, Loures . Source: Authors

Fig. 5.65 Exterior of the supply market of the Lisbon Region (MARL), Loures. Source: Authors

Fig. 5.66 Vendors unloading trucks in the supply market of the Lisbon Region (MARL), Loures. Source: Authors

Fig. 5.67 Fish market, Setúbal. Source: Authors

Fig. 5.68 Food distribution for supermarket in the city centre, Lisbon. Source: Authors

5.6 Trade

Fig. 5.69 Trade: Supermarkets and municipal markets. Source: Authors

Fig. 5.70 Municipal market, Alhos Vedros, Moita. Source: GOOGLE INC. Google Maps

Fig. 5.71 Municipal market, Arroios, Lisbon. Source: GOOGLE INC. Google Maps

Fig. 5.72 Minimarket, Laveiras, Oeiras. Source: GOOGLE INC. Google Maps

Fig. 5.73 Supermarket by the road, Santa Iria de Azóia, Vila Franca de Xira. Source: GOOGLE INC. Google Maps

Fig. 5.74 Fresh vegetables selling at Relógio Street market, Chelas, Lisbon. Source: Authors

Fig. 5.75 Butcher's shop, Lisbon. Source: Authors

Fig. 5.76 Bakery shop, Lisbon. Source: Authors

Fig. 5.77 Itinerant selling of vegetables and fruit, Palmela. Source: Authors

Fig. 5.78 Grocery store, Lisbon. Source: Authors

Fig. 5.79 Rotisserie, Bobadela, Loures. Source: Authors

Fig. 5.80 Açaí ice cream store and wholesaler, Bobadela, Loures. Source: Authors

Fig. 5.81 Interior of municipal market of Ericeira, Mafra. Source: CM Mafra

Fig. 5.82 Butcher's shop, Ericeira, Mafra. Source: Authors

Fig. 5.83 Municipal market of Ericeira, Mafra. Source: Authors

Fig. 5.84 Supermarket, Coina, Sesimbra. Source: Authors

Fig. 5.85 Relógio Street market, Chelas, Lisbon. Source: Authors

Fig. 5.86 Fresh fruits selling at Relógio Street market, Chelas, Lisbon. Source: Authors

Fig. 5.87 Relógio Street market, during the Covid 19 lockdown, Chelas, Lisbon. Source: Authors

Fig. 5.88 Municipal market, Moita. Source: Authors

5.7 Consumption

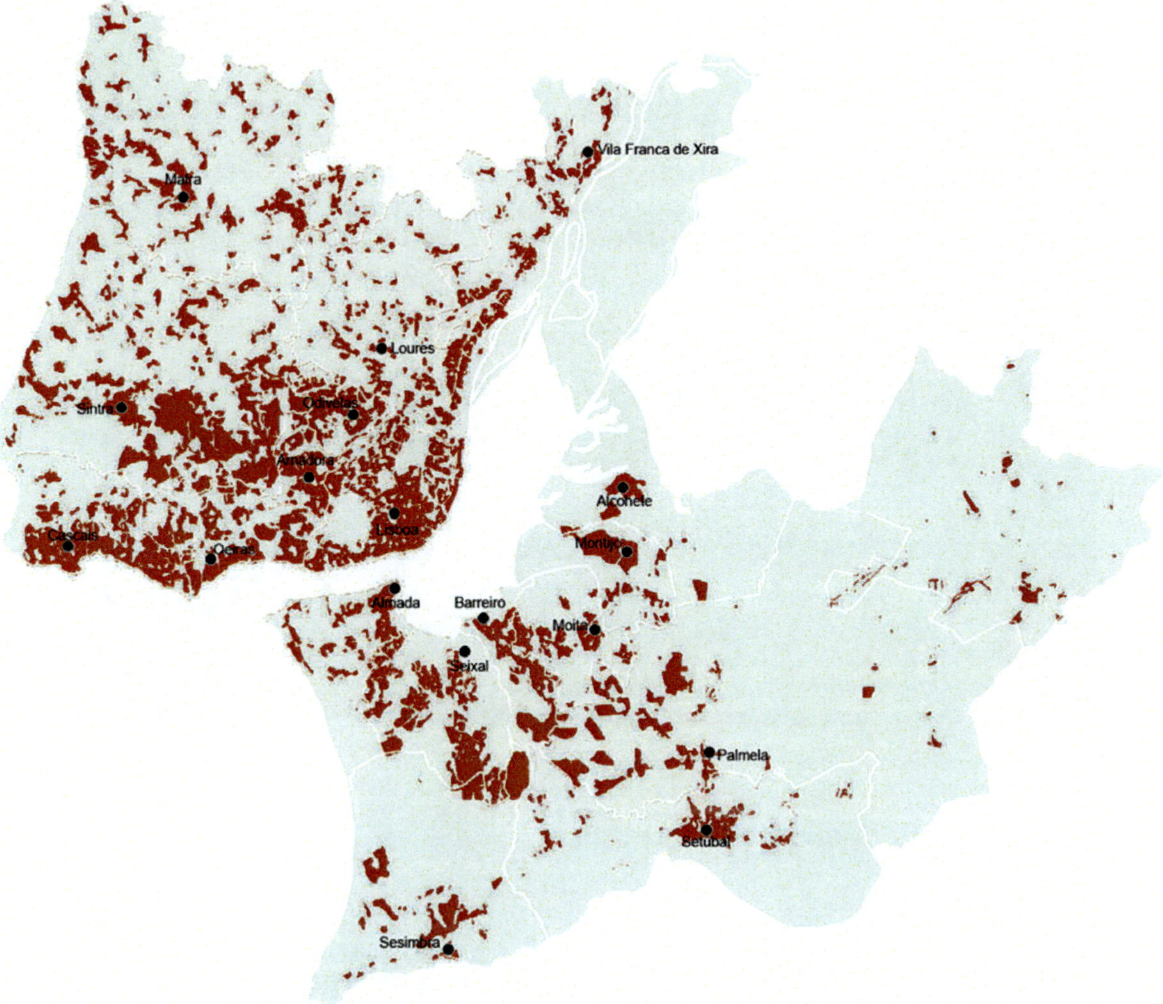

Fig. 5.89 Consumption: Coffeeshops, restaurants and urban areas (*Note* urban areas are surveyed in order to account for household kitchens). Source: Authors

Fig. 5.90 Timeout market, with fast-food restaurants, Lisbon. Source: GOOGLE INC. Google Maps

Fig. 5.91 Town square, with coffees and restaurants, Vila Franca de Xira. Source: GOOGLE INC. Google Maps

Fig. 5.92 Urban areas, Lisbon. Source: GOOGLE INC. Google Maps

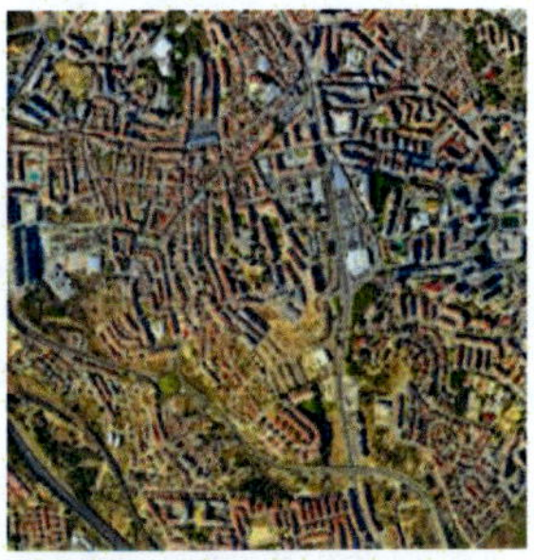

Fig. 5.93 Urban areas, Laranjeiro, Almada. Source: GOOGLE INC. Google Maps

Fig. 5.94 Coffee shop in city centre, Alcochete. Source: Authors

Fig. 5.95 Coffee shop in the 'saloia' village of Arneiro dos Marinheiros, Sintra. Source: Authors

Fig. 5.96 Coffee shops and restaurants in city centre, Cacilhas, Almada. Source: Authors

Fig. 5.97 Restaurants in city centre, Escadas do Duque, Lisbon. Source: Authors

Fig. 5.98 Restaurants in city centre, Rua da Trindade, Lisbon. Source: Authors

Fig. 5.99 Coffee shops and restaurants, Carnaxide, Oeiras. Source: Authors

Fig. 5.100 Rotisserie restaurant, Moscavide, Lisbon. Source: Authors

Fig. 5.101 Restaurant in city centre, EPUL office building, Martin Moniz, Lisbon. Source: Authors

Fig. 5.102 Restaurant at the beach, Olho de Boi, Almada. Source: Authors

Fig. 5.103 Roadside restaurant, Pegões, Montijo Exclave. Source: Authors

Fig. 5.104 Roadside restaurant, Portela de Sacavém, Loures. Source: Authors

Fig. 5.105 TimeOut market, Lisbon. Source: Authors

Fig. 5.106 Coffee shop, Santo António dos Cavaleiros, Loures. Source: Authors

5.8 Waste Disposal

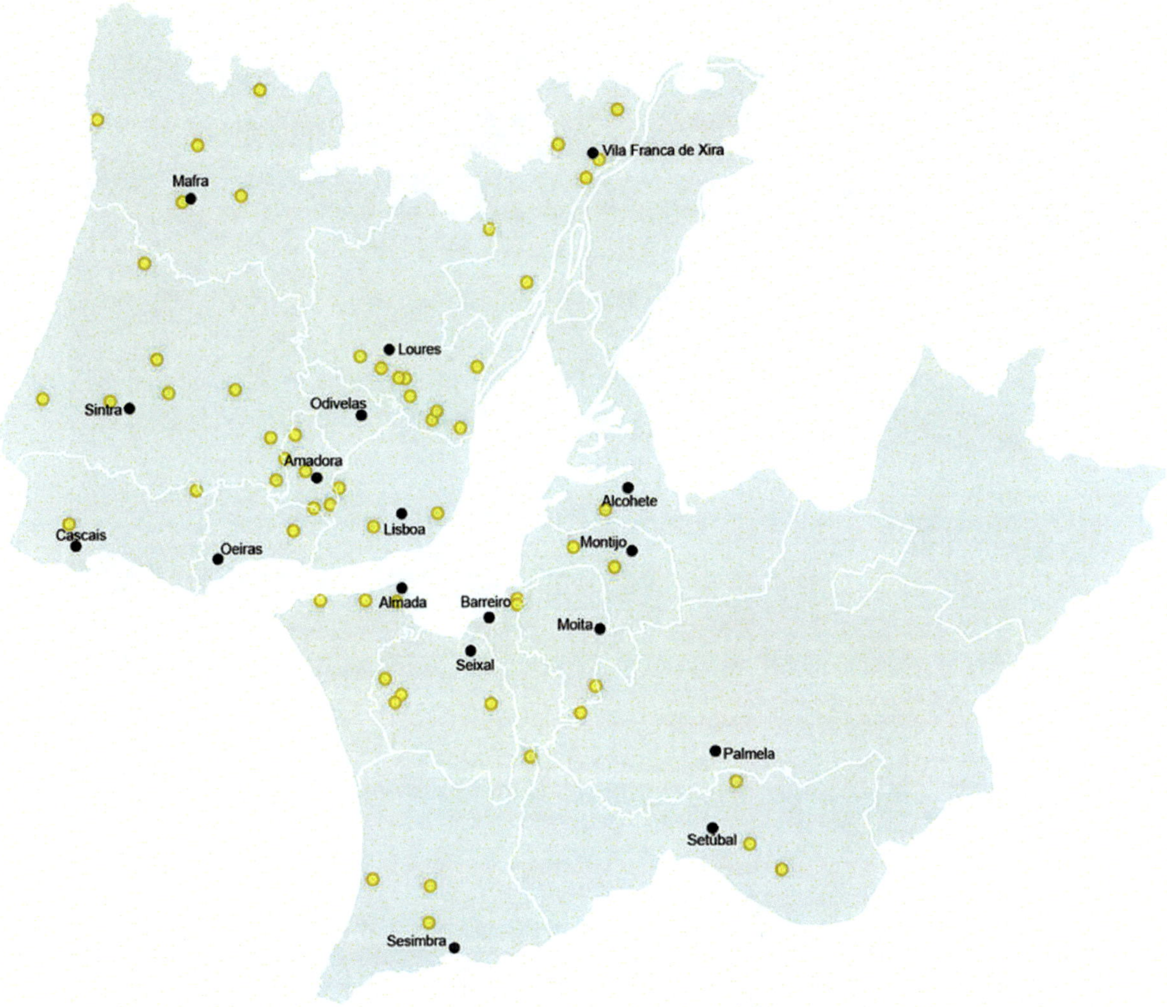

Fig. 5.107 Waste disposal: Water Waste Treatment Plant (WWTP), landfills and recycling centres. Source: Authors

Fig. 5.108 Waste treatment and recovery facility, Brejos das Moita, Moita. Source: GOOGLE INC. Google Maps

Fig. 5.109 WWTP, Lisbon. Source: GOOGLE INC. Google Maps

Fig. 5.110 Recovery and treatment of urban waste, São João da Talha, Loures. Source: GOOGLE INC. Google Maps

Fig. 5.111 WWTP, Cachofarra, Setúbal. Source: GOOGLE INC. Google Maps

Fig. 5.112 WWTP, Casal do Moinho, Mafra. Source: Authors

Fig. 5.113 WWTP, Paio Pires, Seixal. Source: Águas do Tejo Atlântico (Grupo Águas de Portugal)

Fig. 5.114 WWTP, Sentrão, Sesimbra. Source: Águas do Tejo Atlântico (Grupo Águas de Portugal)

Fig. 5.115 WWTP, Lisbon. Source: Authors

Fig. 5.116 Waste Treatment and Recovery facility, Brejos da Moita. Source: Authors

Fig. 5.117 Composting in urban agricultural gardens, Lisbon. Source: Authors

Fig. 5.118 Urban composting in city centre, Lisbon. Source: Authors

Fig. 5.119 Recycling bin for used cooking oil, Lisbon. Source: Authors

References

Anastácio M, Marat-Mendes T (2016) Inquéritos à Arquitectura Popular em Portugal: uma aproximação metodológica. In: Actas do 1° Colóquio Internacional Arquitectura Popular, Município de Arcos de Valdevez, 3–6 April 2013, pp 225–243

Cabrita MA (2008) Território e identidade. MSc Dissertation. Iscte Instituto Universitário de Lisboa, Lisbon

Manning J (ed) (2016) Food and the city. Urban Design 140:14–39

Marat-Mendes T (coord), Mourão J, d'Almeida PB, Niza S (2015) Water and agriculture atlas: Lisbon region in 1900–1940. Atlas da Água e da Agricultura. Região de Lisboa em 1900–1940. Iscte - Instituto Universitário de Lisboa, DINÂMIA'CET-IUL, Lisboa

Marat-Mendes T, d'Almeida PB, Borges JC (2021) Food system photographic surveys: a necessary urban design agenda. In: Bonacho R, Pires MJ, Lamy ECCS (eds) eFood—experiencing food: designing sustainable and social practices. Routledge, London

Niza S, Ferreira D, Mourão J, d'Almeida PB, Marat-Mendes T (2016) Lisbon's womb: an approach to the city metabolism in the turn to the twentieth century. Reg Environ Change 16(6):1725–1737

O'Brien, SA (2020) Uber has lost $5.8 billion in the first three quarters of this year. CNN Business, November 5. In https://edition.cnn.com/2020/11/05/tech/uber-third-quarter-earnings/index.html. Accessed 13 April 2021

Riegl A (1903[1984]) Le culte moderne des monuments. Éditions du Seuil, Paris

Viljoen A, Bohn K (2014) Second Nature Urban Agriculture: Designing Productive Cities. Routledge, London

6 The Food System: A Portrait of Transformations

6.1 Introduction

What food do we produce and what food do we eat? These two questions are perhaps the most determinant aspects of the food system, regardless of the scale we are considering it from. In this chapter, we focus on these questions, pointing out the key dietary habits of the Portuguese—and how these have evolved over time.

In the first moment, we focus on the historical situation of Portuguese eating habits and highlight the characteristic cuisine, the key products utilized as well as their relation to the productive structure of the Lisbon Region during the first half of the twentieth century. As we will see, dietary habits remained relatively stable until the 1960s, when important transformations took place, with dietary habits changing towards more international—and less local—preferences, a transformation which remains to our own day.

Thus, in the second moment, we describe the current food system of the Lisbon Region, based not only on previous studies but also on more recent information made available by the Portuguese National Institute for Statistics (Instituto Nacional de Estatística—INE). Another important source is the recent book by nutritional scientist Pedro Graça (b?) where the key dietary habits of the Portuguese—both historical and current—are described, allowing the observation of consumption and culture as these have developed in the Portuguese diet (Graça 2020).

Considering the emphasis we want to lay on the spatial dimension of the food system, we also seek to establish the key transformations—and the current situation—of land-use within our region. To this end, we compare historical and active PDMs, all approved after the Municipal Planning Law of 1990, discussed in Chap. 3. As we saw, the democratic PDM is the first to order an integral land-use classification of all the territory, hence its central importance for assessing the availability of space for food production.

6.2 The Evolution of the Food System in the Lisbon Region

Dietary habits in the Portuguese territory developed throughout a far-reaching timespan, and developed according to the successive occupations by the Phoenicians, the Greeks, the Catageneses, the Romans and the Arabs, with an important influence arising from contact with the overseas, especially through the Mediterranean Sea (Graça 2020). However, outside influence did not particularly change the eating habits in the inland territories and thus, it is highly likely that the pattern of food habits from the 1940s and 1950s is in continuance with those of the previous centuries, a continuity intensified by belated industrialization and the political isolation imposed by the New State (Graça 2020).

Thus, understanding the food habits of the mid-century in Portugal is, in many senses, to look at a strong and relatively stable cooking tradition, which translates to the eating habits of the inhabitants.

By 1900, the Lisbon city included several agricultural spaces, including olive groves, ploughed lands, woods, vegetable gardens and vineyards—the most extensive agricultural surface was that of ploughed land, but olive groves and vineyards were also particularly relevant (Niza et al. 2016). While ploughed land and vineyards tended to disappear with the advance of urbanization, as we shall see next, the past olive groves were not fully erased, as olive trees are still a common element in the Lisbon landscape, remaining in the gardens of certain neighbourhoods and having even baptized a large new residential area urbanized by the Council in the 1960s, Olivais, which translates to 'Olive Groves'. In 1900, only 16% of Lisbon land was urbanized, while around 40% was signalled as cultivated, a number which could in reality be much greater, considering that the remainder 34% had no indication of land-use (Niza et al. 2016). The data from the 1900s, the transition from the

T. Marat-Mendes et al., *Atlas of the Food System*,
https://doi.org/10.1007/978-3-030-94833-7_6

nineteenth to the twentieth century, suggests that each hectare of the Lisbon city produced between 6 and 11 kg of food produce, although the phylloxera plague in the late nineteenth century decreased the production of wine, and with olive production losing half its produce from 1890 to 1900 and potato production increasing 34% in the same period (Niza et al. 2016).

An interesting aspect of the Portuguese cooking tradition is that while it often makes use of meat and fish, its key elements were indeed vegetables. Until the first half of the twentieth century in Portugal, they were eminently vegetarian by force of circumstances, as generally happened in Southern Europe (Graça 2020). Portuguese cooking was highly based on the vegetable gardens, varying with the seasons, the presence or absence of water and land characteristics, with vegetables mixed with grain, legumes or tubers, almost always cooked with scant use of raw salads (Graça 2020).

Considering the totality of the Lisbon Region, there are important sources from the 1940s which disclose the key patterns of exchange and consumption within the territory, with the whole region organized around a great consumption centre—the Lisbon city—and the surroundings providing different food items, imported in considerable quantities (Marat-Mendes et al. 2015). This situation was consistent with the local hinterland, whereby cities were located close to productive territories, with transported goods conditioned by the availability of river transportation and later by train (Salvador 2019). Within the Lisbon Region, transportation was mostly ensured through animal carriages and later trains, and when the produce originated from the Southern Bank of Tagus, it was transported to Lisbon by boat (Marat-Mendes et al. 2015).

In the 1940s, a study conducted by Maria de Lurdes Santos Pereira pointed out the existence of four 'supply regions' for the Lisbon City, namely the 'Saloia Region' of Oeiras, Cascais, Sintra and Loures; the Southern Bank of Tagus, from Alcochete to Costa da Caparica in Almada; the Setúbal area and the Vila Franca de Xira area—the study included the counting of received produce as well as the key foodstuffs that each 'supply region' provided (Marat-Mendes et al. 2015). The most relevant supply was the 'Saloia Region', providing a total of 26 million kg per year of a great variety of horticultural products (Marat-Mendes et al. 2015). From the Southern Bank of Tagus, a total of 10,679,315 kg was brought to Lisbon in 1946, including savoy cabbage, white cabbage, cauliflower and potatoes for feeding both people and livestock (Marat-Mendes et al. 2015). The Vila Franca de Xira area produced food mostly through vegetable gardens with dry farming, and it supplied in 1946 over 1,776,479 kg in foodstuffs, between kale, sprouts, peas, turnip greens and turnips (Marat-Mendes et al. 2015). The Setúbal area was the least relevant in supplying Lisbon, with 329,124 kg of foodstuffs in 1946 (Marat-Mendes et al. 2015).

From these numbers, we can notice that the most relevant supply area for Lisbon came from its immediate surroundings, mostly in the western vicinities, although Mafra, which remains today a strongly rural territory in the Region, was strangely absent. Moreover, the Tagus does not seem to present a particular barrier to the flow of produce, since the second largest supplier was on the opposite side of the river.

When urban growth started to develop towards a decidedly urban scale in the late 1940s, numerous cultivation spaces within the Lisbon city itself, filled with 'hortas' or vegetable gardens were eliminated in peripheral areas like Lumiar, Campo Grande, Beato, Alto do Pina, Benfica, Olivais. Others would remain in Ajuda, Alto de São João, Alto de Pina, Areeiro, Beato, Benfica, Braço de Prata, Campo Grande, Campolide, Carnide, Chelas, Lumiar, Luz, Marvila, Olivais, Palhavã, Pedrouços, Picheleira, Poço do Bispo, Prazeres, Sete Rios and Telheiras, as well as in borderline suburbs belonging to other municipalities like Algés and Beirolas (Marat-Mendes et al. 2015). Yet most of these cultivations would not last beyond the 1960s and 1970s when the urbanization of the capital reached a new growth dynamic. As we saw in Chaps. 2 and 4, in this era not only is verified a shift in State-promoted urbanization, towards high-density housing but there is also the tide of illegal construction in the city's surroundings. The uneven negotiation between urbanization and the hinterland reaches a critical point—as would dietary habits.

The supply flows from the surrounding areas of Lisbon, as well as some of the dietary habits verified among the regional working-class (Table 6.1) demonstrate that there has been a prevalence of vegetables and grain, and fish (including codfish) had more importance than meat, which was only a relevant consumption item in six out of the 15 municipalities accounted for. Indeed, the Portuguese have a strong tradition of foods based on cooked vegetables and legumes, with soup being a staple of dietary habits, unparalleled in other European countries (Graça 2020). Soups and broths have been a constant presence in the Portuguese food literature as a central dish in the feeding of peasants, accompanied by wine and bread—both of which have an important presence in the Lisbon Region in the 1940s—and sometimes by bacon or codfish chips (Graça 2020), which may explain the use of these foods in the municipalities where they were more consumed. Broths are usually made from horticultural produce harvested on the day of preparation or the days before, added with olive oil, potatoes or beans and occasionally with rice or pasta and reflecting seasonal variation and regional productions (Graça 2020).

However, the use of different produces varied not only according to regions but also according to social classes. Physician António de Almeida Garrett (1884–1961)

Table 6.1 Key dietary elements in the Lisbon Region by 1940

Municipality	Dietary elements of the working-class	Most relevant foodstuffs											
		Bread	Legumes	Vegetables	Bacon	Olive oil	Fish	Codfish	Meat	Wine	Coffee	Sugar	Potatoes
Vila Franca de Xira	Mostly wheat bread alongside some fish or meat	o	o		o	o							
Mafra	Soup and 'cozido' (Portuguese stew)						o		o				
Loures (including Odivelas)	Lunch: fish, bread, wine; dinner: bread soup with meat, wine; supper: fish, bread, wine; short consumption of agricultural produce	o					o		o	o	o		
Sintra	Bread, legumes, short consumption of meat and coffee	o									o	o	
Cascais	Very varied	o						o		o			o
Oeiras (including Amadora)	Soup and 'cozido' (Portuguese stew)						o		o				
Alcochete	Two meals a day	o	o				o		o				
Montijo	No identifyable pattern—includes meat, fish, legumes, farinacious												
Almada	Bread with fish or codfish and legume soup	o					o	o					
Barreiro	Three meals a day		o				o						
Moita	Wheat bread, corn, meat, fish, legumes, vegetables, fruit and wine	o	o		o		o		o	o			
Seixal	Variable	o					o						o
Palmela	Most typical meal is 'cozido' (Portuguese stew)	o	o	o					o				
Setúbal	When there is money, the worker eats as does the middle-class, but does not eat enough when money lacks	o	o				o				o		o
Sesimbra	Very frugal	o					o						

Source Marat-Mendes et al. (2015)

conducted in 1940 a survey on the daily eating habits of the Portuguese, dividing them into four distinct social classes, although with respect to peasants, he considered those from the regions of Minho, Beira and Alentejo—thus, a region from the North, another from the Centre and another from the South of Portugal (Marat-Mendes et al. 2015).

The conclusions reached by Garrett allow us to observe key contrasts on the access different social classes had to food. Transformed products were chiefly consumed by the middle- and upper-classes, including coffee, sugar and pasta. Cheese is a notable case since its consumption was only relevant for the upper-classes. Meat and fish are also important examples to notice: the difference in meat consumption was markedly class-dependent. The city workers and the middle-class consumed similar amounts of fish (60 and 75 g, respectively), but peasants ate much less (10 g) and the upper-classes ate 33 times more fish than the peasants (330 g). Meat was less consumed than fish by all social classes except the middle-class, but the contrasts in quantities were still very pronounced—while peasants and city-workers consumed 10 and 50 g of meat, respectively, the middle-class consumed 250 g and the upper-classes 280 g.

Interestingly, for both the upper- and the middle-classes, the most consumed food item is wine, with a daily 500 ml for both, and while city workers consume the same amount of wine, peasants consume only 150 ml per day. Excluding wine, the most significant food item for the upper-class was potato (350 g) while for the middle-class it was potato and bread (300 g each). For peasants and city workers, the most relevant food item was not wine, but rather bread: 600 g for the latter and 1,000 g for the former.

What do these numbers tell us? First, they testify to a relative proximity between the middle-class and the city worker, with a similar consumption of wine and potatoes and with a protagonism of bread in the overall diet—although the middle-class ate 300 g a day and the urban worker 600 g. The consumption of fish is also very similar in both social classes, although the middle-class consumed five times more meat than the city worker.

Unsurprisingly, the most diversified diets were those of the upper- and middle-classes, while the most limited is that of the peasant. Overall, from peasant diet, milk, cheese, eggs, rice, pasta, flour, bacon, butter, coffee and sugar were excluded—most of them are products that demand transformation and thus will be more expensive and eventually difficult to access. The keys to peasant diet are bread, vegetables and potatoes, with meat and fish as the less relevant elements and consumed in similar quantities (10 g each). Interestingly, the diet of the upper- and middle-classes display the opposite proportions, with meat and fish having a preponderance over vegetables and bread. Other than cheese, it should be noted that eggs were only an element in daily diets for the upper- and middle-classes, the same being true for pasta and flour. With animal fats, the opposite is true—only peasants use it. Considering the importance of fats as a rapid source of energy, it is interesting to note that between animal fat (30 g) and olive oil (20 g), peasants consumed more of it than the remaining social classes, with the upper-classes consuming 45 g, the middle-class 40 g and the city workers 30 g. It should be noted that wheaten bread only became a common food staple among the Portuguese precisely in the 1940s, rapidly becoming the key food item for the Portuguese, a tendency preconized in the role bread had assumed in the diet of the peasantry (Marat-Mendes et al. 2015). As seen in Chap. 1, campaigns for the cultivation of wheat were among the key rural policies in the military dictatorship and the New State, which may explain, at least partly, this tendency (Table 6.2).

The fact that traditional Portuguese cooking is so dependent upon vegetables results from the fact that the majority of the population remained rural until the mid-twentieth century. Indeed, the nature of agricultural work is at the basis of many of the situations verified in Garrett's account of daily diets.

For rural populations, animal protein was relatively scant—bovines were seldom consumed because they were necessary to carry on agricultural labour and because grazing was not sufficient to breed them in large quantities, while pork was more common although not consumed fresh for difficulties in conservation, and game meat was consumed more as a compliment than a food in and by itself (Graça 2020). Fish was a common foodstaple on the coast, but in the countryside its consumption was less significant, being difficult to conserve during transportation (Graça 2020)—hence the fact that in Garrett's account, the peasants consume little fish and only in salted form (Marat-Mendes et al. 2015). Moreover, in the Lisbon Region, as seen in Table 6.1, either fish, codfish or both were a common food item in all Lisbon Region municipalities except for Palmela, Montijo, Vila Franca de Xira, while in Sesimbra, which faces the sea in its western and south borderlines and where the diet of the working-class was classified as 'very frugal', the key foodstaples were fish and bread (Marat-Mendes et al. 2015).

Indeed, it was precisely the difficulty in accessing or conserving protein that has led to some of the most sui generis foods of the Portuguese, including the techniques for conservation in salt and the smokehouse delicacies, whereby meats were minutely harnessed and conserved through smoke in 'chouriços', 'alheiras' and 'morcelas' which display significant variations across the different regions in the country. Portuguese Stew, a very central dish for the Portuguese, and which includes a varied juxtaposition of vegetables, grains, meats and smoked meats—the meats being probably less numerous in days of old—is an example of the creative solutions typical of the Mediterranean in using

Table 6.2 Daily food consumption of Portuguese social classes in 1940

Food items (units of measure)	Peasant/agricultural worker	City worker	Middle-class	Upper-class
Bread (g)	1,000	600	300	250
Milk (cl)	–	100	150	200
Cheese (g)	–	–	–	40
Eggs (unit)	–	–	1	2
Fish (g)	10 (salted)	60 (codfish)	75	330
Meat (g)	10	50	250	280
Beans (g)	40	40	–	–
Potato (g)	400	250	300	350
Rice (g)	–	90	90	90
Pasta (g)	–	–	50	35
Flour (g)	–	–	30	35
Vegetables (g)	600	350	200	200
Wine (cl)	150	500	500	500
Fruit (g)	100	60	150	300
Olive oil (g)	20	30	40	45
Animal fat (g)	30	–	–	–
Bacon (g)	–	15	25	25
Butter (g)	–	–	10	35
Coffee (g)	–	15	25	30
Sugar (g)	–	20	40	50

Source Authors (Adapted from Marat-Mendes et al. 2015)

whatever meats they had to increase the taste of vegetable meals (Graça 2020).

From the pastoral economies of the Portuguese also resulted a great variety of regional cheeses, the majority of which are made from milk by sheep and goats, which could be taken through mountain areas for grazing (Graça 2020). However, as seen above, this form of protein preservation was mostly made available to the upper-classes, and by the 1940s, even the middle-classes did not seem to consume it systematically (Marat-Mendes et al. 2015).

Sugar, a product which was beyond the access of the peasants and which by 1940 was among the least consumed products among all social classes (Marat-Mendes et al. 2015), was however another important trait of Portuguese gastronomy. Indeed, Portugal was one of the first countries in Europe to systematically cultivate sugarcanes, bringing them from Sicily to the region of Algarve, and afterwards to Madeira Island (Graça 2020) which would become emblematically known as the source of the 'white gold' for desserts. Indeed, sugar was produced in Portuguese territory from the early fifteenth century and it was a key cultivation in processes of colonial occupation, but its use remained only accessible to the rich and other social classes in episodic celebrations—only in the twentieth century would it become a key foodstaple (Graça 2020).

Another food product that is important to highlight is rice, although not for its historical importance. It is true that it seems to have been introduced among the Portuguese during the era of Arab occupation, but it did not become an important item for several centuries, mostly because Portugal has a chronic deficit of water while rice is preferably cultivated through irrigated farming (Graça 2020). However, by the 1930s, several paddies emerged in Portugal in strategic areas: the region of Beira Baixa, as well as the floodplains of the rivers Mondego, Tagus and Sado (Graça 2020). Rice cultivation is very demanding in terms of water and of whether, and since favourable conditions are verified in these areas, they remained until today, added by a few sparse wetland paddies the key areas for rice cultivation in Portugal, especially the valleys of Tagus and Sorraia, currently the largest paddy areas in the country (Antunes and Ferreiro 2017). Indeed, in little more than a decade, by 1948, rice was consumed at an average of 8.5 kg per inhabitant each year,[1] rising to 13 kg by the 1960s and continuing to rise until today, making the Portuguese the highest consumers of rice in Europe (Antunes and Ferreiro 2017; Graça 2020).

[1] These numbers include rice produced within Portugal but also imported.

The isolation and relative backwardness of Portugal indeed delayed or even avoided the key transformations that occurred in the food situation of Central Europe, but they also created conditions for the different regions of the country to adapt their dietary and cooking habits to their most convenient production, which ultimately favoured the blossoming of contrasted regional food identities (Graça 2020).

Such a vision is sustained by Modesto (1982), who, as stated before, has conducted the most extensive survey on traditional cooking, highlighting precisely that more than a national culinary identity, Portugal is defined by regional variation. In this sense, Modesto's (1982) findings on the dietary habits of the Portuguese are in tune with the conclusions of the survey on vernacular, popular and traditional architecture conducted by architects in the late 1950s (AAVV 1961). Belated industrialization translated into reduced mobility and a static—rather than dynamic—social setting, which prevented significative influences from afar but also rendered it impossible for food to be preserved during long-distance transformation (Costa et al. 2016; Salvador 2019; Marot 2019).

In Modesto's (1982) book, as in the 1961 Survey on Regional Architecture, the Lisbon Region was included in the Estremadura Region, as it is again in the website repository that has made additional recipes available to the public in 2015. Between these two outlets, 99 recipes were collected for Estremadura: some were generically assigned to this region, while others included a short 'biographical note' that located them more precisely. Excluding those whose origins lay in locations outside the Lisbon Region, we can account for 68 recipes, which are divided among several sections: starters or appetizers (ten recipes), soups (four recipes), fish and shellfish (17 recipes), rice and 'açordas' (three recipes), poultry (four recipes), meats (nine recipes), sweets and puddings (eight recipes), cookies and pastries (seven recipes) and cakes (six recipes). From this brief overview, we can already confirm the prevalence of fish and shellfish dishes over meat and poultry, in accordance not only with the daily diets accounted for by Garrett but also in the main foodstuffs of Lisbon Region municipalities (Marat-Mendes et al. 2015). One of the recipes among 21 were for sweets or desserts, also testifying to the importance of sugary meals in Portuguese cooking habits which we saw rose to preeminence slowly since the fifteenth century (Graça 2020).

The 'Rich Stew', made with 11 types of fishes added with clams, was noticed for being invented by a rugby player from Lisbon, Duarte Amâncio Leal (1921–2018), while the 'Rich Fish Soup', made from several fishes and shellfishes, was an original from the Gambrinus Restaurant in Lisbon (Modesto 1982). In the meat section, there were two similar beef dishes which originated from Lisbon cafés and snack-bars. One of them is the 'Marrare Steak', from a restaurant called 'Marrare das Sete Portas' (Seven Door Marrare) owned by a Galician cook, where bohemian men used to gather. A similar but simpler version is the 'Café Steak', presented as a popular version of the 'Marrare Steak' (Modesto 1982).

More popular dishes were also to be found, including the 'Bulhão Pato Clams' (Fig. 6.4), also known as 'cadelinhas' (little bitches) which are very appreciated in the Southern Bank of Tagus, notably in Barreiro, Montijo, Almada and Caparica. The broad beans used for the 'Broad Bean Soup' were sold on the streets by female sellers who carried their baskets on their heads shouting 'pregões' (announcements) to announce their passage (Modesto 1982). Such female street sellers were common characters in the 1940s novels by Lisbon writer Irene Lisboa (1892–1958) who described life in the popular and poor neighbourhoods of the capital city.

In the Cookies and Pastries section were included some pastries which remained emblematic from the Lisbon Region, namely the 'Pastel de Nata' (Custard Tart) and the 'Queijadas de Sintra' (Sintra Cheese Custard Tarts). At the same time, in the Cakes section was included the 'King Cake', which is a staple cake for Portuguese Christmas in all of the country, and which was thought to be a local adaptation of the 'Galette des Rois' from Bordeaux.

Another important aspect of the recipes collected by Modesto (1982) was the usage of ingredients (see Tables 6.3, 6.4, 6.5, 6.6, 6.7, 6.8, 6.9, 6.10, 6.11, 6.12 and Figs. 6.1, 6.2, 6.3 and 6.4).

Among the appetizers, we encounter snails with a specifically prepared sauce, and indeed, the consumption of snails was one of the deepest marks left in Portuguese cooking from the epochs of scarcity caused by years of bad harvests, which have prompted the people in the countryside to cook and eat snails and mushrooms and those from the coast to catch and eat nearly all kinds of bivalves, all of which remain important items on popular and traditional cooking today (Graça 2020). The use of olive oil as a source of fat is a noticeable aspect, present in half of the appetizers (five recipes), all the rice and 'açordas', four recipes of meat (against two recipes which use lard and one which uses butter), three soups (against one which uses butter) and 13 dishes of fish and shellfish (while two use margarine and two use butter) (Modesto 1982). Only in poultry dishes there is scant use of olive oil—from four dishes, it is only used in 'Gardening Chicken' alongside margarine, while butter is used in two other dishes and the fat from pork meat is used in the last one (Modesto 1982). This is not surprising, since the soil and weather conditions of Portugal made all the territory prone to olive groves cultivation (Graça 2020), with the Lisbon Region not being an exception. As noticed above, olive groves and olive trees are a constant presence, even today, on the regional landscape, while olive oil is consumed

Table 6.3 Appetizers and starters from Estremadura (general) and Lisbon Region locations

		Starters/appetizers									
		Portuguese snails [Caracóis à Portuguesa]	Green eggs [Ovos verdes]	Codfish 'pataniscas' [Pataniscas de Bacalhau]	Garden fishes [Peixinhos da horta]	Pipis'	Rissoles [Rissóis]	Black-eyed peas salad [Salada de feijão-frade]	Meat rolls [Croquetes de carne]	Soft dough pastries [Pastéis de massa tenra]	Pigeon broad beans [Fava de Pombos]
Used ingredients	**Snails**	**2 kg**									
	Olive oil	2 sp		1 sp		2 sp		4 sp			o
	Oregano	o									
	Laurel	o				2 leaves					
	Garlic	2				2 cloves					o
	Onion	1		1	1 sp	2	1 sp	2			
	Salt	o	o	o	o	o	o	o	o	o	
	Pepper	o	o	o	0	o	o	o	o	o	
	Piripiri	o				o					
	Eggs		9	1	1		2 + 2 yolks	1	3	1	
	Margarine		1 sp					1 bunch			
	Parsley		2 sp	o			1/2 sp				
	Flour		o	1 cup	100g		1 cup + 1 sp		2 sp	250 g	
	Vegetable oil		o	o			o		o		
	Codfish			**1 slice**							
	Milk			o			1/2 cup + 2.5 dl		1.5 dl	1.5 dl	
	Lemon			o			2 peel		1	o	
	Green bean				500g						
	Chicken					**1 kg**					
	Paprika					1 sp					
	White wine					1 dl					
	Vinegar					o		2 sp			o
	Water						1/2 cup			1.5 dl	
	Butter						2 sp		2 sp	1 cup	
	Bread						o		o		
	Fish						**1 slice**				
	Nutmeg						o				
	Black-eyed peas							5 dl			
	Cloves							1			
	Potato										
	Broad bean										o
	Lard									**2 sp**	
	Beef								**250 g**	**150 g**	

Source Authors (Adapted from Modesto 1982 and 2015)

by all social classes in Garrett's 1940 account, albeit in different quantities (Marat-Mendes et al. 2015).

Eggs, despite being less noticed in the 1940s data, were a fundamental ingredient in the traditional cooking of Estremadura and the Lisbon Region, being one of the ingredients verified in all sections of Modesto's (1982) book: it is present in three appetizers (out of ten), two rice and 'açordas' (out of three), four meat dishes (out of nine),

Table 6.4 Soups from Estremadura (general) and Lisbon Region locations

		Soups			
		Shrimp cream [Creme de Camarão]	Rich broad bean [Fava Rica]	Rubble soup [Sopa de entulho]	Clams soup [Sopa de Ameijoas]
Used ingredients	Olive oil	2 sp	4 sp	2 sp	
	Garlic		4 cloves		o
	Onion	1		2	1
	Salt	o	o	o	
	Pepper	o	o		
	Piripiri	o			
	Eggs				4
	Parsley				o
	Flour	2 sp			
	Paprika				o
	White wine	0.5 dl			
	Vinegar		o	o	
	Butter				o
	Bread	o			o
	Cloves			2	
	Potato			3	
	Shrimp	**350 g**			
	Tomato	2			
	Broad Bean		5 dl		
	Cabbage			1–2	
	Butter beans			5 dl	
	Pork			**o**	
	Turnip			2	
	Carrot			3	
	Pumpkin			1 slice	
	Shell fish				**o**

Source Authors (Adapted from Modesto 1982 and 2015)

one soup (out of four), five fish and shellfish dishes (out of 17), one of poultry (out of four), in all eight recipes for sweets and puddings and all six cakes, as well as six cookies and pastries (out of seven). Perhaps the frequent use of eggs is one reason why chicken while being a relatively common ingredient was not very present—two poultry recipes and one appetizer, the 'Pipis' (Modesto 1982).

The soups section displays several types of cooking, from that made with expensive ingredients like 'Shrimp Cream' to 'Rubble Soup' (Fig. 6.3) which is made from a dense mix of cut vegetables and grains, with the option of using a ham bone, a clear sign of harnessing all the possible flavour from all the pieces of food, even those which were not directly consumable, such as bones (Modesto 1982).

Another section where the largest social differences can be inferred is in rice and açordas, with a contrast between the 'açordas', which have more ingredients, used in larger quantities, and including shrimp or other shellfishes, usually more expensive—in comparison with the 'Sprout's rice', which uses less quantities with less ingredients, and without any particularly expensive ingredient—this is also one of the recipes that do not use meat, fish or shellfish (Modesto 1982).

Salt and pepper and to a smaller extent paprika, piripiri and white wine are often utilized as a way of seasoning the food, but most of the recipes demonstrate that ingredients are cooked and valued for their taste, without complex sauces or additions—such a characteristic is mostly found on the fish

Table 6.5 Rice and 'açordas' from Estremadura (general) and Lisbon Region locations

		Rices and 'Açordas'		
		Lobster or shrimp 'Açorda' [Açorda de Lagosta ou Camarão]	Shellfish 'Açorda' [Açorda de Marisco]	Sprout's rice [Arroz de Grelos]
Used ingredients	Olive oil	3 sp	3 sp	2 sp
	Garlic	3 cloves	3 cloves	1 clove
	Onion			1
	Salt	o	o	o
	Pepper	o	o	o
	Piripiri	o	o	
	Eggs	4	3	
	Parsley		1 bunch	
	Bread	500 g	500 g	
	Rice			250 g
	Shell fish	**1 kg**	**1.5 kg**	
	Sprout			1/4 of a piece

Source Authors (Adapted from Modesto 1982)

and shellfish section, as well as the meats section in the dishes arising from restaurants and cafés.

From the appetizers section, four of ten dishes are vegetarian, while in the soups there is only one vegetarian dish among four, with the 'Rubble soup' having the option of the ham bone, and again only one in four rice and 'açordas' is without shellfish. From the 21 recipes for desserts, only three are cooked without eggs. Since we know from Garrett's accounting that only the middle- and upper-classes consumed eggs daily (Marat-Mendes et al. 2015), the number of eggs necessary for cooking any of the typical desserts presented in Modesto's (1982) book confirm that most likely, sweets were an episodic habit in the traditional eating habits not only of the Portuguese (Graça 2020) but specifically in the Lisbon Region.

It is also relevant that be it through soups, broths, frying or breading, most elements were always cooked—there are few examples of food consumed raw, confirming this tendency very common among the Portuguese traditional diet.

Meat recipes demonstrate a relatively higher complexity than the remaining ones, most likely because a part of them are really characteristic of middle- and upper-class household as well as from restaurants and snack-bars, belonging to an urban, rather than rural, tradition. Beef is used in six dishes out of nine, in most cases being prepared with sidings or sauces to complexify its taste. Parsley and garlic are also relevant items of meat dishes, for similar reasons. However, there are dishes using pork, a relatively cheaper meat—since pigs are easier and cheaper to feed than cows—being used in three out of the nine dishes: 'Cow Hand Stew', 'Portuguese Broad Beans' and 'Peas with Poached Eggs'. Curiously, these three dishes also make use of 'smoked meats' which, as explained above, were meant to harness the less savoury or fatter parts of meats, mixing them with bread, wine and spices and conserving the result through smoke.

'Smoked meats' are also used in the 'Portuguese Bean Stew', where they are the sole meat utilized alongside black-eyed peas and eggs. A dish that must be highlighted is the 'Cow Hand Stew' which utilizes three different types of meat: beef, pork and 'smoked meats'. Eggs, potatoes, tomatoes, broad beans, white beans, carrots and peas are used as sidings, demonstrating a relative variety in the usage of vegetables and grains in meat dishes.

The fish and shellfish dishes present a great variety of forms of preparation, types of fish utilizes and complexity of recipes. Sardines and codfish are particularly important, having been an emblematic foodstaple of Lisbon to this day. Indeed, from the 17 recipes collected in this section, seven are of codfish and five include sardines. With respect to the latter, the 'Sardinhada', a sort of sardine barbecue, is worth mentioning since the recipe itself presents it as a dish for parties, emphasizing the popular and convivial outline of this rather simple dish, where sardines are indeed the key element, accompanied by a salad of green peppers or cucumber (Modesto 1982).

On the other hand, there some notably rich dishes, namely the cauldron stews—the 'Setúbal stew', the 'Rich stew' the 'Sardine stew'—as well as the 'Rich Fish Soup', all of which imply not only the mix of several types of fishes and also of shellfish, usually meant for large groups meals.

Table 6.6 Poultry from Estremadura (general) and Lisbon Region locations

		Poultry			
		Hunter's rabbit [Coelho à Caçador]	Chicken stew [Frango na Púcara]	Alberto Pimentel's rice [Arroz à Alberto Pimentel]	Frango à Jardineira'
Used ingredients	Olive oil				1 sp
	Laurel	1 leave			
	Garlic	3 cloves			
	Onion	2			
	Salt	o	o	o	o
	Pepper	o	o	o	o
	Eggs			8	
	Margarine				o
	Parsley	1 bunch			
	Chicken		**1**	**2**	**1**
	Paprika				
	White wine	1 dl	1 dl		1 cup
	Vinegar	2 sp			
	Butter		2 sp	100 g	
	Rice			500 g	
	Tomato	1 kg	4		
	Pork	**200 g**	**75 g**		**o**
	Turnip				o
	Carrot				o
	Port Wine		1 cup		
	Alcoholic beverage		1 cup		
	Mustard		1 sp		
	Rabbit	1			
	Scallion		10		
	Raisin		6		
	Brussels sprouts			24 pieces	
	Peas				o

Source Authors (Adapted from Modesto 1982 and 2015)

The 'Rich Fish Soup' in particular, living up to its name, is a mix of up to 12 types of fish—including sardines and shellfish—and is collected from a Lisbon restaurant (Modesto 1982).

Considering the presence of three rivers in the Lisbon Region (Tagus, Sorraia and Sado), as well as the significative coastal areas of Mafra, Sintra, Cascais, Almada, Sesimbra and Setúbal, the importance of fish is not surprising.

In the fish and shellfish section, there are also examples of a popular cooking, such as the Bulhão Pato Clams, believed to have been created by that Portuguese writer (1828–1912) and which, similar to the already mentioned snails appetizer, consists of a simple sauce to accompany the clams (Fig. 6.4).

As noted above, nearly a third of dishes in the Estremadura and the Lisbon Region are desserts. A characteristic of these recipes is that nearly all use eggs and sugar, with only a few exceptions. Considering that both sugar and eggs are not among the most relevant food items in Lisbon Region municipalities nor are they common in all social classes, in Garrett's 1940 accounting even the upper- and middle-class do not consume them in great amounts (Marat-Mendes et al. 2015), and it is likely that most of these desserts were not consumed with great frequency. Such an idea can also be inferred from the fact that some of the desserts are specifically linked with celebrations like Christmas, Easter or weddings. Another important presence is that of the

Table 6.7 Meats from Estremadura (general) and Lisbon Region locations

		Café steak [Bife à Café]	Marrare' steak [Bife à Marrare]	Steak marinade [Bife de Cebolada]	Steak with fried egg [Bife com Ovo a Cavalo]	Baits [Iscas com Elas]	Portuguese bean stew [Feijoada à Portuguesa]	Cow hand stew [Mão de Vaca Guisada]	Portuguese board beans [Favada à Portuguesa]	Peas with poached eggs [Ervilhas com Ovos Escalfados]
Used ingredients	Olive oil			1 sp			2 sp	1 sp	2 sp	
	Laurel			1 leave	1 leave	1 leave	o	1 leave	1 leave	1/2 leave
	Garlic			2 cloves	4 cloves	4 cloves	1 clove	3 cloves	2 cloves	1 clove
	Onion			3			2	2	2	1
	Salt	o	o	o	o	o	o	o	o	o
	Pepper	o	o	o	o	0	o	o	o	o
	Eggs				4		4	1 sp		4
	Margarine									2 sp
	Parsley			1 bunch	1 bunch	1 bunch	o	1 bunch	1 bunch	1 bunch
	Milk	3 sp								
	Lemon	o						1		
	White wine					1.5 dl		1 dl		
	Winegar				1 sp	1 sp				
	Butter	2 sp								
	Black-eyed peas						5 dl			
	Cloves						2			
	Potato				o	o				
	Rice							o		
	Tomato						2	4		2
	Broad bean								4 kg	
	White beans							1 l		
	Pork							**100 g**	**100g**	**100g**
	Carrot							2		
	Mustard	1 sp								
	Lard				**4 sp**	**3 sp**				
	Beef	**200 g**	**200 g**	**500 g**	**4 pieces**	**500 g**		**1 piece**		
	Cream		3 sp							
	"Enchidos" (smoked meat)						**100 g**	**100 g**	**2**	**100 g**
	Sugar								o	o
	Peas									2.5 kg

Source Authors (Adapted from Modesto 1982, 2015)

Odivelas Marmelade, a recipe which has left by one of the last nuns of the Odivelas Monastery, and which makes generous use of sugar—more than 3 kg—thus being a good representation of traditional sweets from feminine monasteries in Portugal, which remain an important tradition among the Portuguese, despite the licentious use of both eggs and sugar in most of these recipes.

Across all the included recipes, Modesto (1982) indeed collected several dishes which clearly reflect the local productions of Estremadura at large, and of the Lisbon Region

Table 6.8 Fish and shellfish from Estremadura (general) and Lisbon Region locations

		Fish and shell fish							
		Setúbal stew [Caldairada à setubalense]	Rich stew [Caldairada Rica]	Sardine stew [Caldairada de Sardinha]	Rich fish soup [Sopa Rica de Peixe]	Bulhão Pato' Clams [Amêijoas à Bulhão Pato]	Cold filled spider crab [Santola Recheada Fria]	Brás' codfish [Bacalhau à Brás]	Raw shredded codfish [Bacalhau Cru Desfiado]
Used ingredients	Olive oil	1.5 dl	2 dl	1 dl	2.5 dl	2 sp		2 sp	3 sp
	Laurel			1 leave	o				
	Garlic			2 cloves	2 cloves	2 cloves		1 clove	1 clove
	Onion	3	2.5 kg	500 g	4			3	1
	Salt	o	o	o	o		o	0	
	Pepper	o	o	o	o		o	o	o
	Piripiri		o				o		
	Eggs						1	6	
	Margarine								
	Parsley			1 bunch	o		I bunch	o	
	Flour								
	Codfish							**1 slice**	**1 slice**
	Milk								
	Lemon					1			
	Paprika			1 cup					
	White wine						0.5 dl		
	Vinegar								1 sp
	Butter				50 g				
	Bread				o				
	Fish	**2.5 kg**	**7 kg**	**1 kg**	**2 kg**				
	Cloves	2					1		
	Potato	800 g						500 g	
	Tomato	4	4 kg	500 g	3				
	Pork								
	Carrot								
	Bell pepper	2	1 1/2		1				
	Pasta	o	125 g						
	Shell Fish		**500 g**		**o**	**1 kg**	**1**		
	Port Wine		1 cup						
	Saffron				1/2 sp				
	Leek				1				
	Picles						1 sp		
	Alcoholic beverage							1 sp	
	Mustard							1 sp	
	Chive							1	
	Olives							o	
	Chickpeas								
	Cucumber	o							

(continued)

Table 6.8 (continued)

		Fish and shell fish							
		Setúbal stew [Caldairada à setubalense]	Rich stew [Caldairada Rica]	Sardine stew [Caldairada de Sardinha]	Rich fish soup [Sopa Rica de Peixe]	Bulhão Pato' Clams [Amêijoas à Bulhão Pato]	Cold filled spider crab [Santola Recheada Fria]	Brás' codfish [Bacalhau à Brás]	Raw shredded codfish [Bacalhau Cru Desfiado]
Used ingredients	Olive oil	1 sp		2 dl	1.5 dl	1 dl	o	o	
	Laurel	2 leaves							
	Garlic	2 cloves	3 cloves	4 cloves	2 cloves	3 cloves			
	Onion	500 g	1		2	3	o	o	
	Salt	o	o	o	o	o	o		
	Pepper	o	o	o	o	o	o	o	
	Piripiri								
	Eggs		3					5	4
	Margarine	1 sp			1 sp				
	Parsley	1 bunch	o			1 bunch		o	
	Flour								2 sp
	Codfish	**4 slices**	**4 slices**	**4 slices**				**1 slice**	**1 piece**
	Milk								o
	Lemon								
	Paprika	1 sp	o		1 sp				
	White wine				o				
	Vinegar		2 sp	o		1 dl	o		
	Butter								
	Bread							o	
	Fish				**1 kg**	**800 g**	**10 pieces**		
	Cloves							o	
	Potato	500 g			750 g		o	o	500 g
	Tomato				300 g		o		
	Pork				**o**				
	Carrot							6	
	Bell pepper			4			o		
	Pasta								
	Shell Fish								
	Port Wine								
	Saffron								
	Leek								
	Picles								
	Alcoholic beverage								
	Mustard								
	Chive								
	Olives								
	Chickpeas		300 g						
	Cucumber						o		

Source Authors (Adapted from Modesto 1982, 2015)

Table 6.9 Sweets and puddings from Estremadura (general) and Lisbon Region locations

		Sweets and puddings							
		Saloio' Sweet Rice [Arroz Doce Saloio]	Farófias'	Royal Slices [Fatias Reais]	Flan Pudding [Pudim Flan]	Orange Pie [Torta de Laranja]	Sonhos à Antiga'	Odivelas Marmelade [Marmelada de Odivelas]	Verdasquinha' Sweets [Doce de Verdasquinha]
Used ingredients	Salt	o					o		
	Eggs	12	4	13	6	6	6	2	8
	Margarine						1 sp		
	Flour						250 g		
	Milk	7.5 dl	7.5 dl		5 dl				
	Lemon	1	1	1					1
	Water						5 dl		
	Bread			12 slices					
	Rice	500 g							
	Potato starch		1 sp			1 sp			
	Sugar	400 g	175 g	750 g	300 g	250 g		3.8 kg	8 sp
	Cinnamon	o		o					1 stick
	Fruit jams			2 sp					
	Orange					1			
	Quince							o	
	Madeira wine								8 cup

Source Authors (Adapted from Modesto 1982, 2015)

in particular, while also mixing cooking of different social classes and belonging to different contexts. The recipes do reflect the diversity of regional cooking, including recipes of a rather simple outline, where there are clear 'protagonists' and a set of seasonings and sidings meant to highlight or intensify the key flavours. Other recipes are more complex, demanding more ingredients and more laborious preparations.

At the same time, the traditional cooking of many recipes is accompanied by the signs of changes which start to process in the eating habits of the Portuguese precisely at the time that Modesto started collecting, comparing and testing recipes received from her public. The transition from a rural plant-based cooking to a more urban and international cuisine where meats and transformed products have a much more important role (Steel 2020) is often present through recipes collected from restaurants and snack-bars, but these dishes would soon become much more homely than they had been before.

By the mid-twentieth century, the rural society of Portugal was about to transform both drastically and rapidly, and the Lisbon Region, with two great centres attracting peasants from the countryside to industrial belts (Lisbon and Setúbal) were to witness this transformation even at a physical level, as discussed in Chap. 4.

During the 1950s, the Lisbon city faced a renewal of its commercial spaces, a transformation which food venues illustrate rather well. In the search for a better and more modern outlook, some of the local tradesmen put aside the traditional customs of the city and opened to the influences of cinema and pop culture: soon Lisbon would see the emergence of the snack-bar, known only through the imagery of North American cinema (d'Almeida 2006, 2017). Architects Victor Palla (1922–2006) and Joaquim Bento d'Almeida (1918–1997) were largely responsible for introducing this international model in Portugal, having even travelled abroad to study the design of this type of food venue. Over the years, this architectural duo designed over

Table 6.10 Cookies and pastries from Estremadura (general) and Lisbon Region locations

		Cookies and pastries						
		Sands [Areias]	Broas de Espécie'	Broas Castelares'	Cascais Walnuts [Nozes de Cascais]	Bean Pastries [Pastéis de Feijão]	Custard Tart [Pastéis de Nata]	Sintra Tarts [Queijadas de Sintra]
Used ingredients	Salt					o	o	o
	Pepper							
	Piripiri							
	Eggs		3	2	6	12	8	4
	Flour	320 g		50 g		250 g	500 g	300 g
	Lemon	1–2		1	1		1	
	Water				1 dl	1.5 dl	3 dl	o
	Butter					2 sp	500 g	1 sp
	White beans					100 g		
	Lard	200 g						
	Cream						5 dl	
	Sugar	150 g	500 g	500 g	125 g	500 g	200 g	350 g
	Cinnamon							o
	Orange			1				
	Sweet potato		750 g	400 g				
	Almonds		80 g	75 g	125 g	25 g		
	Coconut		80 g					
	Manderine		2					
	Honey			0.5 dl				
	Corn flower			225 g				
	Walnuts				20 pieces			
	Cheese							400 g

Source Authors (Adapted from Modesto 1982)

30 examples, among which are counted emblematic ones like Pique-Nique (1952–1955), Pam-Pam (1955–1957), Tique-Taque (1956), Pisca-Pisca (1960), Noite e Dia (1962–1964) and Galeto (1966–1968). Unlike the 'casas de pasto' (eating houses) and delicatessens of traditional meals, snack-bars are markedly cosmopolite, serving fast meals at the counter, sometimes prepared with the customer sitting right in front of the cook (d'Almeida 2006). The first snack-bar designed by Palla and d'Almeida was the Terminus (1949–1950), which successfully attracted many customers from more traditional restaurants (d'Almeida 2006, 2017). Afterwards, the design of the different spaces for both costumers and staff grew more and more complex, with the architects often designing the furnite and the neon signs themselves, and sometimes combining the more fast pace of the snack-bar with other—more traditional and 'slow' forms of food consumption (d'Almeida 2006, 2017). The Snack-Bar Galeto—featured on the IAPXX (Fig. 4.30)—is one of the most emblematic of the snack-bars designed by Palla and d'Almeida, and it attracted a crowd of movie-goers who in there found services of café, take-away, snack-bar and restaurant.

In all their modernity and their cinematic design, these spaces announced a changing city, where urban life was intensifying. Regardless of political oppression, lifestyles and cities were bound to evolve and transform. Interestingly, food was and still is a part of this transformation, for the snack-bar provided a fast meal to professionals in a rush as much as to cinephiles coming out of a film exhibition (d'Almeida 2017). More importantly, the emergence of the snack-bar in the otherwise traditionalist cultural landscape of Lisbon testifies to the powerful ability of art (in this case, both cinema that inspired the snack-bars and the architectural design that brought them to life) to change habits and mindsets.

At this time, the eating habits of the Portuguese were generally similar to the remainder of Southern Europe.

Table 6.11 Cakes rom Estremadura (general) and Lisbon Region locations

		Cakes					
		King cake [Bolo Rei]	Saloio' or wedding cake [Bolo Saloio ou de Casamento]	Folar da Páscoa'	Saloio' or wedding sponge cake [Pão-de-Ló Saloio ou dos Casamentos]	Almond pie [Tarte de Amêndoa]	Rotten cake [Bolo Podre]
Used ingredients	Salt	o	o	o			
	Eggs	4		2	12	2	6
	Flour	750 kg	1 kg	500 g	100 g	200 g	250 g
	Milk		2 dl			0.5 dl	1/2 L
	Lemon	1					1
	Butter	150 g	100 g	100 g		200 g	
	Port Wine	1 dl					
	Sugar		1 kg	3 sp	250 g	300 g	500 g
	Cinnamon		1 sp	1 sp			125 g
	Orange	1					
	Almonds					125 g	
	Fennel			1 sp			

Source: Authors (Adapted from Modesto 1982 and Modesto 2015)

Table 6.12 Typology of land-uses from the Lisbon Region

Type	Aggregated land-uses	Current area in the Lisbon region (km^2)	Current percentage in the Lisbon region (%)
Urban spaces	Urban spaces, Spaces for urbanization, Central spaces, Consolidated urban spaces, Urbanized spaces, Urban spaces for consolidation, Housing areas	476	16.3
Conditioned spaces	Cultural spaces, Spaces for equipment, Public equipment, Military, Conditioned areas, Channel-space/Infrastructure, Channel-space, Infrastructures spaces, Services, Portuary	141	5
Economic activity spaces	Industrial spaces, Urbanizing industrial spaces, Aquiculture spaces, Touristic spaces, Extraction industry, Economic activities/services	153	5
Mixed spaces (urban–rural)	Transition spaces, Unconsolidated urban spaces, Para-urban spaces, Periurban agriculture spaces, Farm built clusters, Residual spaces, Rural settlements, Low-density urban spaces	126	4.3
Rural spaces	Rural spaces, Agricultural spaces, Forestry spaces, Agriculture/Livestock spaces, Agroforestry spaces	1,416	48.4
Ecological spaces	Leisure spaces, Green protection spaces, Green leisure spaces, Urban refurbishing spaces, Urban green spaces, Natural spaces, Environmental requalification spaces	2,925	21

Source Authors (Adapted from Marat-Mendes et al. 2021)

However, between 1960 and 1980, several products which before had only a relatively small importance in the national diet gained increasing importance. The availability of milk grew 181% (71.9–202.3 g per inhabitant daily), as did that of cheese (6.3–12.3 g per inhabitant daily) (Graça 2020). Availability of beer increased 151%, rising from 51.4 g per

Fig. 6.1 Green eggs: ingredients and meal prepared according to Modesto's (1982) recipe. Source: Authors

Fig. 6.2 Garden fishes: ingredients and meal prepared according to Modesto's (1982) recipe. Source: Authors

inhabitant daily to 129.2 g; while pork rose 60% and poultry an impressive 1,379% (Graça 2020). Moreover, margarine rose its availability to 590%, vegetable oils to 441%, butter by 33%, sugar by 63% and beer by 788% (Graça 2020).

While similar changes were taking place all over Europe, and especially in the south, in comparison with the European counterparts, the Portuguese were still consuming less milk and milk products (202 daily g per inhabitant against 480 in Europe), meats (129 daily g per inhabitant against 239 in Europe), eggs (14 daily g per inhabitant against 49 in Europe), margarine (15 daily g per inhabitant against 22 in Europe), sugar (81 daily g per inhabitant against 94.9 in Europe) and beer (110 daily g per inhabitant against 218 in Europe) (Graça 2020). This shows that while the habits of the Portuguese were internationalizing and urbanizing, there was still a significative distance between Europe and Portugal, to which the situation of the country, still living under a dictatorship and with tenuous relationships with outside produce, certainly contributed. Thus, many of the traits of traditional cooking still remained important in this time, especially with regards to fish (51 daily g per inhabitant against 41 in Europe), vegetables (309 daily g per inhabitant against 247 in Europe), legumes (15 daily g per inhabitant against 6 in Europe) and wine (235 daily g per inhabitant against 96 in Europe) (Graça 2020).

Such changes in dietary habits help explain the sustainability issue that was brewing in the country, especially in

Fig. 6.3 Rubble soup: ingredients and meal prepared according to Modesto's (1982) recipe. Source: Authors

Fig. 6.4 Bolhão Pato Clams: ingredients and meal prepared according to Modesto's (1982) recipe. Source: Authors

the Lisbon Region. On the one hand, we had an increasing urban population with peasants arriving in the vicinities of Lisbon, and with a demographic growth accompanied by changing food habits that relied even more on outside trade. On the other hand, we had a shrinking regional hinterland where cultivations were displaced to make way for industrial estates and residential neighbourhoods, decreasing the productive capacity of the region. While the Lisbon Region had never been self-sufficient, the more it urbanized—and the more it changed its diet—the more dependent it became from the produce of others.

The democratic regime, which had the integration in Europe as one of its chief goals, continued in the pathway for a new, globalized food system, whereby the Portuguese became increasingly free to eat as their European counterparts did, that is, free from the restraints of local production. From 1980 to 2000, the availability of beef increased from 28 daily g per inhabitant to 33 g, pork from 26 daily g to 58, poultry from 32 daily g to 50, while milk availability rises from 166 daily g to 247 (Graça 2020). Availability of eggs, a source of animal protein which had a key role in the cuisine of Estremadura (Modesto 1982), also rose from 14 daily g to

20 until 1997, fish from 54 daily g to 66, vegetables from 171 daily g to 237, fresh fruits from 93 daily g to 179 and citrines from 25 daily g to 46 (Graça 2020).

In these dietary habits, local cultivations or livestock seem to become unimportant—eating becomes uniformized, one could eat in Lisbon as one did anywhere else, gastronomy lost a sense of place while gaining a sense of limitlessness. Not all changes were positive, obviously, since the consumption of fats also grew exponentially: butter rose from 2 daily g per inhabitant to 4, margarine from 14 daily g to 19, vegetable oils from 30 daily g to 35 and olive oil from 12 daily g to 15, all transformations which favoured a loss of health and prompt the increase in diet-related diseases (Graça 2020).

By the 1990s, there was a reframing of the Portuguese traditional diet: cereals and rice, which in the 1970s accounted for 38% of consumed energy, represented in the 1990s only 30% while meat and animals products rose from 12 to 18% (Graça 2020).

A new significant period of change starts in 2011, in the aftermath of the 2007 Financial Crisis, which hits Portugal the hardest between 2011 and 2013, causing a reduction in the consumption of certain products and altering specific consumptions, although not the general dietary patterns. Thus, beef was replaced by poultry and there was a rise in self-label brands and on-sale products (Graça 2020). Pork consumption reduced in comparison with poultry, since these were cheaper and more accessible, with fish consumption also declining (Graça 2020). Vegetables, whose availability was 237 daily g in 1997, rose to 296 in 2016, while the availability of fresh fruits, legumes and vegetable oils remained stable throughout this period (Graça 2020).

6.3 The Evolution of Land-Uses in the Lisbon Region

As discussed in Chap. 3, it was only with the democratic planning policy, whose cornerstones are laid originally in 1982 and confirmed in 1990, that the whole of the national territory was determined to be covered by planning instruments and namely by an integral scheme of land-uses. Although in the case of the Lisbon Region the 1964 PDRL had proposed a full land-use scheme for the regional territory, this plan was never approved, and even at the time it was designed, it was already outdated, ignoring many of the ongoing tendencies verified in the suburbs of the Lisbon city.

Given their integral nature, PDMs allow a relatively stable basis for the comparison of land-use schemes. It must be noted that the land-use classifications managed through the PDMs do not reflect the actual usage of the territory, but rather its dominant potential activity. In other words, a land parcel being classified as rural and categorized as agricultural does not mean that it is cultivated, but rather that cultivation is the usage more suited to its characteristics. Obviously, this implies a long-standing problem with planning instruments, whereby urban land is classified according to a social value attributed to the soil with basis on social interests, while rural soils are classified according to their physical, geological or ecological characteristics (Cavaco 2010). However, if we seek to assess how productive soils are distributed within the region, at least according to the evaluation conducted by planning professionals and municipal decision-makers, then PDMs are our strongest clues.

Not that they are without problems. Focusing specifically on the Lisbon Region, there are two problems that can be highlighted: the first is that many municipalities have failed to comply with the rule whereby PDMs must be revised, updated and republished every decade. In the 18 municipalities of the Lisbon Region, eight have PDMs approved in the 1990s—more specifically between 1994 and 1999—and which remain active, in principle, since they never underwent a full revision or updating. These correspond to the municipalities of Amadora (1994), Almada (1994), Sesimbra (1998), Barreiro (1994), Montijo (1997), Alcochete (1997) and Palmela (1997). Sintra and Setúbal are currently in the process of approving the revision of their PDMs which had originally been published in 1999 and 1994, respectively. The fact that some of these PDMs have been without updating for nearly 30 years confirms the endemic weakness that geographer João Ferrão (b1952) points out to the Portuguese planning policy as well as to the problem of lack of social recognition of the importance of spatial planning (Ferrão 2011). Indeed, the problem is not just the nonchalant attitude of municipal decision-makers that causes the delay and obsolence of planning instruments but also the lack of social pressure to demand a properly planned territory.

In 2015, the Portuguese General Directory of the Territory (*Direcção-Geral do Território*—DGT) conducted a survey on the active PDMs, synthesizing all the active documents in a single drawing, which had been kindly made available to the SPLACH Project. This drawing also included seven of the ten retired PDMs which have eventually been replaced by updated versions—namely for the municipalities of Lisbon, Loures, Vila Franca de Xira, Oeiras, Cascais and Seixal.

While the data collected from these drawings allowed a closer inspection of the land-use schemes instated by the PDMs of the Lisbon Region, a second problem arises from the lack of uniformity in the land-use schemes. The several Planning Acts approved during the democratic era have all, as seen in Chap. 3, suggested a key partition of the territory between urban and rural, but municipalities have relative autonomy to create more complex land-use classes or to create a series of functional categories which allow a more nuanced distinction between different types of urban land and different types of rural land. While such autonomy is important to make sure that municipalities have the ability to properly render the characteristics and potentials of their territory, this inevitably makes the comparison between the land-use schemes of different municipalities difficult to attain.

The SPLACH Project has proposed a typology for the aggregation of land-use classes and categories, uniformizing those in existence in the Lisbon Region for comparison sake (Marat-Mendes et al. 2021). Precisely to avoid the oversimplification implied in the urban versus rural schism, we propose a more layered typology (Table 6.12), which includes urban and rural, and also highlights the specific spaces for economic activities and those conditioned by a public easement. Ecological spaces, including forestry parks and infrastructural corridors, which are in principle non-productive were also detached from productive soils. Finally, special attention was paid to all land-uses which included typically urban or typically rural spaces, as well as those resulting from old farms, from periurban cultivation or from unfinished urbanization processes which may be harnessed for food-related activities.

Because they presuppose a residential structure allied with productive space, rural settlements were also included in this type of mixed spaces. Because the new PDMs from Setúbal and Sintra were only created after the DGT survey, they are not included in the following maps, although we discussed the new Setúbal PDM in Chap. 3.

Focusing specifically on the parts of the territory that are more prone to food cultivation, we can notice that, in the municipalities where a comparison between active and retired PDMs is possible, there are some changes. In Lisbon there is a clear increase in mixed spaces, while rural spaces are non-existent, in the past as now. In Loures there is a slight decrease and also a slight fragmentation of rural spaces, especially in the Trancão Floodplain, while mixed spaces, which were scant in the 1993 version, become practically residual in the 2015 PDM. In Vila Franca de Xira, alterations in rural land were slight but were more pronounced in mixed spaces, which become more dispersed. In Cascais, rural soils from the first PDM virtually disappeared in the newest version, although this municipality has put great effort into the promotion of urban agricultural spaces. However, the decrease in mixed spaces is also very visible: overall, Cascais registers a very drastic change from one PDM to the next, with spaces with cultivation potential decreasing in both typologies. In Mafra changes are also not drastic, with rural soils being generally maintained and mixed spaces becoming more disseminated throughout the whole municipal territory. Finally, in Seixal there is a slight decrease in rural land, and mixed spaces also decreased.

Lisbon and the urban areas of Vila Franca de Xira are, in the Northern Bank of Tagus, the municipalities where mixed spaces are more relevant in active PDMs, while in the Southern Bank there is a significant amount of these spaces in Almada, Moita, Palmela and Setúbal—a contrast with the bygone PDMs where Cascais and Seixal were also relevant municipalities with these types of spaces.

With regards to rural space, some territories necessarily stand out. In the Southern Bank of Tagus, the Montijo Exclave, Palmela and Alcochete as well as the Great Wetland of Tagus, stand out for being extensive areas almost uninterruptedly rural. On the Northern Bank, the interior of Vila Franca de Xira, Loures, Mafra and Sintra stand out for their rural areas, although these are more fragmented and marked by villages and settlements more visibly than their Southern Bank counterparts. In this regard, there is less change in comparison with the bygone PDMs, except for Cascais which lost the majority of its rural soils.

6.4 The Contemporary Food System of the Lisbon Region

To make sense of the meaning of the patterns of land-use predicted in the PDMs, we must also try and assess what was produced in this territory. Data from 2015 notices that in the 18 municipalities of the Lisbon Region, and despite the high degree of urbanization, there are still important productive resources: indeed, the region produced 740 thousand tons of agricultural biomass, including livestock, having imported an additional 3.7 million tons from Portugal and 5.4 million from the rest of the world, mostly Europe (Niza 2017). At the same time, 4.7 million tons of agricultural biomass are exported from here to the country and 1.8 million for the rest of the world (Niza 2017). Overall, the consumption of food in the Lisbon Region seems to be of about 3.3 million tons of agricultural biomass (Niza 2017) (Figs. 6.5, 6.6, 6.7 and 6.8).

Obviously, the regional food system is not determined solely by the existence of rural or other forms of productive spaces, since it includes other phases and other activities—and thus necessarily other spaces, such as canteens, municipal markets, wholesale, retail and supply trade, restaurants, transformation industries (Oliveira et al. 2014) as well as domestic kitchens, street markets and fairs and artisanal workshops, all of which are much harder to highlight or isolate in the land-use schemes of PDMs.

Luckily, the Portuguese INE has made some important statistics available that allow us to observe with some more detail the contents of the rural land of the Lisbon Region. Two documents stand out, the first being the Agricultural Statistics (INE 2018) and the second the most recent Agricultural Census conducted in 2019 (INE 2021). One key problem with these two documents is that while the former isolates the Lisbon Region from the remainder of the territory, in the Census, Lisbon is included in the 'Ribatejo e Oeste' (RO) region, making it impossible to break down the numbers corresponding to the several items surveyed in the Census. However, from the presented maps, it is possible to extract the distribution of productive activities in the territory of the Lisbon Region, allowing us to take some conclusions on the key aspects of the organization of food production—and indirectly of other food-related activities—across the metropolitan space.

From the Agricultural Statistics, it is possible to conclude that, being the largest metropolitan conurbation of Portugal (Portas et al. 2007), Lisbon is also the least productive of the five existing agrarian regions of the country, although all the national productions have a presence in its rural areas, regardless of how small (Table 6.13). Vineyards for wine are its most significative use of useful agricultural area (UAA), followed by rice, tomato, potato and forage whey. However, in terms of produced tons of foodstuffs, forage whey and tomato are the most relevant productions. The dimension of its utilized agricultural area (UAA) for industrial tomato plantations is the second largest in the country, while its production of sunflowers is the second-highest in the country. Except for rye, which is non-existent in the region, beans, almonds and chestnuts are its least voluminous productions, while beans, kiwis, almonds and chestnuts are the least significative uses of UAA. In general, it is clear that fruits are one of the least significant produces of the Lisbon Region, both in terms of its use of UAA and the resulting produced tons—indeed, in nearly all fruits except for apples, pears, cherries and grapes, the Lisbon Region is the smallest national producer, and even in those three exceptions, it is only the second smallest producer.

While the comparison established by the 2017 Agricultural Statistics and published in the following year (INE 2018) are not necessarily the most adequate for the Lisbon Region—whose area is significantly smaller and more urbanized than most other regions as they are organized in this document—it is also true that, as we saw, the productive space of the Lisbon Region decreased for several decades, between urbanization and rural abandonment.

The distribution of regions in the 2019 Agricultural Census (INE 2021) is more balanced, although the inclusion of the Lisbon Region in the agrarian region of 'Oeste e Ribatejo' does eliminate much specificity regarding the productive structure of the capital region. However, it is still a highly relevant document, which exposes some important aspects of agricultural businesses, particularly suggestive for the study of the food system.

The Census notices that in Portugal, the abandonment of agricultural activities has slowed down since the last Census in 2009, while the average size of agricultural holdings has increased 13.7% (INE 2021). As we saw in Chap. 3, the new Planning Act of 2014 inhibited land-use conversions and eliminated spaces for urbanization. Although it is impossible to state if there is a causal relation between these two phenomena, it is possible that the underlying intensification of the urban–rural schism may have prompted owners of rural estates to make their land productive. Entrepreneurialization of agriculture also rose, with agricultural enterprises managing 36% of UAA and more than half of livestock units (INE 2021). Each agricultural holding generates about 23.3 thousand euros of total standard output, an increase in relation to 2009 (INE 2021). Specialized farms increased 7%, and the total area of holdings increased by more than 400 hectares, reaching 55.5% of the territorial area (INE 2021). The UAA itself increased 8.1% and its usage transformed: a decrease of 11.6% in arable land was overcompensated by a 24.6% increase in permanent crops and of 14.9% in permanent pastures (INE 2021).

Areas under temporary grasses and grazing crops grew 12%, while areas for grain cereals decreased 32.2% and for potatoes decreased 28.6% (INE 2021). Permanent cultures were marked by the establishment and modernization of olive groves and orchards, especially for berries, subtropical fruits and almonds (INE 2021). Permanent pastures increased 14.9%, now occupying more than half of the UAA (INE 2021). Irrigated surface increased 16.6%, with 69.7% of fresh fruit orchards counting on irrigation, as do 11.5% of nut groves, 31.7% of olive groves and 27.8% of vineyards (INE 2021).

With respect to livestock, bovine holdings decreased by 27.8% but the animals increased by 10.6%, pigs increased by 15.7%, sheep have remained stable, goats decreased 11.5% although large herds faced a growth of 27.6% and poultry increased 1.5 times (INE 2021) (Table 6.14).

Mechanization and precision farming more than quadrupled, organic farming agricultural holdings tripled, but the agricultural labour force decreased 14.4% (INE 2021). However, the report says nothing on the severe problem of precarious migrant labour which has caused a series of polemics in the Portuguese press recently due to the inhumane conditions that some proprietors and contractors arrange for temporary workers to live into.

A contrast is visible also among the heads of agricultural business, for while sole holders are 67.1% male with an average of 64 years old, agricultural enterprise managers are in average 13 years younger and have high academic and professional qualifications, while 46.3% of sole holders have only the first level of basic education (INE 2021).

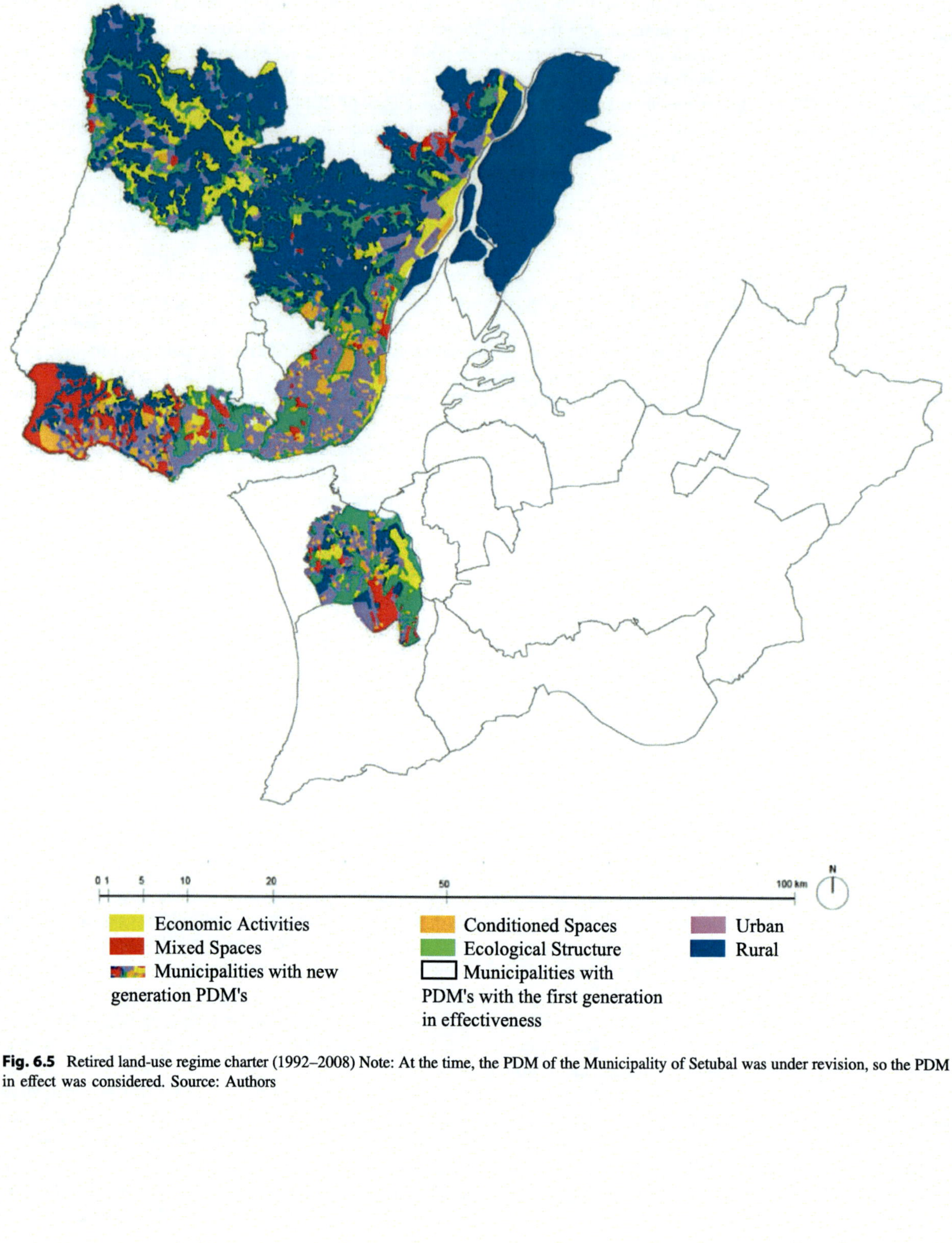

Fig. 6.5 Retired land-use regime charter (1992–2008) Note: At the time, the PDM of the Municipality of Setubal was under revision, so the PDM in effect was considered. Source: Authors

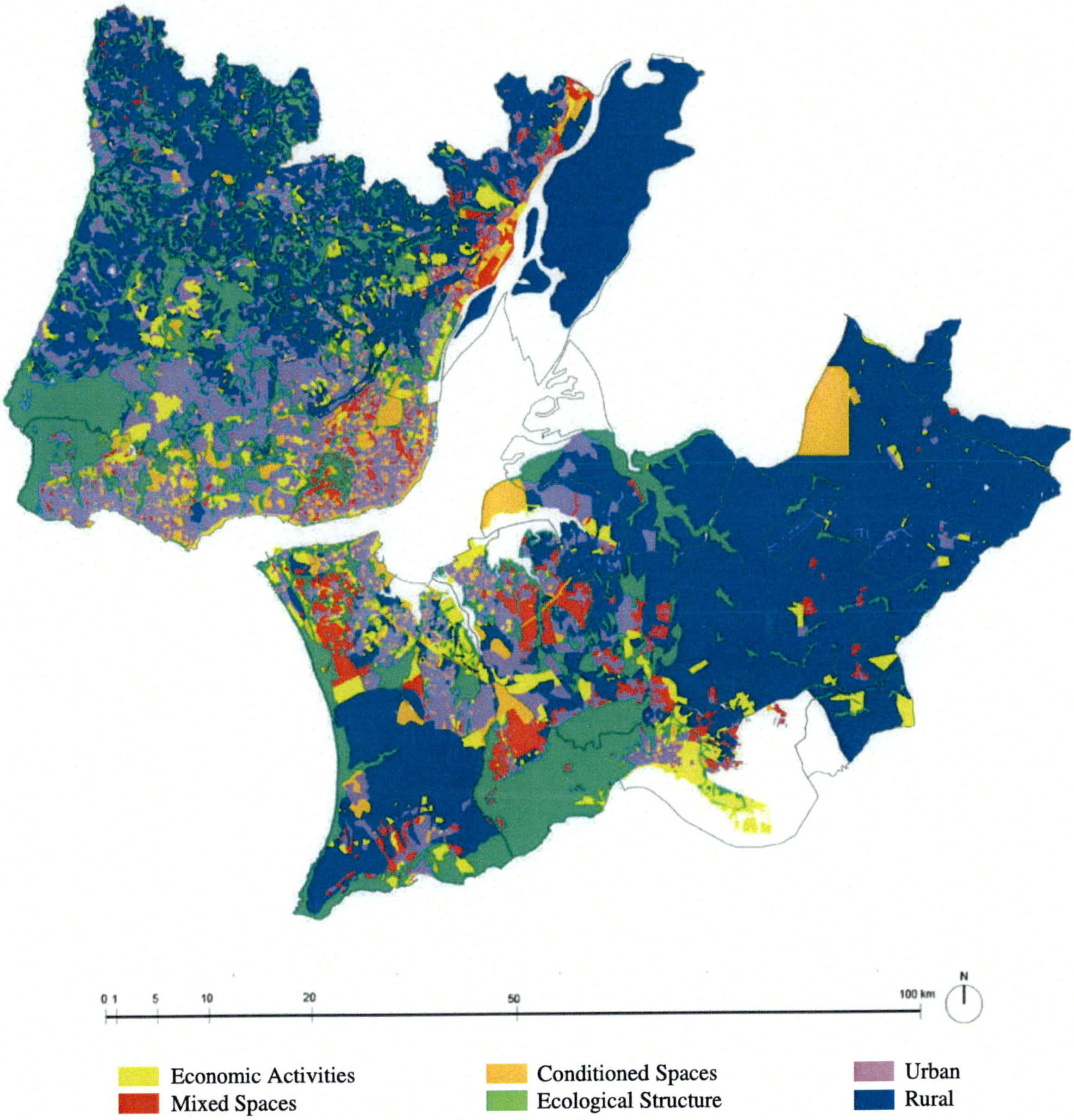

Fig. 6.6 Active land-uses in the Lisbon Region. Source: Authors

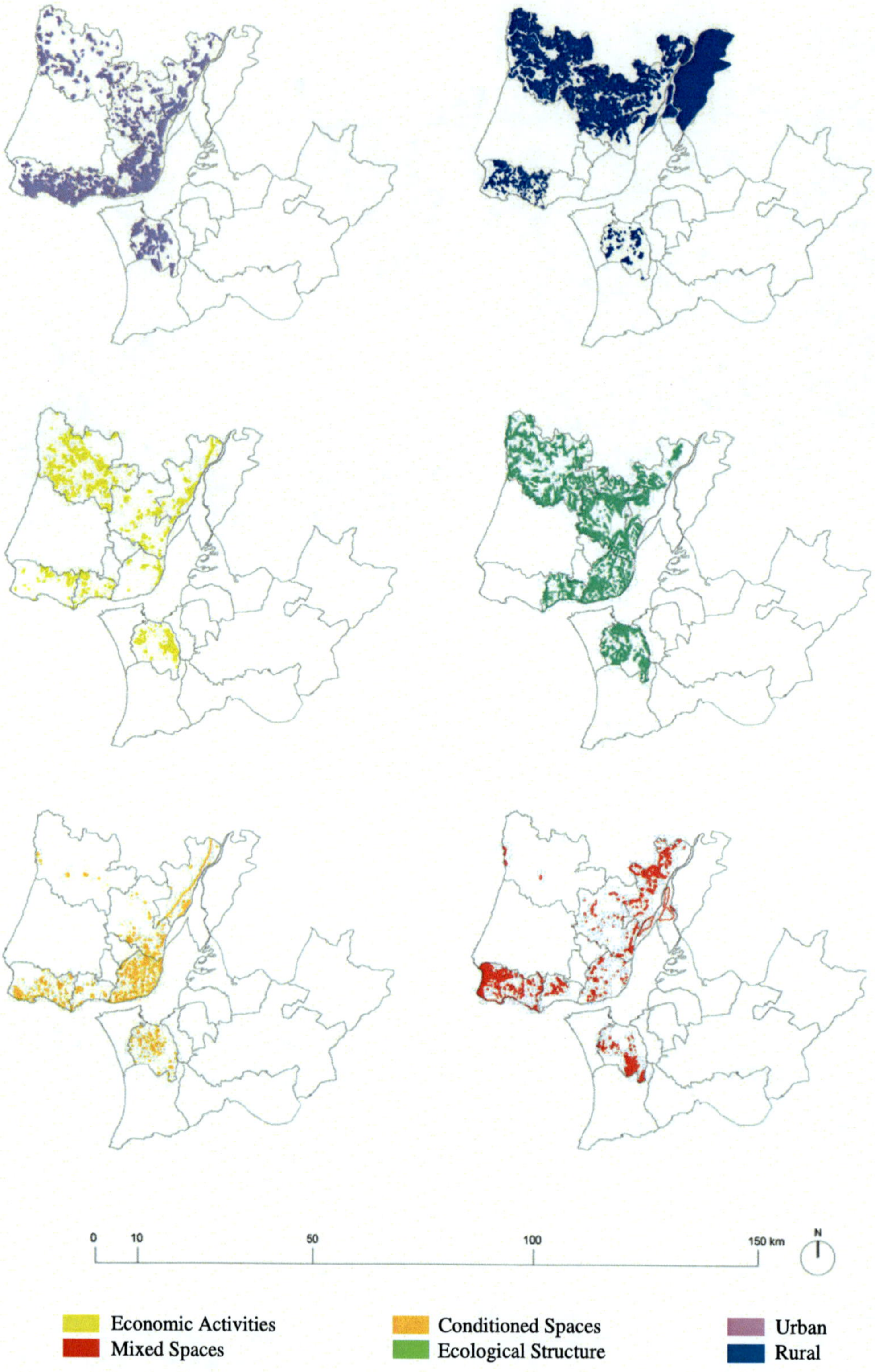

Fig. 6.7 Retired land-use regime charter, by typology. Source: Authors

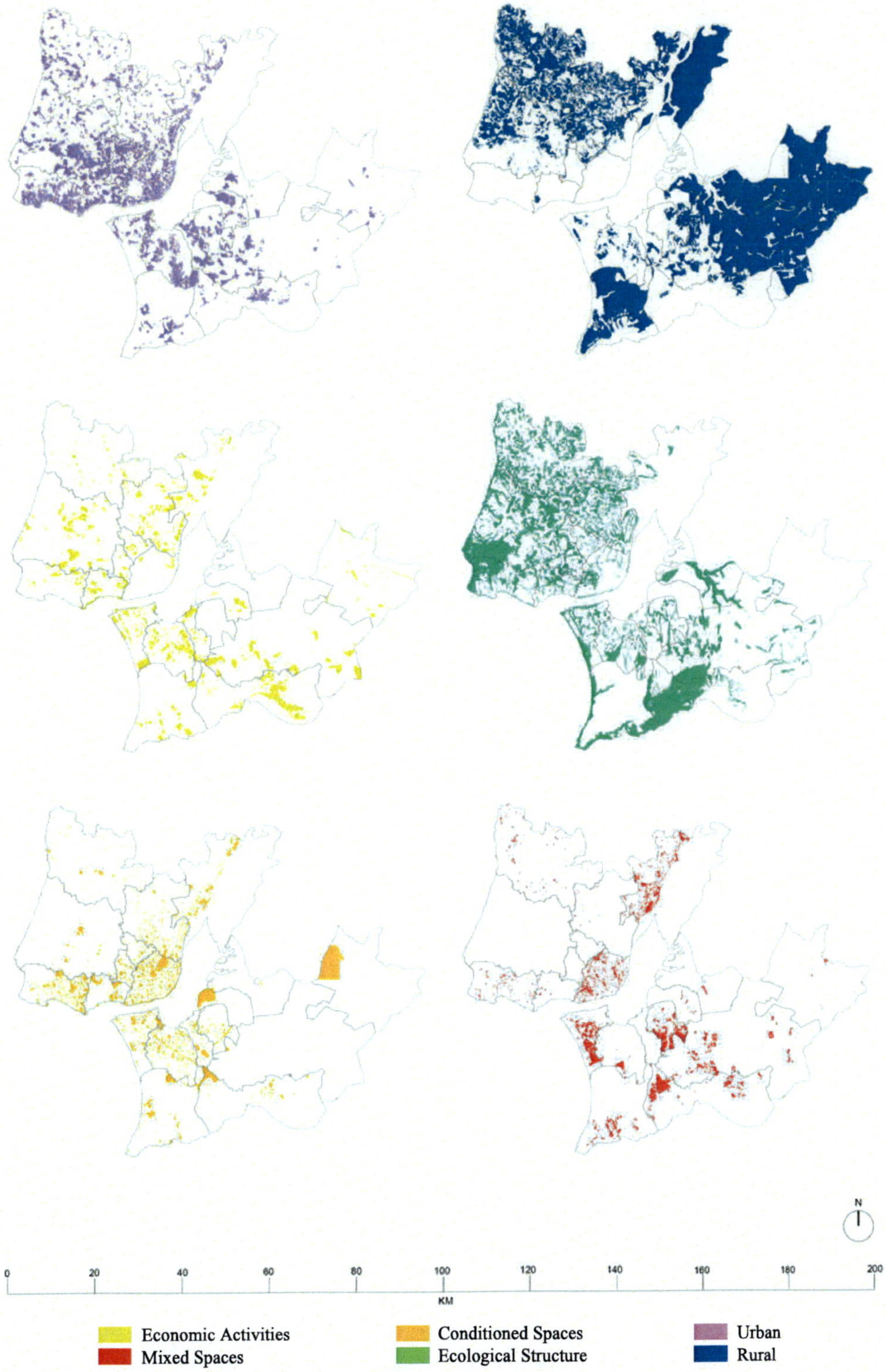

Fig. 6.8 Active land-use regime charter, by typology. Source: Authors

Table 6.13 Food production in the Lisbon Region

Produce	Area in the UAA (ha)	Produced tons	Continental production (ha)	National context—UUA	National context—production
Durum wheat	395	1,092	59,536	5th from 5	4th from 5
Common wheat	379	1,056	50,190	5th from 5	4th from 5
Maize for grain	1,355	18,517	744,577	4th from 5	4th from 5
Rye	0	0	14,439	5th from 5	5th from 5
Rice	5,269	33,080	179,777	3rd from 5	3rd from 5
Whey	78	90	45,856	5th from 5	5th from 5
Barley	285	774	47,862	4th from 5	3rd from 5
Beans	2	2	2,323	5th from 5	5th from 5
Chickpea	34	32	1,506	4th from 5	4th from 5
Potato	2,402	101,954	473,018	4th from 5	3rd from 5
Tomato (industrial)	3,847	336,693	1,650,429	2nd from 5	3rd from 5
Sunflower	436	1,432	20,814	3rd from 5	2nd from 5
Forage maize	1,241	80,881	2,466,828	4th from 5	4th from 5
Forage whey	291	416,058	954,579	5th from 5	5th from 5
Apple	173	2,684	327,503	4th from 5	4th from 5
Pear	87	1,196	201,928	4th from 5	4th from 5
Peach	94	746	41,617	5th from 5	5th from 5
Cherry	8	29	19,267	5th from 5	4th from 5
Plum	75	941	29,173	5th from 5	5th from 5
Kiwi	2	25	35,254	5th from 5	5th from 5
Orange	294	2,504	315,935	5th from 5	5th from 5
Tangerine	28	240	37,204	5th from 5	5th from 5
Almond	5	4	20,139	5th from 5	5th from 5
Chestnut	5	6	29,640	5th from 5	5th from 5
Nut	22	31	4,569	5th from 5	5th from 5
Olive for eating	48	17	17,802	5th from 5	5th from 5
Olive for olive oil	596	18,077	858,413	5th from 5	5th from 5
Grapes	130	873	21,657	5th from 5	4th from 5
Grapes for wine	8,024	69,005	868,635	4th from 5	4th from 5

Source Authors (Adapted from INE 2018)

Table 6.14 Livestock production in the Lisbon Region

Livestock	Total	Continental production	National context production
Pigs	221,000	2,131,000	3rd from 5
Sheep	45,000	2,218,000	5th from 5
Goats	9,000	326,000	4th from 5
Cows	88,000	1,388,000	4th from 5
Poultry	189,359	211,385,805	4th from 5

Source Authors (Adapted from INE 2018)

Overall, Portugal has a larger UAA than the European average—43% of the national territory against 39%—and while the majority of UAA of Europe is used for arable land, in Portugal predominance goes to permanent pastures (INE 2021). However, the size of agricultural holdings is on average smaller in Portugal than in Europe by nearly 3 ha (INE 2021).

Beyond the general considerations about agriculture in Portugal, which seems to face a period of revival and reintensification—although this is not entirely good news—the Agricultural Census proceeds to break down its results through the nine agrarian regions it establishes, including RO where the Lisbon Region is included.

Thus, from 290,229 agricultural holdings in the national territory, RO is home to 34,486, being the fourth region with more numerous explorations—11.9% of the national total (INE 2021). The total UAA reached 3,963,945 ha, 409,095 (10.3%) located in RO (INE 2021), having the second-largest average UAA per agricultural holding—11.9 ha (INE 2021). In comparison with 2009, the number of agricultural holdings in the country decreased by 4.9% and in RO by 13.5%, but the UAA increased in the country by 8.1% and by 4.6% in RO (INE 2021). RO also has the second-largest increase in ha per holding, 21%, against a national average of 13.7% (INE 2021).

While the majority of agricultural holdings is still managed by sole holders, 94.5–50.6% of agricultural enterprises are concentrated on RO (INE 2021), affirming this region as a platform for the renovation that seems to be underway in agricultural businesses.

Regarding the composition of the UAA, arable land decreased in Portugal by 11.6% but it increased slightly in RO by 1.8%. Non-perennial farming represents 141,114 ha of the RO arable land, while fallow land represents 28,193 ha (INE 2021). Familiar horticulture has decreased in Portugal since 2009 by 18%, but in RO this decrease reaches 37.7%, totalizing now 1,211 ha from the UAA (INE 2021). Permanent farming increased by 17.1% in Portugal but in RO it decreased by 7%, counting now only 76,100 ha of land. Permanent pastures have increased by 14.9% in Portugal but in RO only 7.8% (INE 2021). Overall, the UAA increased nationally 8.1%, but only 4.6% in RO (INE 2021).

Within non-perennial farming, cereals for grain decreased by 32.2% in Portugal, but only by 12.5% in RO, while legumes increased 41.2% nationally and 36.8% in RO (INE 2021). Non-perennial pastures and forage cultures increased 12% in Portugal, but 24% in RO, while potatoes decreased by 28.6% nationally, but only decreased 6.5% in RO. Industrial cultures decreased in the country by 57%, but in RO they have increased by 39%, almost the opposite proportion (INE 2021).

Fruit production seems to have been a particular area where the Portuguese invested. Production surface of fresh fruits increased by 14.1% in the country but only by 0.8% in the RO (INE 2021). Citrines production surface increased by 15.8% in the country and decreased by nearly the same amount, 15.9% in RO (INE 2021). Subtropical fruits, however, show a national increase of 152.7%, while in RO the increase is of 3,570.5% (INE 2021). Nuts production space increased 98.6% in the country and 100.7% in RO (INE 2021). Olive groves and vineyards, which we have seen to be traditionally present in the landscape of the Lisbon Region, show a different tendency: olive groves increased in the country by 12.3% while in RO they have decreased by 11.5%, whereas vineyards decrease in both cases, but in the country (−2.6%) much less than in RO where it decreased by 11.5% (INE 2021). Permanent pastures in open fields have increased their surface by 29.2% nationally, while in RO the increase is 32.5%; however in woods or forestry, it has decreased by 4.4% in the country and by 9.3 in RO; and finally, under permanent farming areas, the pasture surface has increased by 1.2% in Portugal but decreased by 2.2% in RO (INE 2021) (Figs. 6.9, 6.10, 6.11, 6.12, 6.13, 6.14, 6.15, 6.16, 6.17, 6.18, 6.19 and 6.20).

With respect to irrigation farming, such was the system of 46.4% of agricultural holdings and 15.9% of the UAA. From the irrigated land, 46.8% comprised arable land, 43.3% permanent cultures and 9.9% permanent pastures (INE 2021).

Mechanization was also an important change in the most recent Agricultural Census, and it demonstrates that the use of agricultural machinery has increased radically: while Portugal seems to be using it less, having a decrease of 0.5%, the RO region demonstrates an opposite tendency, with a rise of 4.2% (INE 2021). Agricultural holdings with their own machinery are more numerous, with a 4% increase in the country and a 4.6% increase in the INE (2021).

On the other hand, agricultural labour force has decreased in the national total by 14.4% and in the RO by 11.7% (INE 2021). The decrease in familiar labour force—of 27.3% in Portugal and 35.7% in RO—is less significant than non-familiar labour—which decreased by 30.7% nationally and 27.7% in RO (INE 2021). Service hiring however has escalated, with an increase of 159.9% in Portugal and of 485.6% in the RO (INE 2021).

Finally, biological agriculture has registered a very positive evolution since 2009, witnessing an increase of 214% in Portugal and of 319.4% in RO, registering an increase occupation of UAA of 112% in Portugal and 132.5% in RO (INE 2021). Breaking it down into its several sectors, non-perennial biological farming increased 78.8% in Portugal but only 26.6% in RO (INE 2021). Permanent biological farming increased by 134.9% in Portugal but fell 74.7% in

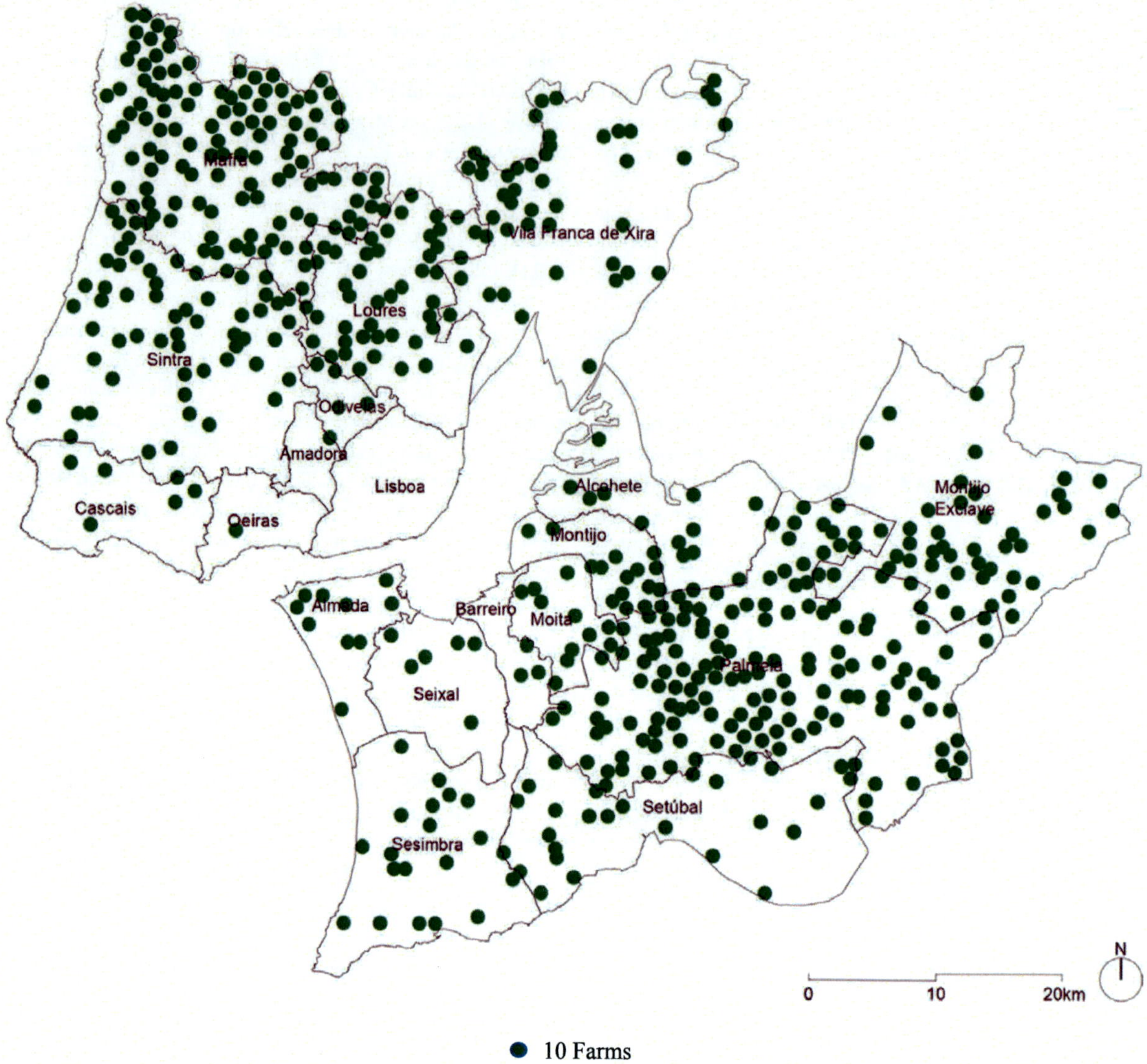

Fig. 6.9 Agricultural holdings (2019). Source: Authors (Adapted from Agricultural Census, 2019)

RO. Permanent biological pastures increased by 113.8% in the country and by 195.6% in RO (INE 2021). Cow biological breeding grew 127.4% in Portugal and 217.2% in RO (INE 2021). The biological production of pigs disappeared from RO, while in the country it just decreased by 24.2% (INE 2021). And biological sheep breeding increased 22% nationally and 83.2% in RO (INE 2021).

In nearly all aspects accounted for in the Agricultural Census, there is a clearly formed rural crown in the Lisbon Region. The rural territories of Palmela, the Montijo Exclave, Vila Franca de Xira (mostly the Great Wetland of Tagus, but not only) and Mafra, as well as the northern areas of Sintra and Loures, stand out for their clearly developed agricultural activities. Palmela, Mafra and Loures have the most agricultural holdings, although the largest properties are, unsurprisingly, to be found in the Montijo Exclave and in the Great Wetland. Grazing areas seem to be stronger in the Southern Bank of Tagus—again with a particular weight in Palmela and the Montijo Exclave, while these two territories, alongside Mafra seem to have the greatest variety of livestock types.

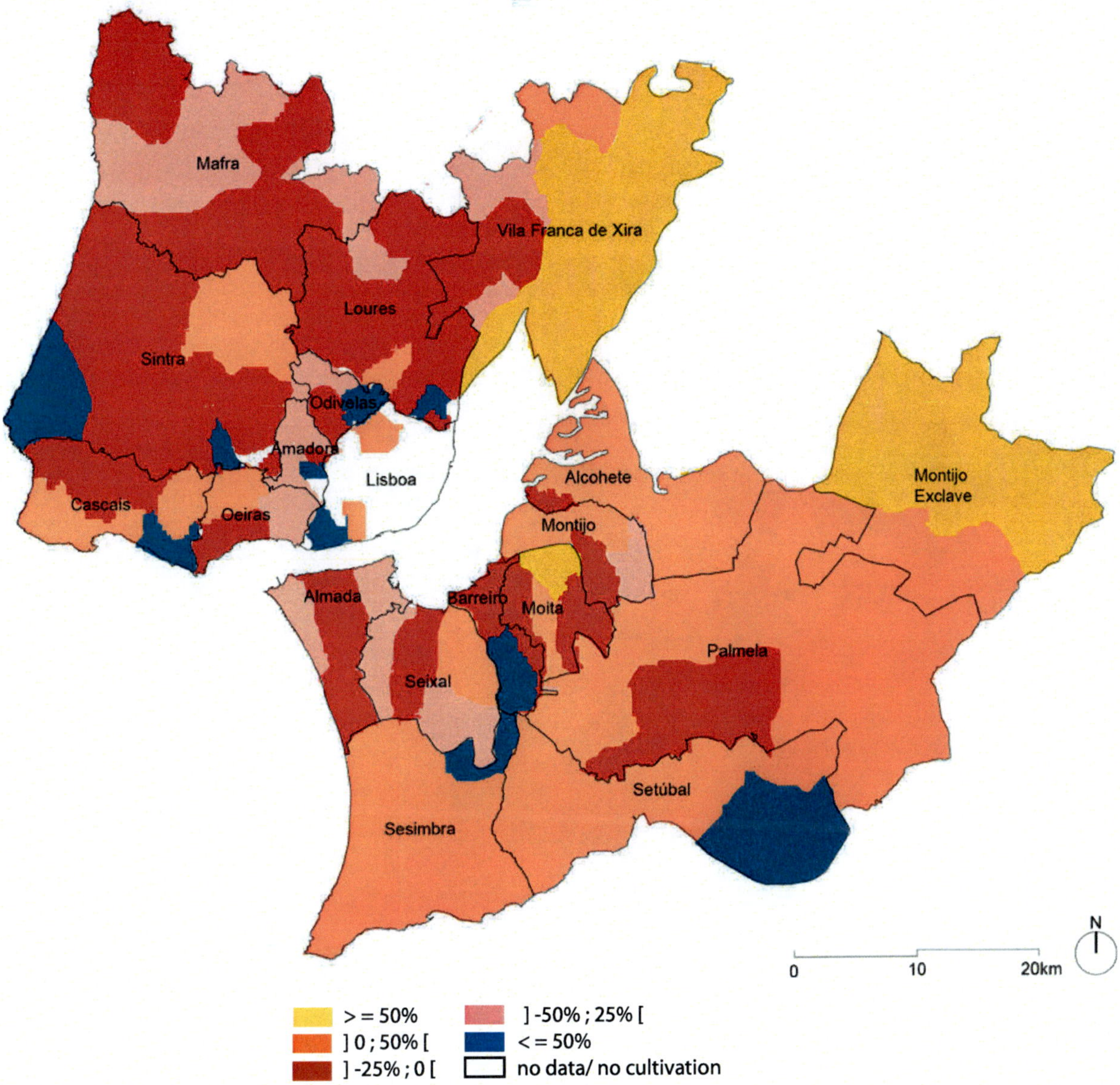

Fig. 6.10 Average size of agricultural holdings (2019). Source: Authors (Adapted from Agricultural Census, 2019)

However, Loures, Sintra and Sesimbra have considerable amounts of livestock, with a special emphasis on sheep, goats and herbivores, which are indeed the dominant type of livestock in the Lisbon Region. With respect to non-perennial farming, arable crops are dominant on the Southern Bank of Tagus (particularly strong in Palmela and the Montijo Exclave, but also in the Montijo urban territory and in Moita and Alcochete) as well as in Vila Franca de Xira. On the rest of the Northern Bank of Tagus, intensive horticulture is dominant—confirming the contrasts in regional production described long ago in the survey on regional architecture (as quoted in the Introduction of this Atlas). Agricultural machinery is also more significant in these areas. While biological agriculture is not yet a particularly relevant aspect of the regional food production structure, there are some examples already, mostly of perennial farming, which cluster in Sesimbra, the Montijo Exclave but mostly in Vila Franca de Xira. With regards to diversity, it is precisely Vila Franca de Xira that displays the most diverse sample of biological agriculture, showing the endless productive potential of this territory.

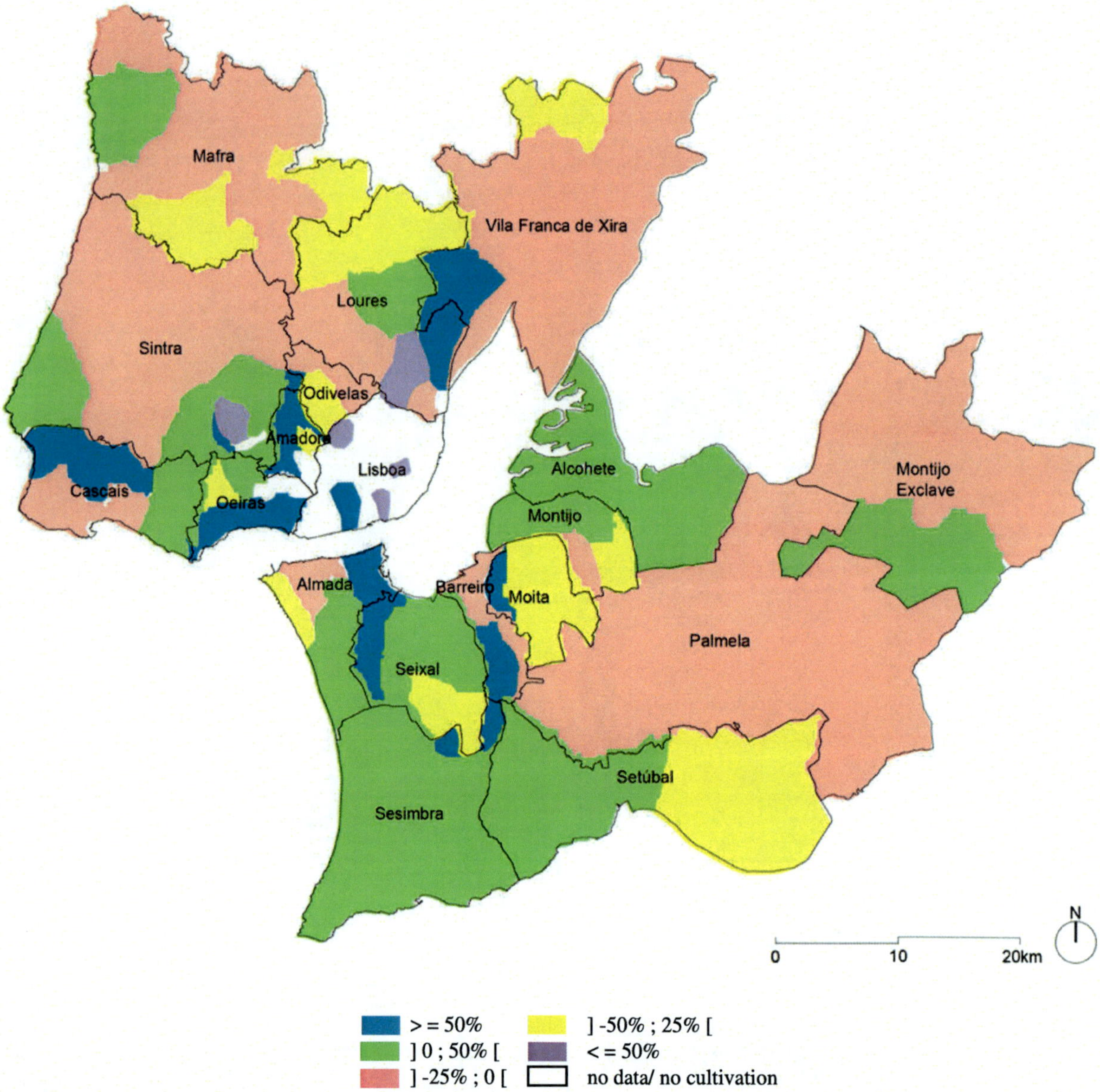

Fig. 6.11 Variation in agricultural holdings (2019). Source: Authors (Adapted from Agricultural Census, 2019)

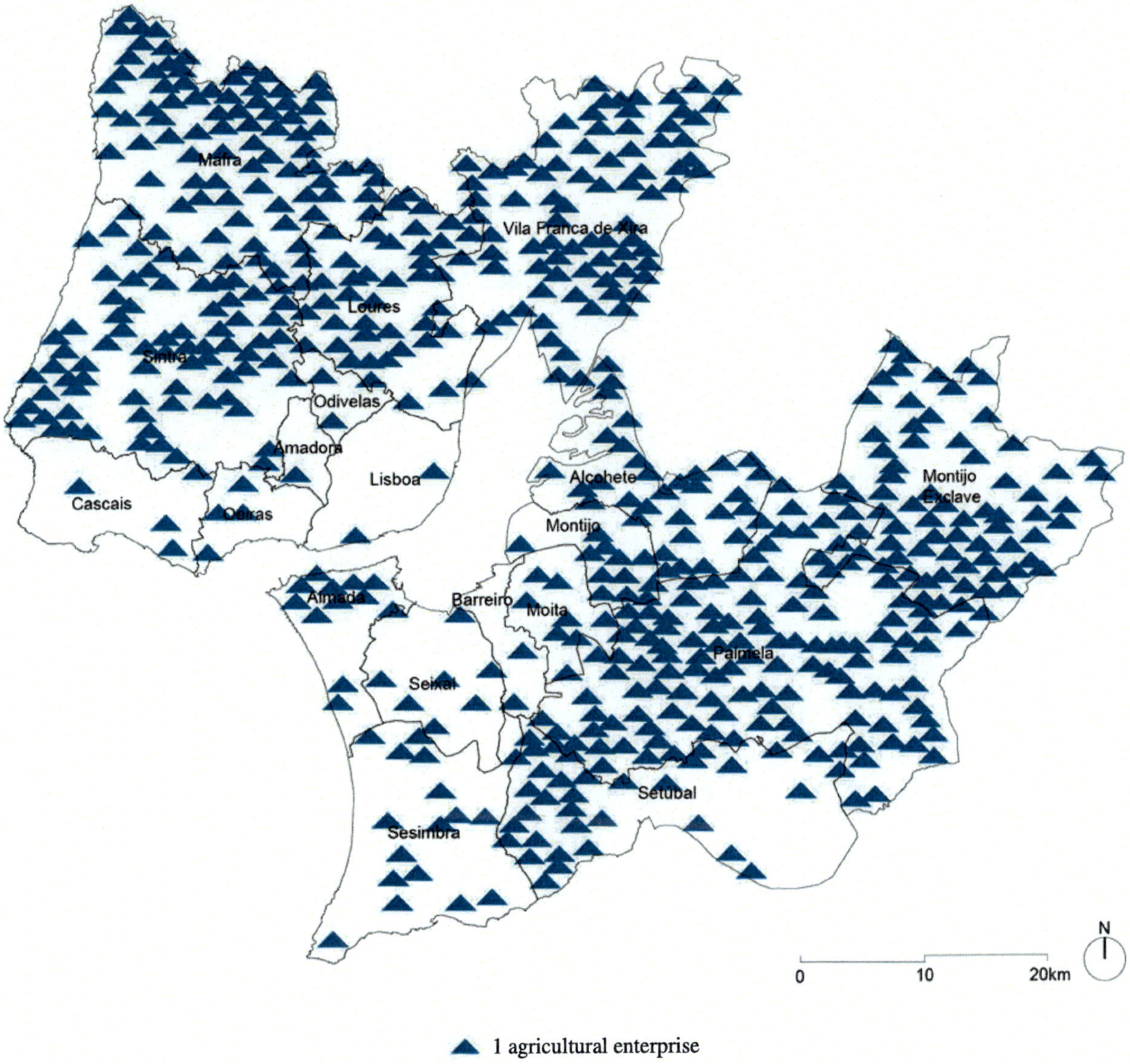

Fig. 6.12 Agricultural enterprises (2019). Source: Authors (Adapted from Agricultural Census, 2019)

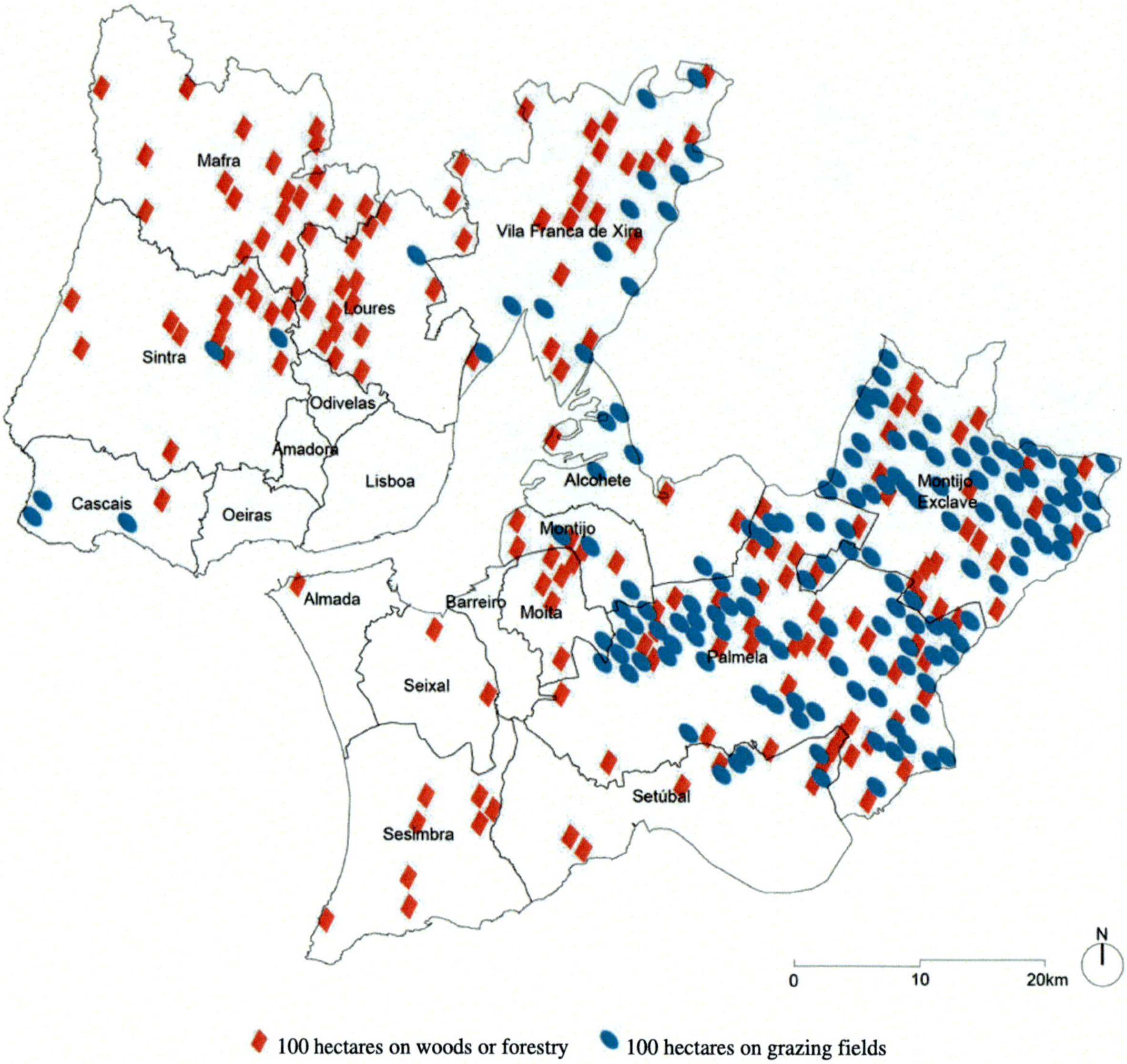

Fig. 6.13 Permanent pastures (2019). Source: Authors (Adapted from Agricultural Census, 2019)

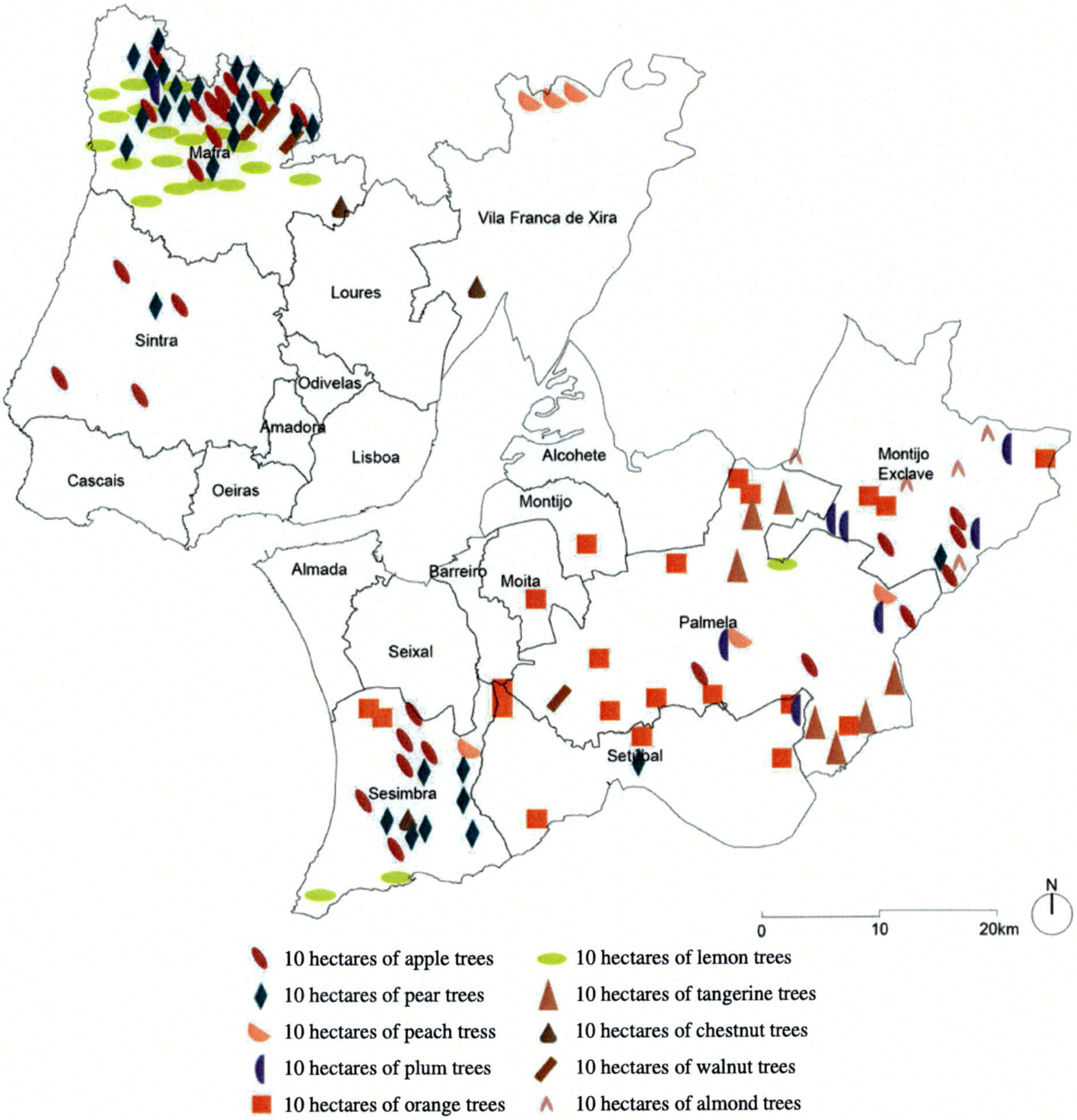

Fig. 6.14 Key fruit production in trees (fresh fruit, citrus fruit, nuts) (2019). Source: Authors (Adapted from Agricultural Census, 2019)

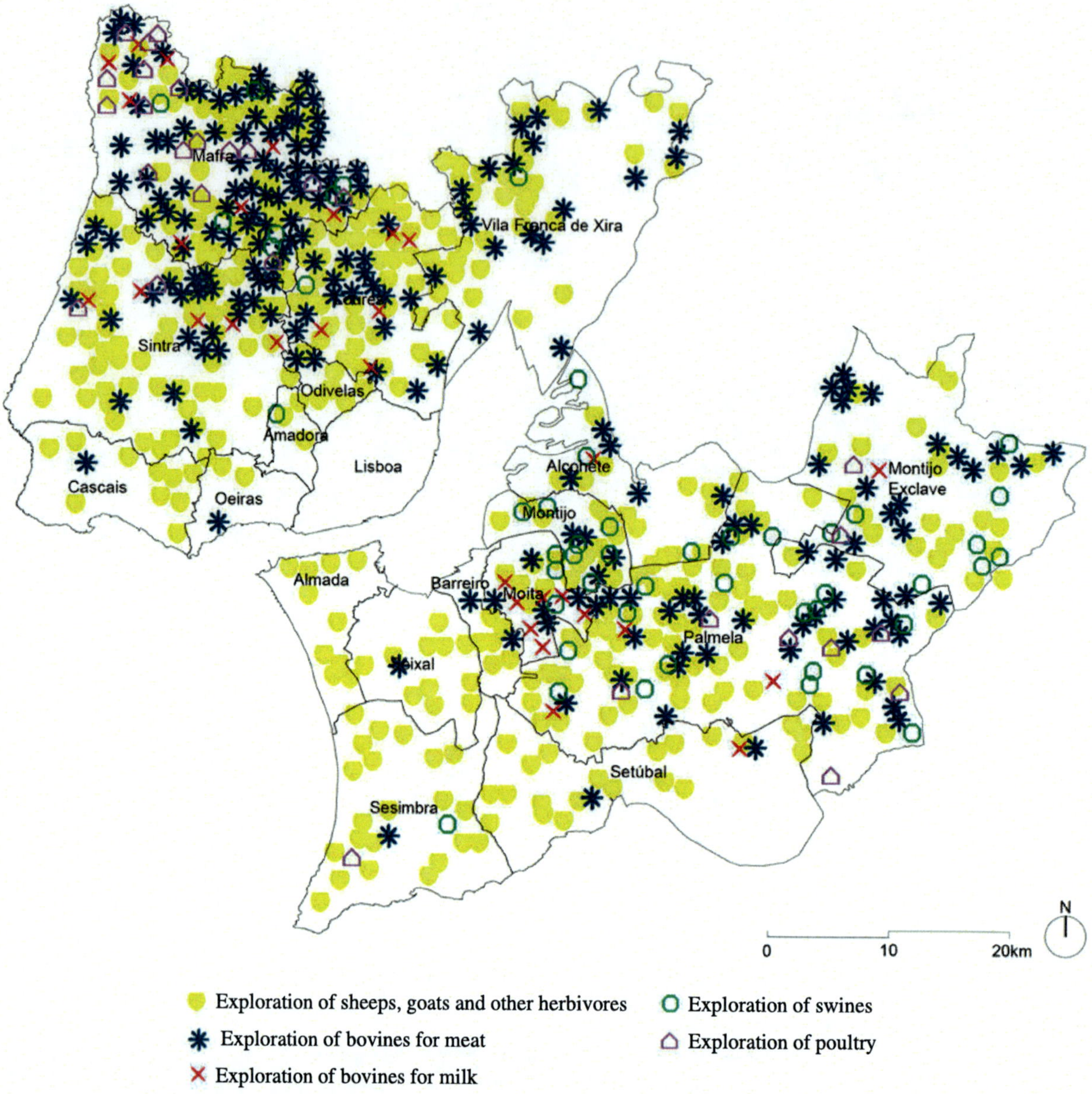

Fig. 6.15 Main types of livestock (2019). Source: Authors (Adapted from Agricultural Census, 2019)

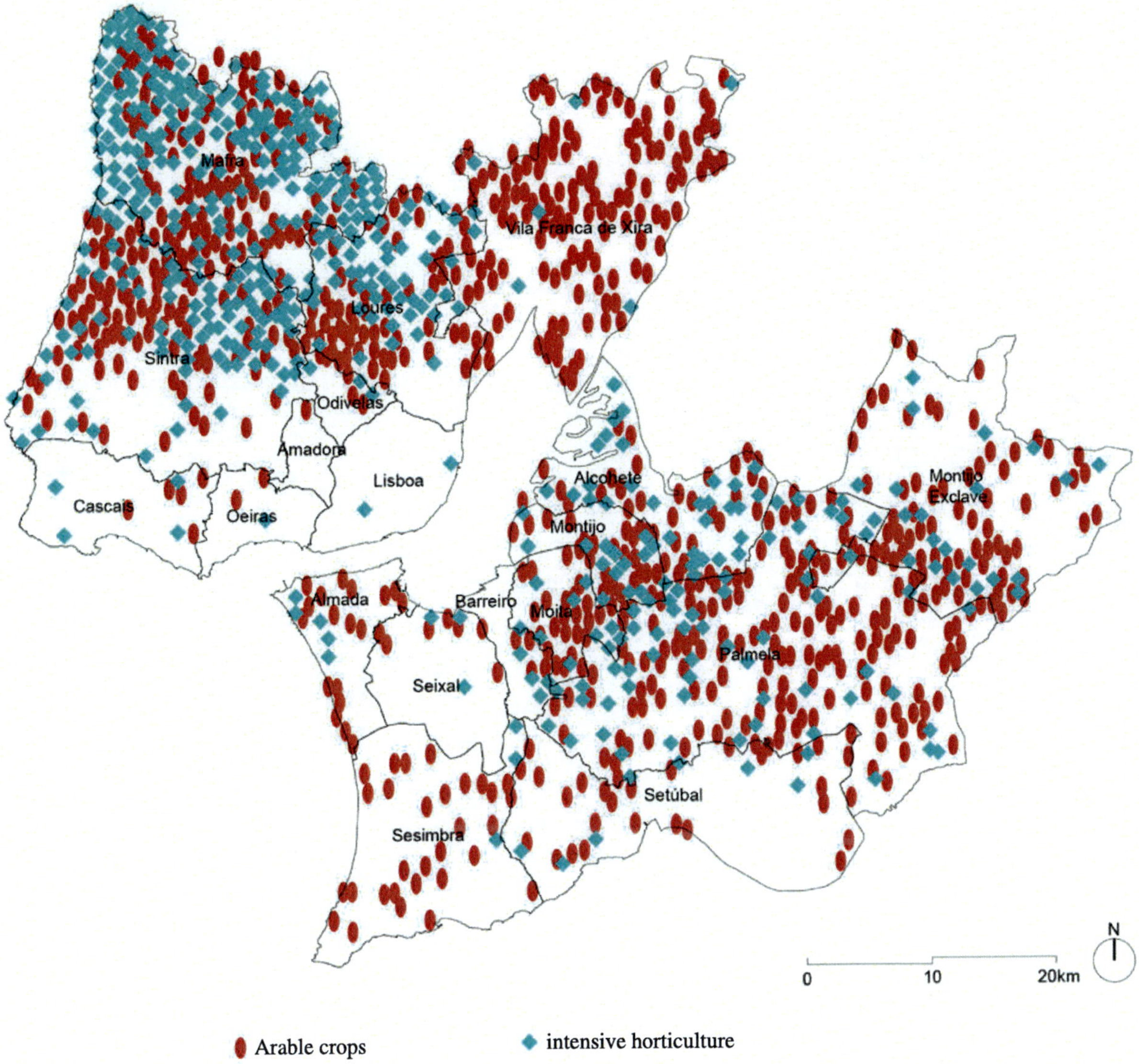

Fig. 6.16 Types of non-perennial farming (2019). Source: Authors (Adapted from Agricultural Census, 2019)

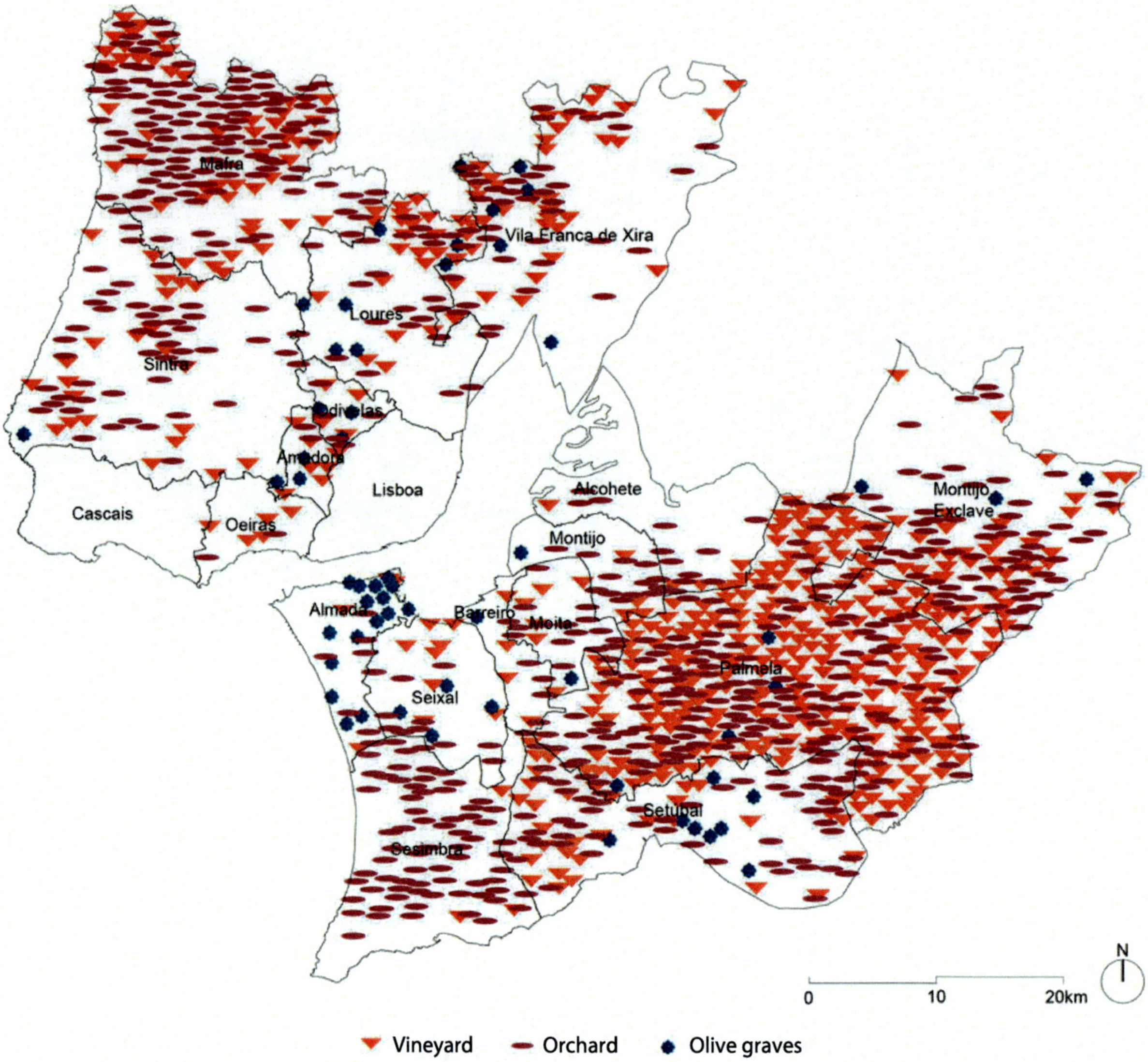

Fig. 6.17 Main types of perennial farming (2019). Source: Authors (Adapted from Agricultural Census, 2019)

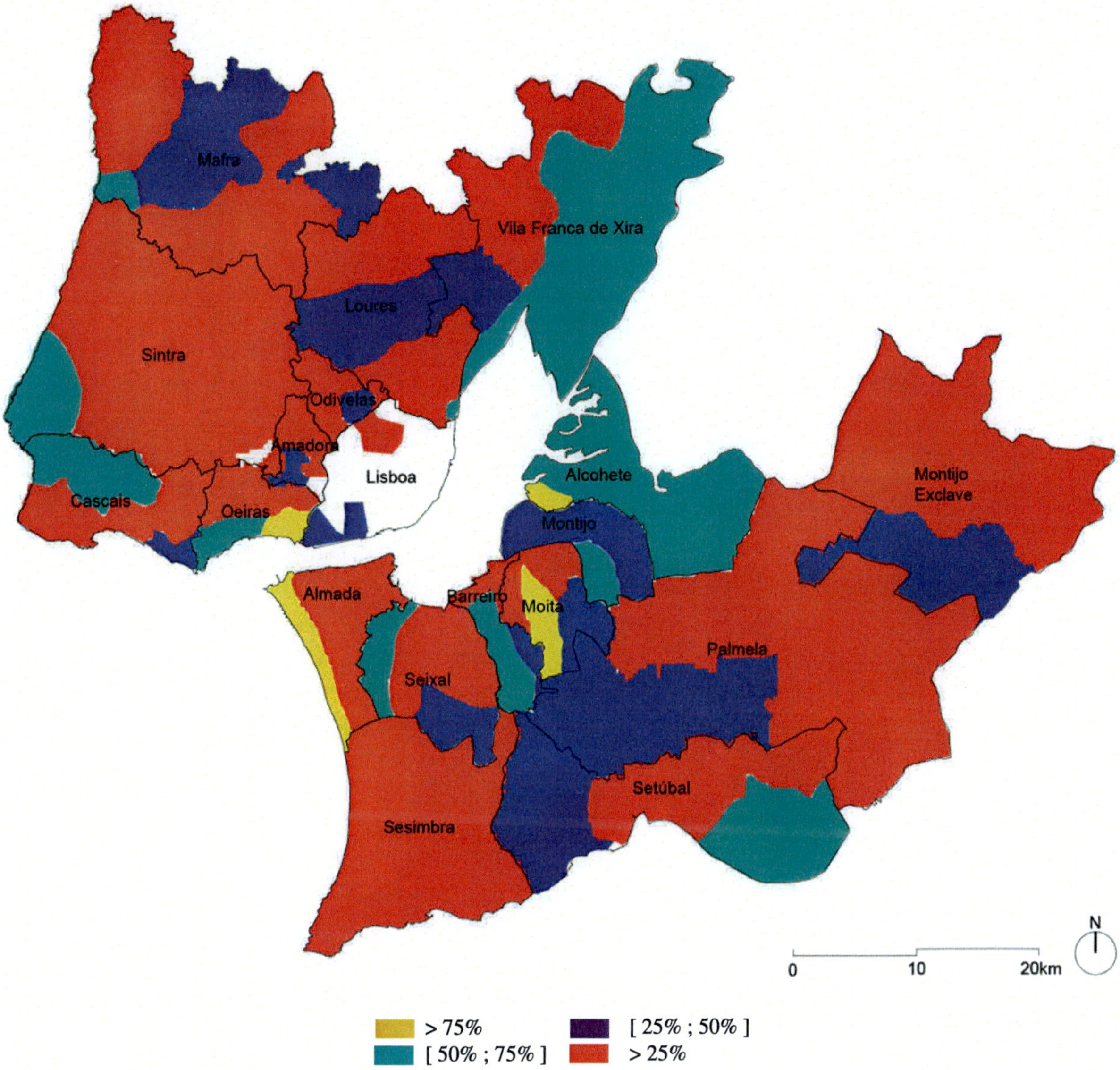

Fig. 6.18 Irrigated surface from the useful agricultural surface (2019). Source: Authors (Adapted from Agricultural Census, 2019)

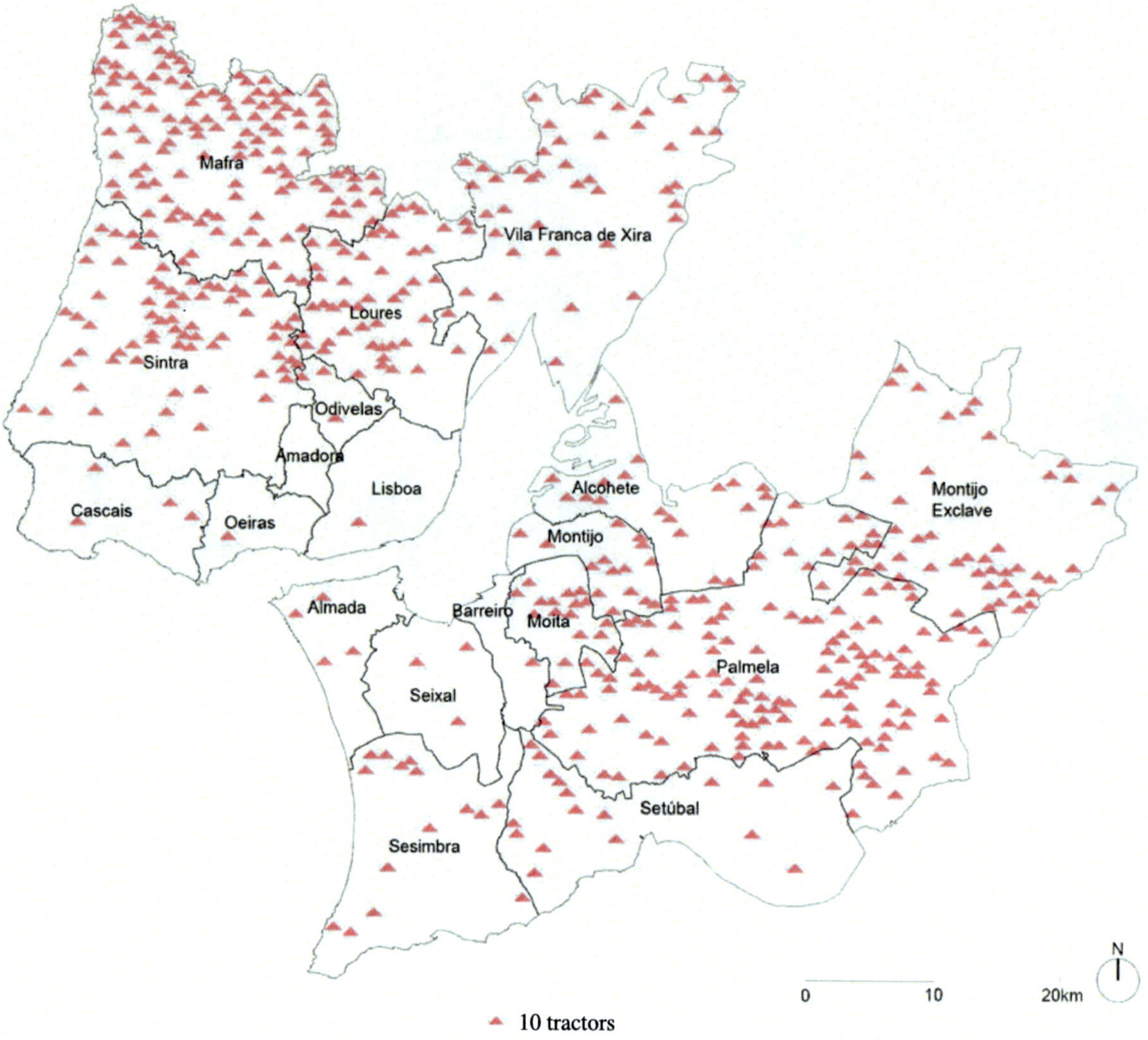

Fig. 6.19 Agricultural machinery (2019). Source: Authors (Adapted from Agricultural Census, 2019)

References

AAVV (1961 [2004]) Arquitectura Popular em Portugal. Ordem dos Arquitectos, Lisbon

Antunes HI, Ferreiro MF (2017) Segurança alimentar e sustentabilidade: o caso do setor do arroz no Vale do Tejo e Sorraia - perceção e práticas. Working Paper 2017/03. https://doi.org/10.15847/dinamiacet-iul.wp.2017.03

Cavaco CS (2010) Formas de habitat suburbano. Tipologias e modelos na área metropolitana de Lisboa. PhD Thesis. FAUL, Lisbon

Costa S, Fox-Kämper R, GoodR SI, Treija S, Atanasovska JR, Bonnavaud H (2016) The position of urban allotment gardens within the urban fabric. In: Bell S, Fox-Kämper R, Keshavarz N, Benson M, Caputo S, Noori S, Voigt A (eds) Urban allotment gardens in Europe. Routledge, London, pp 201–228

d'Almeida, PB (2006) Victor Palla e Bento d'Almeida. Obras e Projectos de um Atelier de Arquitectura, 1946–1973. MSc dissertation, Faculdade de Ciências Sociais e Humanas da Universidade Nova de Lisboa, Lisboa

d'Almeida PB (2017) Victor Palla e Bento d'Almeida ou Bento d'Almeida e Victor Palla. In: d'Almeida PB (coord), Martins JP (eds) Victor Palla e Bento d'Almeida: Arquitectura de outro tempo. Caleidoscópio, Casal de Cambra, pp 40–105

Ferrão J (2011) O ordenamento do território como política pública. Fundação Calouste Gulbenkian, Lisbon

Graça P (2020) Como comem os portugueses. Fundação Francisco Manuel dos Santos, Lisbon

INE - Instituto Nacional de Estatística (2018) Estatísticas Agrícolas 2017. INE, Lisboa. https://www.ine.pt/xurl/pub/320461359. ISSN: 0079-4139, ISBN: 978-989-25-0445-2

INE - Instituto Nacional de Estatística (2021) Recenseamento Agrícola 2019. Análise dos principais resultados. INE, Lisboa. https://www.ine.pt/xurl/pub/437178558. ISBN: 978-989-25-0562-6

Marat-Mendes T (coord), Mourão J, d'Almeida PB, Niza S (2015) Water and agriculture atlas: Lisbon region in 1900–1940. Atlas da Água e da Agricultura. Região de Lisboa em 1900–1940. Iscte - Instituto Universitário de Lisboa, DINÂMIA'CET-IUL, Lisboa

Marat-Mendes T, d'Almeida PB, Borges JC (2021) Concepts and definitions for a sustainable planning transition: lessons from

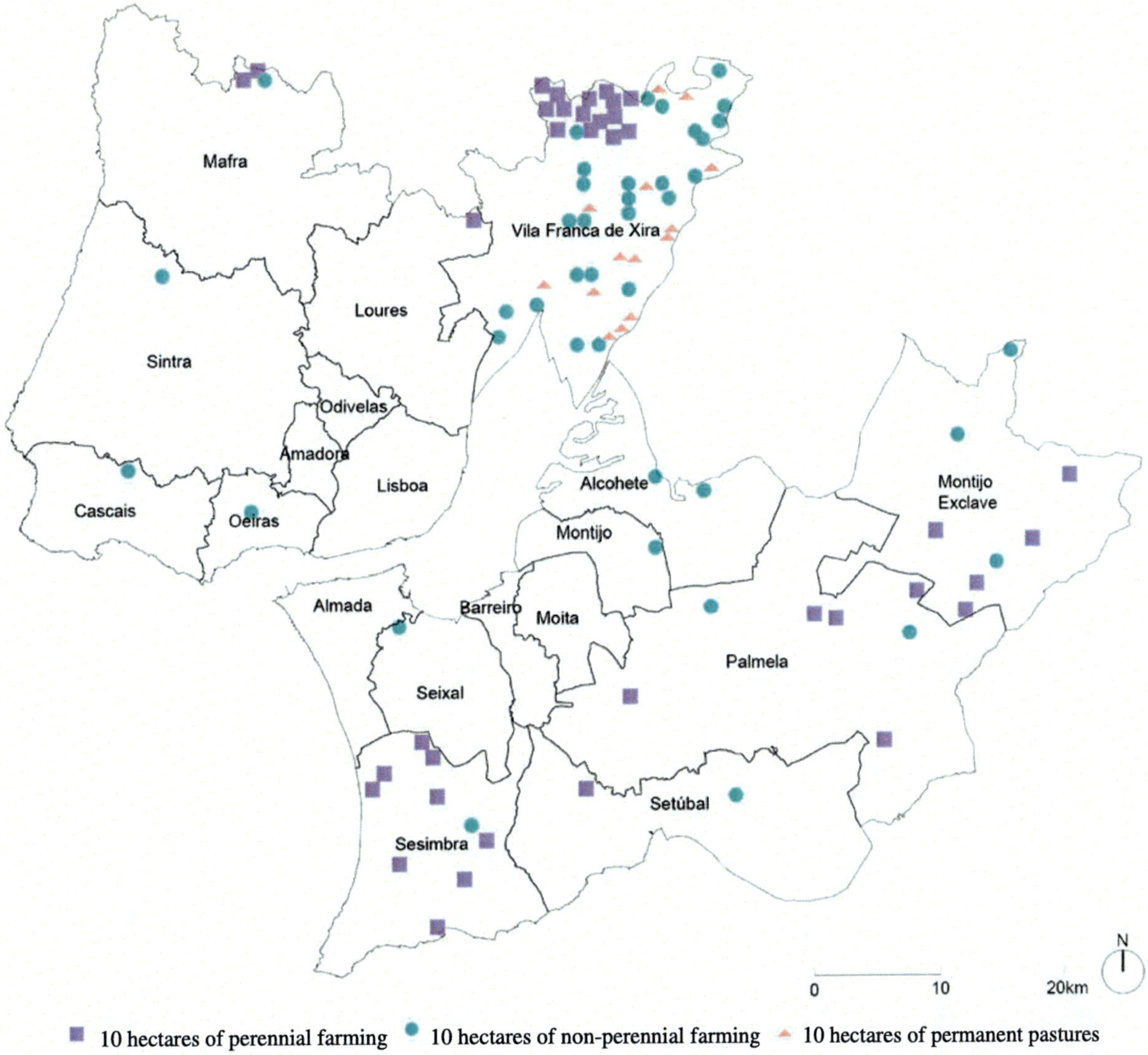

Fig. 6.20 Biological agriculture (2019). Source: Authors (Adapted from Agricultural Census, 2019)

moments of change. European Planning Studies 1–23. https://doi.org/10.1080/09654313.2021.1894095

Marot S (2019) Taking the Country's side: agriculture and architecture. Polígrafa & Trienal de Arquitectura de Lisboa, Lisbon

Modesto ML (1982) [1990] Cozinha tradicional Portuguesa. Verbo, Lisbon

Modesto ML (2015) Acervo Maria de Lourdes Modesto. https://acpp.pt/maria-de-lourdes-modesto

Niza S (2017) O sistema alimentar no contexto do metabolismo urbano da Área Me Lisboa. In: Oliveira R, Amâncio S, Fadigas L (eds) Alfaces na avenida. Universidade de Lisboa, Colégio Food, Farming and Forest, Lisboa, Estratégias par cidade, pp 42–47

Niza S, Ferreira D, Mourão J, d'Almeida PB, Marat-Mendes T (2016) Lisbon's womb: an approach to the city metabolism in the turn to the twentieth century. Reg Environ Change 16(6):1725–1737

Oliveira R, Morgado MJ, Martinho R, Maurício I (2014) O sistema alimentar urbano da Área Metropolitana de Lisboa—análise e diagnóstico. FCSH, Lisboa

Portas N, Domingues A, Cabral J (2007) Políticas urbanas—tendências, estratégias e oportunidades. Fundação Calouste Gulbenkian, Lisboa

Salvador MS (2019) Shaping the city through food: the historic foodscape of Lisbon as case study. Urban Design International 24:80–93

Steel C (2020) Sitopia. How food can save the world. Chatto & Windus, London

Part III

Urban Agriculture and Its Role in the Territory

As the planning situation of the Lisbon Region and key elements of its food system are already presented, in this part, we focus on one of the subjects that has received more attention in scholarly literature on the relationship between food and urbanism, namely, the subject of urban agriculture. Either acknowledged as an important activity to counteract the artificial land-uses that take predominance in the city (Dias, 2018), or as a form of heritage specific to the urban periphery (Parham, 2015); either as an activity for leisure and promotion of urban well-being, either as a form of negotiation between the competing interests of different social classes and between the system and those left at its margin (Cabannes & Raposo, 2013; Hardman & Larkham, 2016), either as a link between urban and rural (Viljoen et al, 2005, 2014) or as an use to be integrated in urban buildings to promote the greening of the city itself (Rodriguez, 2016; Orsini et al, 2017), urban agriculture has been discussed from several perspectives. Because it synthesizes such different functions and cuts across the occupational preferences of such different people, it is already an activity worth taking seriously as a form of city usage. But the practice of urban agriculture even questions the limits of urban and rural as self-sufficient and opposing realities.

This part explores the situation of urban agriculture in the Lisbon Region. The first chapter is a brief review of definitions and key ideas associated with this practice and its importance in urban territories. In the second chapter, we present a survey conducted at the SPLACH project on urban agriculture in the 18 metropolitan municipalities. Finally, a selection of case studies is presented, to demonstrate the diversity of forms that the phenomenon of cultivating in the city assumes in this territory.

7 Urban Agriculture from a Historical Perspective

7.1 Introduction

Cities have become parasites on the landscape—huge organisms draining the world for their sustenance and energy, relentless consumers, relentless polluters (Rogers 1998, 27).

The contention quoted above is by architect Richard Rogers, in his memorable attempt at constructing a sustainable model for urbanism, over 20 years ago. The city as a parasite may be a rather poetic and pessimistic reiteration of a very central problem Rogers was identifying. Namely, that cities have become arenas of consumerism, an individualistic consumerism that largely exceeds the capacities of local production and thus dependent upon global systems for distribution of goods (Rogers 1998). The future, as architectural historian Sébastien Marot (2019) notices, predicates upon a paradox whereby urbanization seems inevitable because we accept it as an integral part of history (even if in reality it is a rather recent artefact) but at the same time impossible because of the environmental and social threats it poses, which the recent pandemic of COVID-19 has proved all too real (Kordshakeri and Fazeli 2020; Abusaada and Elshater 2020; Honey-Rosés et al. 2020).

Debates over the city as a place of consumption are not new. Clyde Weaver (1984) synthesizes the key critique to the capitalist spatial outlook: that separation between city and countryside directly translates the organization of production and consumption in the interest of profit rather than the local community. A similar idea, more explicitly or less, underlies the theories of early regional planners, like Patrick Geddes, Ebenezer Howard and geographers like Piotr Kropotkin (1842–1921). To be sure, cities had long been distinct from the countryside insofar as their key activity was trade and not agriculture (Weber 1966). Yet, until modernity, most trade-based cities were reliant upon a productive hinterland whose key function was to ensure food provision for urbanites—many of whom, in some cases, had their private backyards for cultivation (Weber 1966; Steel 2008; Salvador 2019; Marot 2019). Having a '*villa suburbana*', out of the city and combining cultivated leisure with work, civility with rurality, was not uncommon among the Romans, whose words for such practices remain in many contemporary languages—*hortus*, *silva*—and such habit was revived in the Renaissance, afterwards providing a key for 'regimenting' the countryside by enclosing former agrarian commons (Steel 2008; Marot 2019). These practices, not so distant from us in history, testify to how the food system provides a unique point of view from which to observe territorial transformation, both within and around cities.

With modernity and its separation of production and consumption, both the territory and the food system transform greatly. As rural land started to be urbanized, witnessing the sprouting of suburbs, new cities and towns or industrial belts, the food system reached ever-growing scales, with foodstuffs being transported across great distances—until this process peaks with today's global-scale food systems suited to the interests of economies of scale but with less convincing advantages for the local community or the local landscape (Steel 2008).

There were attempts at bridging the gap between urban and rural. Fourier's Phalanstéres were surrounded by agroforestry, to ensure the diet of the residents (Barthes 1971). Cerdá's plan for Barcelona stipulated that only two of the four fronts of each urban block would be occupied with buildings, the remainder being open as green spaces to create a rural city (Marat-Mendes 2002). Ebenezer Howard's (1902) Garden Cities had large agricultural estates serving as a foodshed for the city and its satellites. Constant's utopian vision of a New Babylon (1959–74), Kisho Kurukawa's Agricultural City (1960), Brazni's Agronica (1993–94), although all of these merely consider the rural landscape as a backdrop into which urban protagonists are inscribed (Marot 2019) and all were conceived as 'utopian' conceptual projects, often designed to express, as a radical gesture, an idea of how the city might be changed, evolved or overcome.

In practice, few attempts at building Phalanstéres, notably André Godin's (1817–1888) Familistére, included no agroforest comparable with what Fourier imagined. Cerda's

T. Marat-Mendes et al., *Atlas of the Food System*,
https://doi.org/10.1007/978-3-030-94833-7_7

blocks eventually became fully urbanized, and Barcelona is no urban–rural residential city. And as for Howard's Garden City, most of its iterations were inspired by the works of his followers Raymond Unwin and Barry Parker, who emphasized Howard's lessons on the construction of suburban neighbourhoods, and generally not including the integral hinterland that Howard's original vision included (Lôbo 1995; Pothukuchi and Kaufman 2000; Steel 2008).

So it would seem like agriculture has vanished from at least the majority of cities in the Global North. But has it really?

The practice of food growing has always been a frequent aspect of the city's perceived or physical edge, and its erasure and substitution with urbanization is but a recent phenomenon, that in many cities is still underway (Parham 2015). However, the opposition between urban and rural has become a highly defined one. Typically, when one thinks of agriculture, large, cultivated fields in rural landscapes come to mind—immense explorations, sometimes with a single produce, places like those described by Portuguese novelist Alves Redol (1911–1969), landscapes of "only flatland and sky—sky and flatland"[1] (Redol 1971: 27).

Considering how much the world's population is growing —and how that population is overwhelmingly concentrated in cities—large-scale agriculture is undoubtedly necessary to ensure food for all. In that sense, agriculture may be difficult to articulate with the typical residential density of cities. But may that incompatibility result from preconceived notions of both agriculture and the city?

7.2 Defining Urban Agriculture

Sustainability concerns, which have been absorbed by national and international political agendas, from the Brundtland Report (UN 1987) to the recent 2030 Agenda for Sustainable Development (UN 2015) and Habitat III (UN 2017), suggest that our conceptions of the city must change —confirming Richard Roger's (1997) parasitic metaphor. In the specific proposals of the FAO (2019) in the 'Urban Food Agenda', the impacts and meanings of the urban–rural dichotomy are directly challenged. Indeed, such a schism, while taken for granted in planning policies (in Portugal and beyond) does overlook how systems for provision of basic human needs run across both types of territories (Ibañez and Katsikis 2014). The Milan Urban Food Policy Pact (MUFPP 2015) urges the public sector (as well as other actors) to rethink food systems within the perspectives of resilience and sustainability, acknowledging their entanglement with other issues faced by municipalities, including poverty, health, land-use, hygiene, and transportation.

Significantly, in the past years, there has been a blooming recognition from professionals and scholars of architecture, urban design and urban morphology that agriculture and architecture must be reconnected as concerns and as modes of living, in conjunction with one another, seeking alternatives to industrial agriculture and market economy, whose typical spaces are "the parallel dead-ends of metropolitan congestion and monocultural deserts" (Marot 2019: 8).

Countering such assumptions, urban agriculture is developed from scattered grassroots initiatives, official or illegal (Cabannes and Raposo 2013; Delgado 2018; Dias 2018; Dias and Marat-Mendes 2020), and often lacking interest or support from decision-makers and urban designers. In North America (Ashe and Sonnino 2013; Napawan 2016), Canada (Koc et al. 2008; Komisar et al. 2009), Australia (Edwards and Mercer 2010), among others, few, but relevant initiatives, have prompted urban agriculture into municipal planning (Deakin et al. 2016).

While urban agriculture eventually designates a very vast plethora of situations, with a high degree of adaptability into different physical and cultural contexts (Smit et al. 2001), it is still possible to provide a general definition going beyond the mere practice of food production within urban settlements, namely that.

> Urban agriculture entails the production of food for personal consumption, education, donation, or sale and includes associated physical and organizational infrastructure, policies, and programs within urban, suburban, and rural built environments. From community and school gardens in small rural towns and commercial farms in first-ring suburbs to rooftop gardens and bee-keeping operations in built-out cities, urban agriculture exists in multiple forms and for multiple purposes (Hodgson et al. 2011: 2).

Within this description, urban agriculture can encompass varying types of economic activity, categories and subcategories of products, be located in several contexts, from intra-urban to periurban areas, destination and provenance of products, their marketing, scale and production system may also vary (Mougeot 2000). Despite variations, there are two classic forms of urban agriculture activity, namely allotment vegetable gardens and community gardens (Iaquinta and Drescher 2010; Costa et al. 2016).

Allotment gardens usually refer to individual parcels of land allocated to individuals or families to be worked independently, although the allotment may be contiguous with others, with the participating individual households organized into self-governing associations (Flavell 2003; Iaquinta and Drescher 2010). On the other hand, community gardens are maintained by a group of individuals or households, producing collectively on a piece of land primarily for

[1] This is Alves Redol's description of the landscape of the Tagus Wetland, whose southern end is the Great Wetland in Vila Franca de Xira.

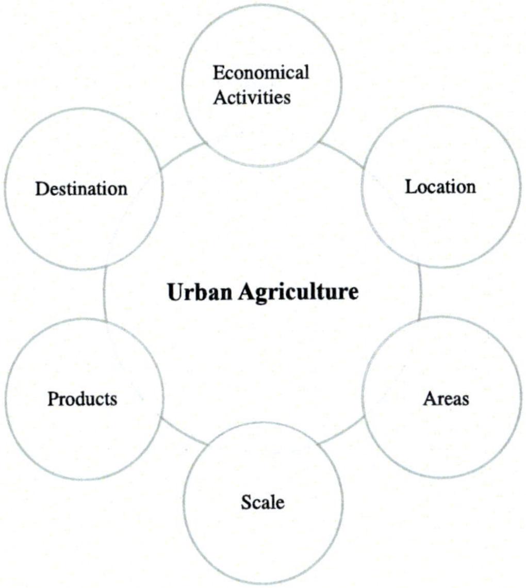

Fig. 7.1 Variables in the practice of urban agriculture. Source: Authors (adapted from Mougeot, 2000)

their own consumption (Drescher et al. 2006; Iaquinta and Drescher 2010) (Fig. 7.1).

In metropolitan regions, urban agriculture sometimes resembles periurban agriculture, meaning the production, distribution and marketing of food, including in community and school gardens, backyard and rooftop horticulture and innovative food production methods for maximizing production in a small area, as well as practices on physical edges like farms, community-supported agriculture and family farms in metropolitan greenbelts. Obviously, these are complex activities, impacting the community's food security, patterns of neighbourhood development, environmental sustainability, land-use management, and preservation (Dias 2018, Marat-Mendes et al. 2021).

While urban spaces, whenever planned, have tended to include green spaces, directed towards the leisure of the population and the visual quality of the landscape, urban agriculture implies a specific type of garden, which for long has been excluded from planning instruments (as seen with regards to the Lisbon Region) but whose recognition is increasing. The vegetable garden, whose succinct definition would have it as a.

> Plot of land made available for individual, non-commercial gardening or growing food plants. Such plots are formed by subdividing a piece of land into a few or up to several hundred land parcels that are assigned to individuals or families (Cambridge Dictionary).

In this apparently simple definition lie the key contradictions that the vegetable garden rises with regards to traditional assumptions about urban space. Indeed, its relatively small dimensions—in comparison with rural farmland—leave it close to the usual green leisure space of any residential neighbourhood. But its specific use to the production of food challenges the very exclusion of this activity that has defined, for over 100 years, policies of urbanism. Vegetable gardens have often been seen regarded as 'other' spaces, precisely because they challenge traditional understandings of urban and rural, but also of public and private (Costa et al. 2016).

Historically, the phenomenon of urban agriculture is not new. Even in Europe, the practice of having allotment gardens dates back to the nineteenth century, but in African colonial cities, vegetable patches had their roots in ancient communal practices, and in Mexico the farming system known as chinampas predates the arrival of Columbus (Smit et al. 2001). Indeed, it is unclear if urban agriculture was developed by the first urban settlers as a systematic source of food for the local community, or if it resulted from incremental transformations in food-producing practices as urbanization took form (Smit et al. 2001).

While congestion, compaction and even conurbation may be considered as key features of contemporary cities, these are by no means ordered. Indeed, landscape architects Rob Aben and Saskia DeWit consider objectness to be another key characteristic, whereby cities, suburbs, agriculture etc., are juxtaposed and separated by residual space and tied together by infrastructure (Aben and De Wit 1999). Urban agriculture often occupies the thresholds of these unordered objects, calling out attention to the space that knits together these juxtaposed objects, showing that borderlines have a spatiality and area capable of welcoming a life of their own (Kullman 2017).

Urban gardening is a pivotal form of making connections to nature, establishing a relationship where humans directly transforming nature, refashioning, regulating and denaturalising nature to suit specific needs (Hardman and Larkham 2014). Often, the creation of urban agricultural gardens implies an extensive work of adapting cultivations to the size and form of available space, which in many cases is not sympathetic to such use (Costa et al. 2016). Indeed, within a certain perspective, urban agriculture can be the best image to demonstrate that there is no 'natural' landscape, but rather natural spaces that are transformed by human labour and social history (Telles 1997).

Since nature in its raw form can be perceived as uncivilised, wild, and untamed, in need of mastering, the control over nature can be seen everywhere in cities, from the trimming of trees to the maintenance of street lawns (Hardman and Larkham 2014). While many of these tasks are often performed by municipal workers charged with the treatment of landscape elements, urban agriculture often presents itself as a form of landscape transformation for which the required tasks are performed by local people themselves, who take the leading role (Cabannes and Raposo 2013; Costa et al. 2016).

Indeed, countering the acceleration of urban activities, gardens may be seen as 'oases of peace' (Aben and DeWit 1999, 12) and urban agriculture in particular questions this acceleration and even growth (in the economic sense) by physically occupying the spaces urbanization is supposed to grow towards. Moreover, urban agriculture proposes an alternative use for urban space which, however insignificant from an economic perspective, constitutes a great contribution towards urban sustainability, far superseding the eventual 'green' technologies that seem to offer further urbanization without further environmental degradation (Marat-Mendes 2020). Such advantages include the promotion of biodiversity and links between city and countryside, a contribution towards healthier lifestyles and the provision of recreating spaces also appropriate for sustainable mobility—like pedestrian and cycling corridors—also physically linking different (urban) areas within and outside the city (Costa et al. 2016).

Considering the sustainability potentials of urban agriculture, a key aspect of its defence by either scholars and decision-makers is that it promotes the consumption of local food. However, this notion of local food has also been noticed critically, since in many cases, the environmental or economic cost of local production can indeed supersede the disadvantages of bringing it from the outside (Hardman and Larkham 2014). Another aspect that must always be considered, although it remains obvious, is the sheer impossibility of feeding a metropolitan city—or even a metropolitan region—solely through produce from urban agriculture, whatever its scale. The 'Urban Food Agenda' (FAO 2019) calls for articulations between local and regional production with economies of scale, a solution which obviously contains a compromise with the existing system, but which remains realistic in face of the exponential growth of urban areas, expected to increase in the following decades. Moreover, from the perspective of urban planning, urban agriculture has the potential to propose alternative to existing land-use schemes, legitimizing existing de facto uses of the territory, to promote sustainable and long-term local solutions by empowering communities, to generate jobs and increase income and to strengthen multi-level governance through city participation and collaboration (Delgado 2018).

The FAO, in association with other international organisms, created the City Region Food System Program (CRFS) to analyze specific urban food systems and highlight the role of civil society in them. Through specialized teams in loco, data was collected to allow the observation of the most relevant initiatives carried on between 2014 in Toronto (Canada), Utrecht (The Netherlands), Colombo (Sri Lanka), Lusaca (Zambia), Kitwe (Zambia), Medellin (Colombia) and Quito (Ecuador). The SPLACH Project (Marat-Mendes et al. 2021e) analysed the results of FAO reports and grouped the key parameters in five areas, namely, local food production, food legislation, food waste management, health and food habits and food safety. Interestingly, urban agriculture can be integrated with several of these FAO parameters, or even all of them, depending on the type of cultivation practised.

References

Aben R, de Wit (1999) The enclosed garden. History and development of the Hortus Conclusus and its reintroduction into the present-day urban landscape. Nai010 publishers, Amsterdam

Abusaada H, Elshater A (2020) COVID-19's challenges to urbanism: social distancing and the phenomenon of boredom in urban spaces. J Urban. https://doi.org/10.1080/17549175.2020.1842484

Ashe LM, Sonnino R (2013) At the crossroads: new paradigms of food security, public health nutrition and school food. Public Health Nutr 16(6):1020–1027

Barthes R (1971 [2009]) Sade Fourier Loyla. Points, Paris

Cabannes Y, Raposo I (2013) Peri-urban agriculture, social inclusion of migrant population and Right to the City—Practices in Lisbon and London. City 17(29):235–250. https://doi.org/10.1080/13604813.2013.765652

Costa S, Fox-Kämper R, GoodR, Sentić I, Treija S, Atanasovska JR, Bonnavaud H (2016) The position of urban allotment gardens within the urban fabric. In: Bell S, Fox-Kämper R, Keshavarz N, Benson M, Caputo S, Noori S, Voigt A (eds) Urban allotment gardens in Europe. Routledge, London, pp 201–228

Deakin M, Diamantini D, Borrelli N (eds) (2016) The governance of city food systems: case studies from around the world. Fondazione Giangiacomo Feltrine, Milan

Delgado C (2018) Contrasting practices and perceptions of urban agriculture in Portugal. Int J Urban Sustain Develop 10:170–185

Dias AM, Marat-Mendes T (2020) The morphological impact of municipal planning instruments on urban agriculture. Cidades, Comunidades & Territórios 41. http://journals.openedition.org/cidades/2991

Dias AM (2018) The shape of food: an analysis of urban agricultural shapes in Lisbon's Greater Area. M.Sc. dissertation, Iscte—Instituto Universitário de Lisboa, Lisbon. https://hdl.handle.net/10071/19300

Drescher A, Holmer R, Iaquinta D (2006) Urban homegardens and allotment gardens for sustainable livelihoods: management strategies and institutional environments. Tropical Homegardens, Springer, The Netherlands, pp 317–338

Edwards F, Mercer D (2010) Meals in Metropolis: mapping the urban foodscape in Melbourne, Australia. Local Environ 15:153–168. https://doi.org/10.1080/13549830903527662

FAO (Food and agriculture organization) (2019) FAO framework for the Urban Food Agenda. FAO, Rome. https://doi.org/10.4060/ca3151en

Flavell N (2003) Urban allotment gardens in the eighteenth century: the case of sheffield. Agric History Rev 51(1):95–106. http://www.jstor.org/stable/40275844

Hardman M, Larkham PJ (2014) Informal urban agriculture. The secret lives of guerrilla gardeners. Springer, London

Hodgson K, Campbell MC, Bailkey M (2011) Urban Agriculture: growing healthy, sustainable places. Report Number 563. American Planning Association, Chicago

Honey-Rosés J, Anguelovski I, Chireh C, Daher VK, Van den Bosch CK, Litt JS, Mawani V, McCall MK, Orellana A, Oscilowicz E, Sánchezv U, Senbel M, Tan X, Villagomez E, Zapata O,

Nieuwenhuijsen MJ (2020) The impact of COVID-19 on public space: an early review of the emerging questions. Cities Health. https://doi.org/10.1080/23748834.2020.1780074

Howard E (1902) Garden cities of tomorrow. Swan Sonnenschein & Co, London

Iaquinta D, Drescher A (2010) Urban agriculture: a comparative review of allotment and community gardens. In: Aitkenhead-Peterson J, Volder A (eds) Urban ecosystem ecology. American Agronomy Society, USA, pp 199–226

Ibañez D, Katsikis N (eds) (2014) Grounding metabolism. New geographies 6. Harvard University, Graduate School of Design, Harvard, Cambridge, Massachusetts

Koc MR, MacRaeR DE, Roberts W (2008) Getting civil about food: the interactions between civil society and the state to advance sustainable food systems in Canada. J Hunger Environ Nutr 3(2–3):122–144. https://doi.org/10.1080/19320240802243175

Komisar J, Nasr J, Gorgolewski M (2009) Designing for food and agriculture: recent explorations at Ryerson University. Open House Int 34(2):61–70. https://doi.org/10.1108/OHI-02-2009-B0007

Kordshakeri P, Fazeli E (2020) How the COVID-19 pandemic highlights the lack of accessible public spaces in Tehran. Cities Health. https://doi.org/10.1080/23748834.2020.1817690

Kullmann K (2017) Concave Worlds, Artificial Horizons: Reframing the urban public garden. Stud History Gardens Designed Landscapes 37(1):15–32

Lôbo MS (1995) Planos de urbanização – a época de Duarte Pacheco. FAUP Edições, Porto

Marat-Mendes T (2002) The sustainable urban form. A comparative study in Lisbon, Edinburgh and Barcelona. Ph.D. Thesis, University of Nottingham, Nottingham UK

Marat-Mendes T (2020) Designing for sustainability: retrieving a systemic role for urban form. Urban Morphol 24(2):235–238

Marat-Mendes T, Borges JC, Dias A, Lopes R (2021) Planning for a sustainable food system. The potential role of urban agriculture in Lisbon Metropolitan Area. Journal of Urbanism: International Research on Placemaking and Urban Sustainability 14(3):356–386. https://doi.org/10.1080/17549175.2021.1880960

Marat-Mendes T, Henriques JM, Perestrelo M, Pinto TC, Costa P, Pereira MM, Borges JC, Lopes SS, Henriques CN (2021e) Sistema Alimentar e Sustentabilidade: aferição de oportunidades e iniciativas à escala local. In: Pinho (ed) Compêndio de Políticas Urbanas. FEUP-ISCTE-UA, Porto

Marot S (2019) Taking the Country's side: agriculture and architecture. Polígrafa & Trienal de Arquitectura de Lisboa, Lisbon

Mougeot J (2000) Urban agriculture: definition, presence, potentials and risks. Growing cities, growing food: Urban agriculture on the policy agenda. (Thematic Paper 1, RUAF), pp 1–42

MUFPP (Milan Urban Food Policy Pact) (2015) Milan urban food policy pact. Accessed November 24. http://www.milanurbanfoodpolicypact.org/wp-content/uploads/2016/06/Milan-Urban-Food-Policy-Pact-EN.pdf

Napawan NC (2016) Complexity in urban agriculture: the role of landscape typologies in promoting urban agriculture's growth. J Urban Int Res Placemaking Urban Sustain 9(1):19–38. https://doi.org/10.1080/17549175.2014.950317

Orsini F, Dubbeling M, de Zeeuw H, Gianquinto G (eds) (2017) Rooftop urban agriculture. Springer, Switzerland

Parham S (2015) Food and urbanism—the convivial city and a sustainable future. Bloomsbury, London

Pothukuchi K, Kaufman JL (2000) The food system. J Am Plann Assoc 66(2):113–124. https://doi.org/10.1080/01944360008976093

Redol A (1971) Gaibéus. Europa-América, Amadora

Rodriguez O (2016) Towards an indigenous, sunlit, rooftop food production. Urban Des (Food City) 140:24–27

Rogers R (1998) Cities for a small planet. Basic Books, London

Salvador MS (2019) Shaping the city through food: the historic foodscape of Lisbon as case study. Urban Des Int 24:80–93

Smit J, Nasr J, Ratta A (2001[1996]) Urban agriculture: food, jobs and sustainable cities. UNDP, New York

Steel C (2008[2013]) Hungry city—how food shapes our lives. Vintage, London

Telles GR (1997) Plano Verde de Lisboa. Colibri, Lisboa

UN (United Nations) (2015) Transforming our world: the 2030 agenda for sustainable development. United Nations, New York

UN (United Nations) (2017) New Urban Agenda—Habitat III. United Nations, New York

UN (United Nations) (1987) Our common future. http://www.un-documents.net/our-common-future.pdf

Weaver C (1984) Regional development and the local community. Wiley, New York

Weber M (1966) The city. Free Press, New York

8 A Survey of Urban Agriculture in the Lisbon Region

8.1 Introduction

In Portugal, within the democratic planning scheme, as seen in Chap. 3, vacant areas within urban perimeters may integrate the Municipal Ecological Structure, but they are not acknowledged their agricultural potential. The same conceptual understanding excludes vacant urban land from agricultural policies, including with regards to their integration in the RAN.

Throughout the twentieth century, and even during the democratic era, when spatial planning started to apply to the totality of the territory, the spatial planning policies rest upon two important dichotomies. The first is the 'urban', for consumption, versus the 'rural', for production. The second is the schism between 'agricultural spaces', located outside urban settlements, with restricted building regulations to safeguard food production, versus 'natural spaces or with cultural and landscape value', which exist in both urban and rural spaces, and legally safeguarded not for food production, but for the sustainability of territorial ecological structures (Marat-Mendes et al. 2021a, b).

The dichotomy of ecological versus agricultural relevance reflects another expressive difference: while agricultural and forestry areas correspond to private land, natural spaces located in urban settlements are usually of public domain, and thus often utilized as parks for social, recreational and environmental purposes. Only recently has such perception somehow changed, with urban parks finally being considered in a more multifunctional perspective and urban vegetable gardens gaining greater visibility in public programmes and occasionally (at least in Portugal) in planning instruments. However, even when PDMs acknowledge the existence of urban agriculture this is most often subordinated to socio-environmental functions. Horticulture is framed as a form of leisure, with the food production potential being, either practically or symbolically, undermined.

In spite of instruments for spatial planning, in the territory of the Lisbon Region one is often confronted with forms of land-use that do not conform with official stipulations, as the level of non-compliance is very considerable in the region as a whole (Abrantes et al. 2017) and even from a historical perspective, an important part of current urbanized areas underwent such process illegally (Pinto and Guerra 2019).

Several studies conducted in the past clarify not only that urban agriculture has been practiced in the Lisbon Region and in the Lisbon city for decades now (Castel-Branco et al. 1985; Cabannes and Raposo 2013; Costa et al. 2016; Harper and Afonso 2016; Delgado 2017, 2018; Messina and Mourato 2018; Dias 2018; Marat-Mendes and Borges 2019b; Dias and Marat-Mendes 2020; Marat-Mendes et al. 2021a, b).

The presence of this practice is naturally more visible in urban and periurban areas, and its origin is often informal, while the practice of municipalization has become more and more expressive in the last decade. The benefits, in terms of domestic economy, social cohesion and health have justified its blossoming and its acceptance, overturning existing preconceived notions installed in the local culture (Paizinho and Ferreiro 2017; Hespanhol 2019).

The case of urban agriculture also represents an important example of non-compliance with official land-uses within the contemporary context, since they do not merely contradict official land-uses, but most importantly they challenge the key assumption that urban land excludes is distinct from rural land insofar as it does not include food production as an activity (Weaver 1984). Moreover, while the spree of illegal construction raised problems regarding infrastructural integrity and a unbalanced pattern of suburbanization, often resulting in conurbation, with urban agriculture the case may be the opposite. In principle, and even considering the relatively low scale of development and of economic organization in most urban agriculture initiatives in Portugal and specifically in the Lisbon Region (Delgado 2017), urban agriculture, both official and grassroots, is often countering an excessive extension of urbanized land, especially in areas of high-density development.

While the practice of urban agriculture has been noticed for long, it is only recently that it gained official recognition from municipal authorities, perhaps accompanying an international

T. Marat-Mendes et al., *Atlas of the Food System*,
https://doi.org/10.1007/978-3-030-94833-7_8

tendency to recognize this practice as a legitimate, even desirable one. While for a long time, urban agriculture in the Lisbon Region was chiefly associated with peripheral neighbourhoods, often with low-income populations, in the aftermath of the 2007 Financial Crisis, interest in it blew up, including among the urban middle-class (Cabannes and Raposo 2013; Costa et al. 2016; Delgado 2017).

Even considering this rise in interest, urban agriculture cannot be considered a key aspect of the Lisbon Region food system. Since nearly a third of the Portuguese population currently lives in these 18 municipalities, it would be difficult for urban agriculture to attain such importance. However, it represents a fundamental practice when considering what changes can be pursued towards sustainability. If the ability to scale-up positive practices is key to sustainable territorial transitions, then urban agriculture must surely be assessed, since its scaling-up would have at least two advantages: first, the retention of more space towards food production instead of further soil artificialization and second, a positive practice of use of vacant space with benefits for local residents, even those who do not partake themselves in cultivation practices.

8.2 The Survey and Its Methodology

Between 2019 and 2021, a survey on urban agriculture was conducted by the SPLACH Project. It surveyed 315 examples of urban agricultural spaces from all 18 municipalities of the Lisbon Region, collecting examples with different forms of management (informal, municipal, collective, associative), and also of their form and their usage (size, shape, circulation spaces) and only private kitchen gardens were excluded (Marat-Mendes et al. 2021b). Some municipal councils already acknowledge urban agriculture as a legitimate land-use and have included specific land-use categories in urban space when revising their PDMs. Yet urban agriculture can be verified in all the remaining ones, including those that have not amended their PDMs—leaving the question of whether it will be included when the upcoming revision arrives.

The methodology for this survey on urban agriculture was divided into three phases: survey, mapping and analysis (Fig. 8.1).

The first phase consisted in identifying the typical patch of urban agriculture in online mapping instruments (Google Maps and Google Earth), and additionally enlisting all cases known to the researchers, all of whom reside in the Lisbon Region. Information and data made available by municipal offices was also added to this identification. All possible urban agriculture examples were collected, regardless of their background or legal status (Fig. 8.2).

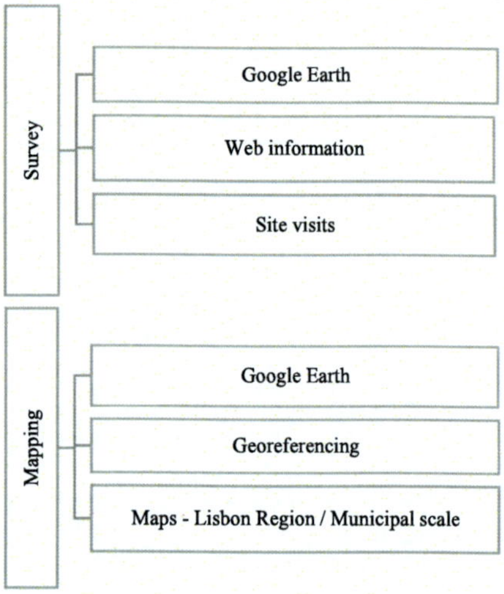

Fig. 8.1 Survey methodologies. Source: Authors

From this largest inventory (Fig. 8.3), 109 were selected for in-loco visit. These visits were documented through photographic survey and the design of two-dimensional (2D) schemes of the morphological features of the vegetable garden, including its plot division and level of use. For completing the mapping phase, information was collected in KMZ format georeferencing, allowing for specifying the location and the outline shape of each urban vegetable garden. Maps were made for each municipality, further including a characterization index card for visited cases. An upcoming publication by the SPLACH Project will make the full results of this survey available to the public, exposing with more detail the particularities of visited vegetable gardens (Marat-Mendes et al. forthcoming).

Finally, a portion of territory was selected by each of the 18 municipalities representing the diversity of vegetable gardens identified. Taken as case studies, these samples allow an assessment of urban agriculture in Lisbon Region.

To ensure the plausibility of any comparisons between different examples of urban agriculture, each case studies was studied for criteria other than spatial, namely regarding:

Land-use in the active PDM—most cases, but not all, are located on land classified as urban or for urbanization;

Management form—several types of management were encountered, and we sought to include all forms in the set of case studies;

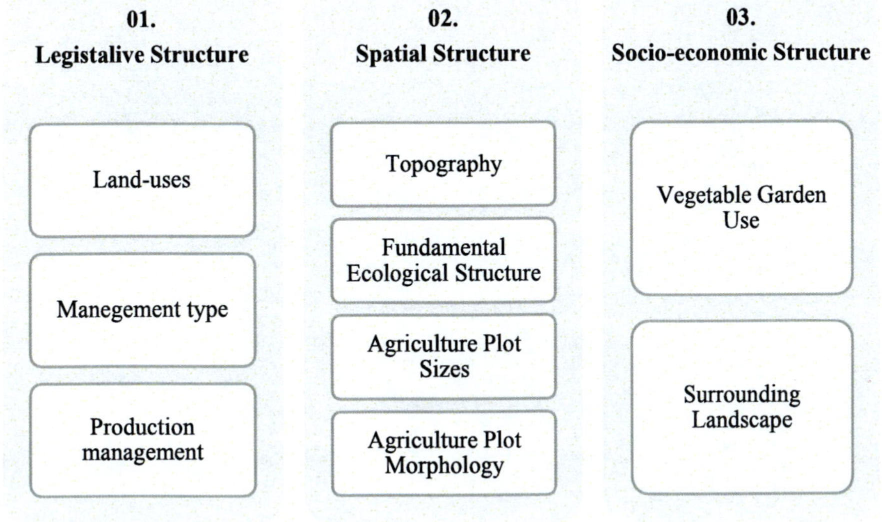

Fig. 8.2 Indicators of the structures. Source: Authors

Surrounding areas—the characteristics of inception of vegetable gardens, accounting not only for sui-generis cases but also for recurrent situations, such as location on a fringe-belt area or close to social housing estates;
Exceptions—in municipalities with low number of examples, at least the two first criteria are met.

These criteria demonstrate the diversity of physical and management forms which urban agriculture assumes in the Lisbon Region. Only by assessing the extent and the physical characteristics of such existing practices can urban agriculture be better understood and integrated into urban planning or design solutions, either for improving or creating new vegetable gardens adequate to the specific needs of the local community.

We interpret the urban vegetable garden as a manifestation of human action which explicitly implies the design of productive spaces. That these may play a key role in a sustainability strategy that follows some of the cornerstones of the 'Urban Food Agenda' (FAO 2019) provides a unique opportunity for their acknowledgement as legitimate and desirable elements of the contemporary city.

From the urban agriculture survey, beyond an inventory of 315 examples, different vegetable gardens, resulted a systematized summary of the recurrent typologies of vegetable gardens in the Lisbon Region, while reflecting upon the combination of legislative and spatial factors, as well as with the local context, topography and vegetation. Elements from this survey have informed several publications from the SPLACH Project (Marat-Mendes et al. 2020, 2021a, b).

Our analysis made it possible to identify the types of vegetable gardens practiced in the 18 municipalities, verifying which types recur the most and to observe their patterns of adaptation to conditions specific to their site. Furthermore, we could assess the legislative and management structures that support the identified spatial structure. The following scheme identifies the indicators of the structures analysed and defines the sub-indicators and recognizes the typologies observed (Tables 8.1, 8.2 and 8.3, Figs. 8.2, 8.3, 8.4 and 8.5).

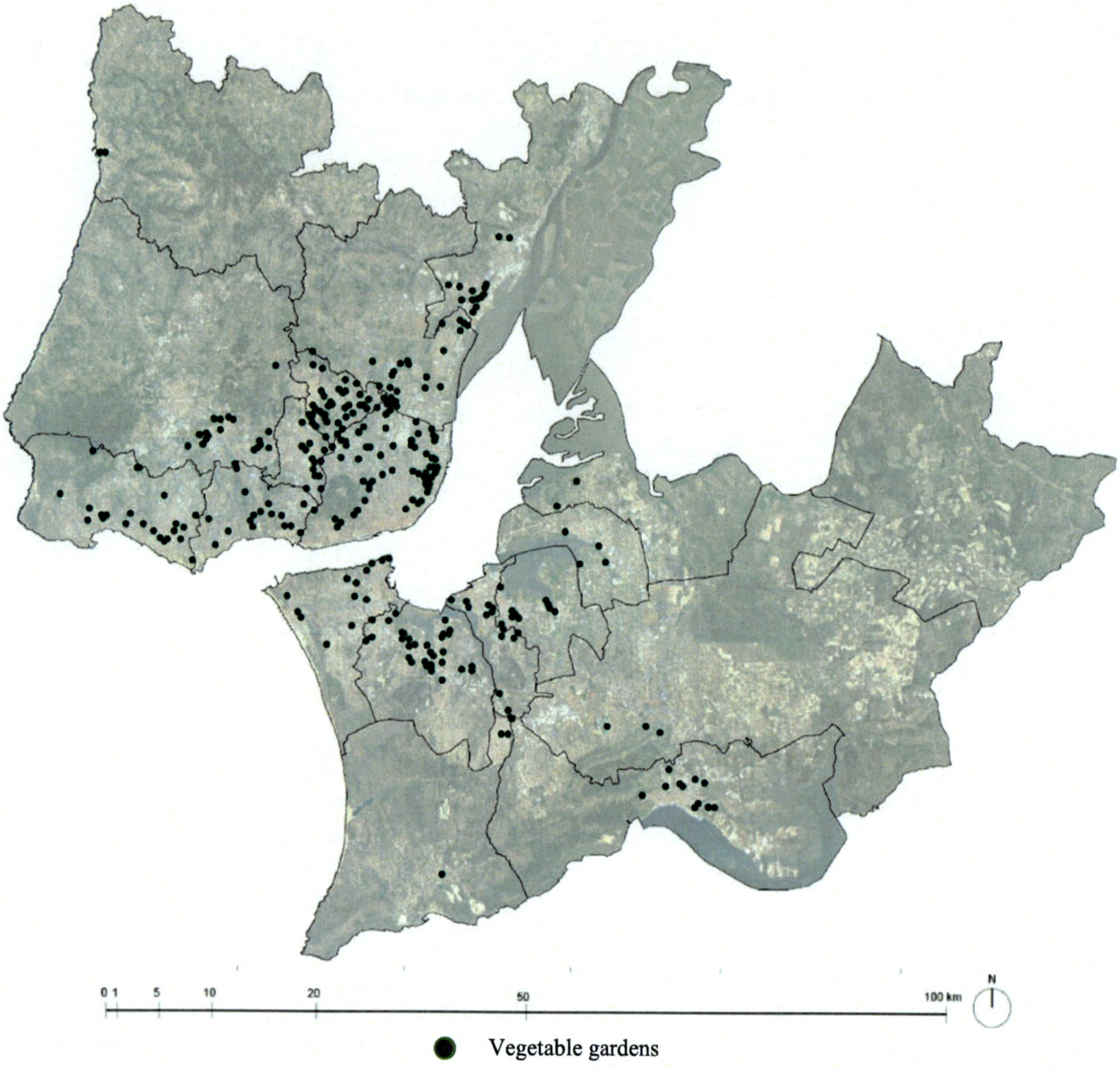

Fig. 8.3 Location of urban agriculture examples. Source: Authors

Table 8.1 Quantities of urban agriculture examples

Municipalities	a	b	c	d	e	f	g	h	i	j	k	l	m	n	o	p	q	r
Total	2	13	13	13	24	74	20	2	8	4	33	23	3	34	3	13	20	13

a) Alcochete
b) Almada
c) Amadora
d) Barreiro
e) Cascais
f) Lisboa
g) Loures
h) Mafra
i) Moita
j) Montijo
k) Odivelas
l) Oeiras
m) Palmela
n) Seixal
o) Sesimbra
p) Setúbal
q) Sintra
r) Vila Franca de Xira

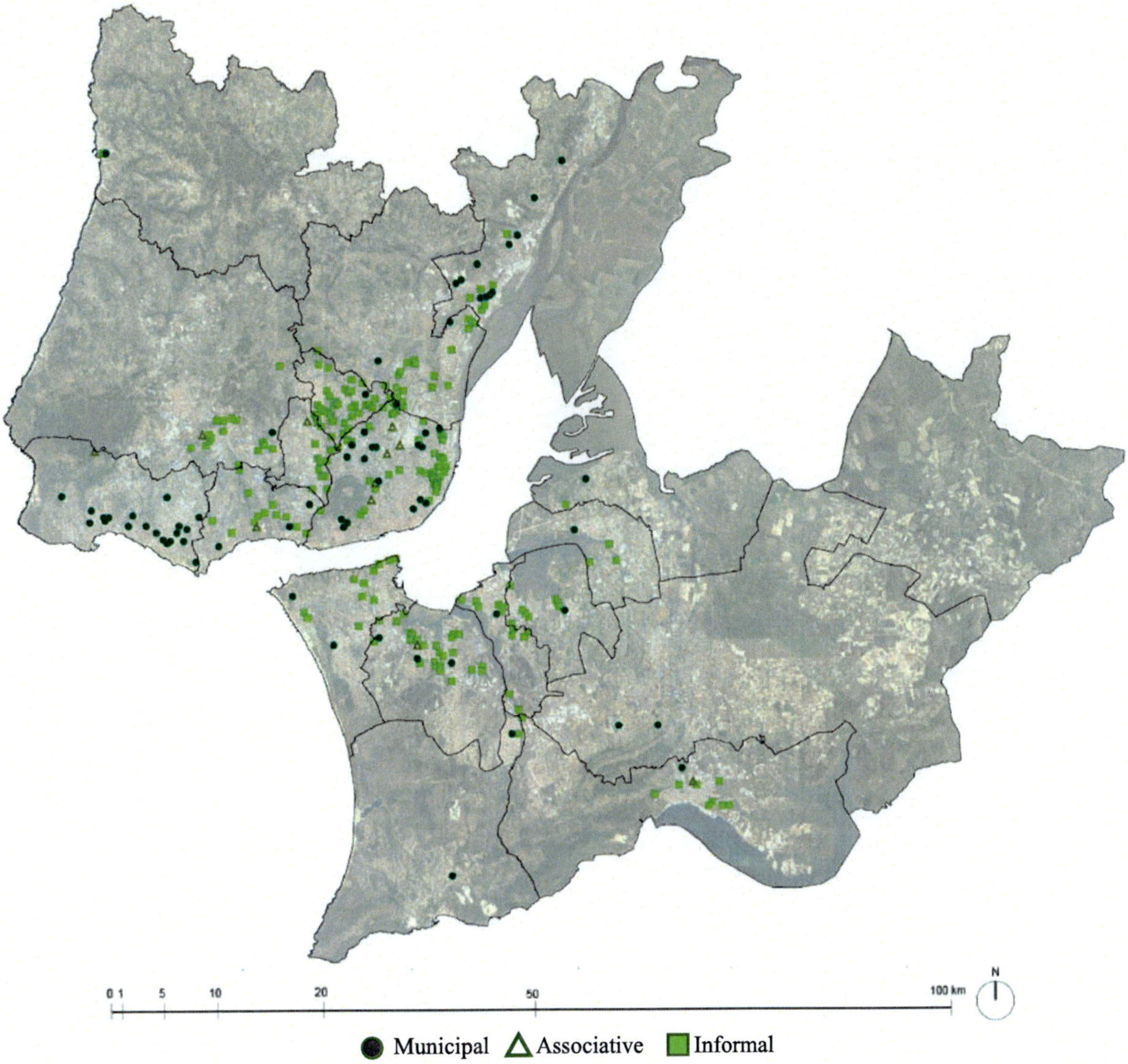

Fig. 8.4 Vegetable Gardens by Management type. Source: Authors

Table 8.2 Quantities of vegetable Gardens by Management type

Municipalities	a	b	c	d	e	f	g	h	i	j	k	l	m	n	o	p	q	r
Formal	1	2	2	0	22	21	2	1	2	1	2	3	2	5	2	3	4	5
Associative	0	0	1	3	1	4	0	0	0	0	2	0	1	0	0	0	0	0
Informal	1	11	10	10	1	494	18	1	6	3	31	17	2	28	1	10	16	8

a) Alcochete
b) Almada
c) Amadora
d) Barreiro
e) Cascais
f) Lisboa
g) Loures
h) Mafra
i) Moita
j) Montijo
k) Odivelas
l) Oeiras
m) Palmela
n) Seixal
o) Sesimbra
p) Setúbal
q) Sintra
r) Vila Franca de Xira

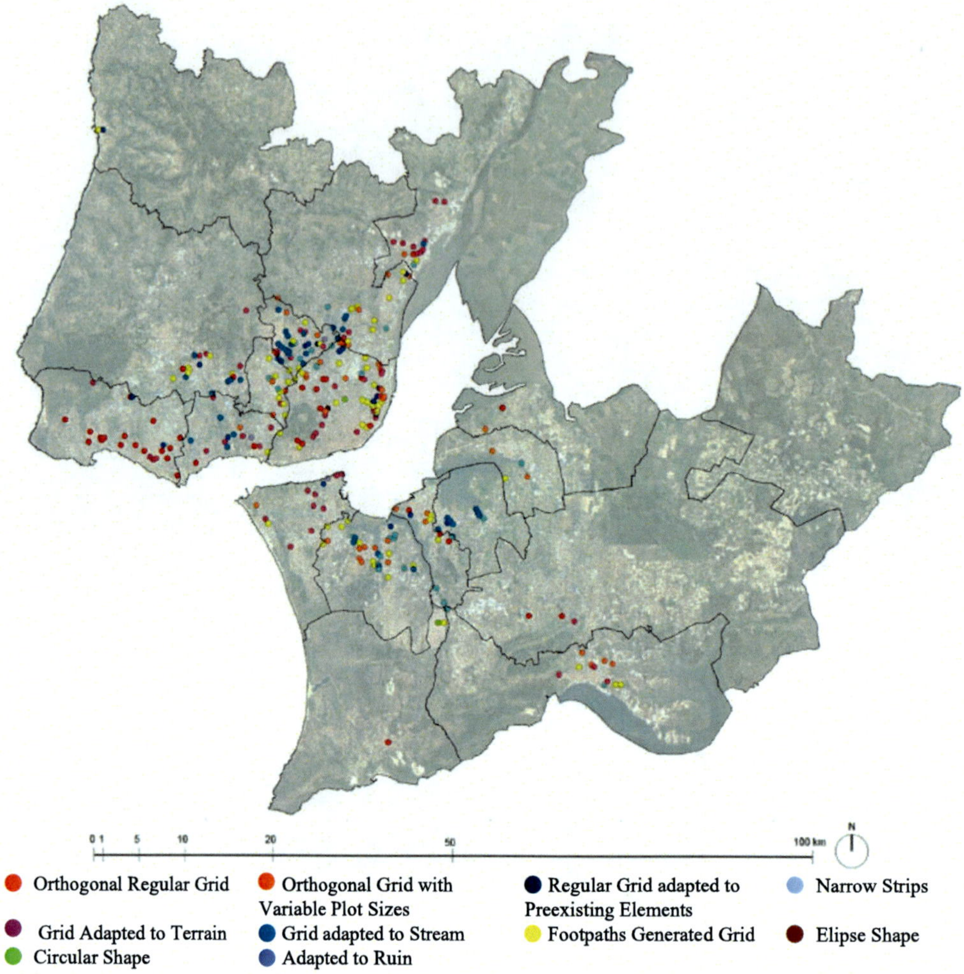

Fig. 8.5 Vegetable Gardens by Typo-morphologies. Source: Authors

Table 8.3 Quantities of vegetable Gardens by Typo-morphologies

Municipalities	a	b	c	d	e	f	g	h	i	j	k	l	m	n	o	p	q	r
	1	2	1	1	19	12	0	0	0	1	0	1	2	0	1	1	1	2
	1	0	0	6	0	10	5	0	0	1	4	5	0	8	0	3	1	1
	0	5	1	0	3	16	2	0	0	0	11	5	1	1	0	4	4	8
	0	0	2	0	0	0	2	0	6	0	9	8	0	6	0	0	7	0
	0	2	0	1	2	10	1	1	1	0	2	1	0	1	0	0	1	0
	0	4	7	2	0	20	8	1	0	1	4	2	0	11	1	3	6	1
	0	0	2	3	0	4	2	0	1	1	2	0	0	7	0	2	0	1
	0	0	0	0	0	1	0	0	0	0	0	0	0	0	0	0	0	0
	0	0	0	0	0	1	0	0	0	0	0	0	0	0	1	0	0	0
	0	0	0	0	0	0	0	0	0	0	1	1	0	0	0	0	0	0

a) Alcochete
b) Almada
c) Amadora
d) Barreiro
e) Cascais
f) Lisboa
g) Loures
h) Mafra
i) Moita
j) Montijo
k) Odivelas
l) Oeiras
m) Palmela
n) Seixal
o) Sesimbra
p) Setúbal
q) Sintra
r) Vila Franca de Xira

8.3 Results

Urban agricultural spaces are naturally concentrated along the more continuously urbanized areas of the Lisbon Region, and especially those from the Northern Bank of Tagus. A cluster can be found across the territory of Costa do Sol, extending from Monsanto in Lisbon through Algés towards Cascais. A second cluster contours the Lisbon northwest borderline, extending through Amadora towards Sintra. The borderlines of Lisbon with Amadora, Odivelas and Loures also present a significative amount. Chelas and the eastern end of the Lisbon city represent an important cluster, which will be continued through the riverside of Loures and Vila Franca de Xira. On the Southern Bank of Tagus, the largest concentration runs between the eastern end of Almada and all the riverside areas of Seixal.

Unsurprisingly, municipalities with large rural zones like Mafra, Palmela, Montijo, Alcochete and Sesimbra display less examples of urban agriculture. This is probably due to the greater availability of land to cultivate, as well as to the characteristics of housing, which often include outside areas. A somewhat different situation is that of Sintra, Loures and Vila Franca de Xira, which despite belonging to the 'Saloia' Region and being indeed municipalities where rural territory is extensive, also have settlements with high residential density—it is in many of these that vegetable gardens sprout.

On the other hand, it must be noted that in the case of Vila Franca de Xira and, to a lesser extent, the Loures riverside, urban agriculture is intertwined with other forms of periurban agriculture, including farms of different dimensions and olive groves.

There is a clear disproportion between the type of management and its territorial distribution. The number of informal gardens is significantly higher than municipal ones, with associative gardens being, in comparison, a relatively residual case. Cascais is a notable exception, having municipalized all urban agricultural spaces in the municipality or replacing them with alternatives. Indeed, our survey found no examples of informal or associative urban agriculture, but instead a network of diversified and well-treated urban agricultural gardens, often functioning as effective public spaces in residential neighbourhoods, especially those with multi-family buildings. A program for promoting urban agriculture in the municipality was launched in 2009 by the Council, with an administration elected by the PSD and which has included other forms of food production, for instance, in balconies (Dias 2018).

In Palmela and Sesimbra, all vegetable gardens encountered are municipal, although this refers, respectively, to two and one examples. Mafra only has two examples, both of them located in Ericeira, a coastal town which has experienced great urban growth due to its touristic advantages. Alcochete has also two examples, located in the outskirts of the salt pans area between the main town and Samouco. On both cases, one of the examples is municipal, the other informal.

Vila Franca de Xira, despite the great availability of rural land, has many examples of vegetable gardens in its urban cores, and a significant amount has been municipalized, although at Póvoa de Santa Iria and Alverca, informal examples subsist.

Sintra, Odivelas, Amadora and Loures are municipalities where urban agriculture seems to have found little acceptance, with the overwhelming majority of existing examples being informal. In Almada, Seixal, Barreiro, Moita and Setúbal, the majority of examples is of an informal nature. However, Seixal is promoting a programme for urban agriculture where this activity is promoted as leisure and as food production, to each corresponding different physical conditions, while in Setúbal, the most recent PDM (analysed in Chap. 3) acknowledges and promotes the practice of urban and periurban agriculture.

Regarding the typology, 'Footpaths generated grid', where the circulation trails do not necessarily follow topographical conditions, but still determine the limits of different vegetable gardens, this is one of the most common, being a dominant type in Lisbon, Amadora, Loures and Seixal, and having a strong presence in Sintra. In Mafra, this is the typology of one of the two existing urban agricultural spaces —the other is a set adapted to pre-existing elements.

Grids adapted to a water stream are also a relevant typology, the most common in Oeiras, Sintra, Odivelas and Moita, and being relevant too in Lisbon.

Orthogonal grids are the preferred typology in Cascais, which is not surprising, considering that this is the preferred typology in municipal horticultural parks, followed in many examples where a process of officialization was promoted (Dias 2018; Marat-Mendes et al. 2021a, b). In Sesimbra, the only existing horticultural park has an orthogonal grid, as have two of the three existing in Palmela—the other has a grid adapted to the terrain. In Montijo and Setúbal is predominant a variation of the orthogonal grid, whereby the individual horticultural plots are of different dimensions. The two urban agriculture spaces in Alcochete are in orthogonal grid, one regular, the other with varying sizes.

Barreiro constitutes a very particular case, because while it does not have a very large sample of examples, these show an even distribution of typologies, with footpaths generated grids (two examples), orthogonal grids (two examples), orthogonal grids with different dimensions (two examples) and another, rarer, typology, where cultivation spaces are developed in long narrow strips (two examples) (Figs. 8.6, 8.7, 8.8, 8.9, 8.10, 8.11, 8.12, 8.13 and 8.14).

Fig. 8.6 Green house in informal urban agriculture in Alto do Zambujal, Amadora. Source: Authors

Fig. 8.7 Improvised walls in informal urban agriculture in Alto do Zambujal, Amadora. Source: Authors

Beyond the morphological aspect, with informal examples presenting more variety than municipal horticultural parks, further differences regard the types of materials used for structuring cultivation space, the secondary uses the vegetable garden shelters and the presence of fruit trees. Fruit trees are excluded from most municipal gardens, since the relatively small size of each allotment means the trees may cast shade upon the contiguous allotments. However, in many instances, there are zones within horticultural parks with some fruit trees, usually belonging to the whole

Fig. 8.8 Informal urban agriculture in Tapada do Mocho, Paço de Arcos (Oeiras). Source: Authors

Fig. 8.9 Informal urban agriculture in Arrentela, Seixal. Source: Authors

Fig. 8.10 Informal urban agriculture in Queluz, Sintra. Source: Authors

Fig. 8.11 Informal urban agriculture in Odivelas. Source: Authors

park—this happens frequently in Seixal, Oeiras and Cascais, for instance. Flowers are usually not allowed on municipalized examples (except for the limiting ridges), but in informal and associative examples, they are very frequent. Another example is the creation of improvised greenhouses, which are frequent in informal and associative examples, but non-existent in municipal ones.

In informal gardens, improvised sheds are often found, both to keep produce and tools, but also with tables and chairs, probably meant for meals and social conviviality

Fig. 8.12 Informal allotment gardens in Quinta do Conde, Sesimbra. Source: Authors

Fig. 8.13 Flowers in informal allotment gardens in Quinta do Conde, Sesimbra. Source: Authors

Fig. 8.14 Informal community gardens in Quinta do Conde, Sesimbra. Source: Authors

Fig. 8.15 Municipal horticultural park in Espargal, Oeiras. Source: Authors

(Fig. 8.9). This is most likely due to the greater liberty horticultors have in informal settings, which allow them to decide on an individual basis what they want to have within the confines of their vegetable garden, but also to the availability of more space. However, some municipal examples already include convivial areas, notably in Cascais, Almada or Seixal, where barbecues and meals prepared with produce from the vegetable gardens are often promoted.

Finally, a key aspect of official municipalization is that the materials used for support equipment and division of allotments within a park are standardized and limited, whereas in the informal examples, there is a predominance of recycled materials, often pieces of home furniture, as well as fridges and bathtubs used for harvesting water. In municipalized horticultural parks, net fences with wooden structures are usually preferred—often disappearing under hedgerow—and water is provided by the municipalities, thus

Fig. 8.16 Municipal horticultural park nearby an old windmill in Espargal, Oeiras. Source: Authors

eliminating the improvised elements for its harvest. While access to water is indeed a fundamental aspect for improving the conditions under which agriculture is practiced, and while arguably fridges and bathtubs may not be very visually appealing, structures for harvesting rain water are not frequent in municipal examples.

Only in a few examples—the Yellow Estate in Almada, or in Arrentela, Seixal—did we find livestock, usually sheep. However, vegetable gardens of all kinds are clearly attractive to other animals, from bees and birds to dogs and mostly cats. In some horticultural parks, cat houses (wood large boxes with entrances and plates for pet food) are being introduced (Figs. 8.15, 8.16, 8.17, 8.18 and 8.19).

The survey conducted in 2019–2021 obviously does not exhaust the practice of urban agriculture in the Lisbon Region.[1] First, because many examples still escaped our scanning of the territory through Google Earth (whose images are captured at different times) and *in loco* visits, mostly by walking and by public transportation. Second, as the several studies conducted since the 1980s demonstrate, this is a phenomenon that is naturally defined by its adaptability and malleability, with gardens being abandoned at the same time others are coming to existence, due to climatic or financial conditions, or because their site has been redeveloped.

As is often the case with urban agriculture, the disappearance of vegetable gardens does not result from abandonment in a proper sense, but rather by the occupation of the land, usually with a different function than that which is officially assigned.[2] Since this is a problem that afflicts many examples of urban agriculture, its resolution is key to the maintenance of this activity.

One important solution would be the inclusion of these spaces in the Municipal Ecological Structure, which includes spaces whose biophysical or cultural features make them important for the protection and conservation of the environment of spaces both urban and rural.[3] Specifically, in urban land, this includes Urban Green Spaces, areas for environmental balance, for landscape protection and leisure.

This is another example where Portuguese legislation follows the ideas of landscape architect (and former Minister) Gonçalo Ribeiro Telles, who since the 1980s developed several studies for the ecological structure of both the Lisbon city and the Northern Bank of Tagus. In the resulting 'Green Plan for Lisbon', Telles (1997) defines an Urban Green Structure, as a sequence of spaces, continuous or discontinuous, with their own identity, formed from cultural and

[1] A more personal approach, where we would conduct small interviews with horticultors was not possible either, due to the restrictions of COVID-19 which not only advised the practice of social distancing but also, for several periods of time, even prohibited circulation between different municipalities, significantly conditioning the way we conducted our in-loco visits.

[2] One of our case studies, the informal horticultural gardens in the southern area of the Chelas Valley, in Lisbon, have been destructed since we conducted our survey, to make way for the construction of the Hospital de Todos-os-Santos. In Bobadela, Loures, the happened to make way for a new urbanization.

[3] See Regulation 15/2015, published in Diário da República 161/2015 – Série I, 19–8-2015.

Fig. 8.17 Municipal agricultural Space of Monte Sião, Seixal. Source: Authors

Fig. 8.18 Hortas de São João, Almada. Source: Authors

Fig. 8.19 Quinta do Texugo Horticultural Park, Almada. Source: Authors

landscape values of natural and urban space, based on the primordial telluric values of the site. Another important notion Telles (1997) advances is that of *Continuum Naturale*, a continuous system of natural occurrences supporting wildlife and the maintenance of genetic potential and contributes to the balance and stability of the territory. This will to articulate the characteristics of the landscape with its history and its identity means accepting different types of green spaces, which often include productive gardens. Indeed, Telles (1997) includes several horticultural spaces in his 'Green Plan for Lisbon'.

Part of the 'Green Plan' was absorbed by the 2012 Lisbon PDM. In their real complexity and reach—and not as merely decorative spaces as the Lisbon PDM seems to have them—horticultural spaces in urban cores were a further opportunity to realize the importance and the role of Ecological Municipal Structures and of Urban Green Structures. Their position with relation to spaces of rural architecture also provided an opportunity to think of urban–rural links which, while suggested by the Urban Food Agenda (FAO, 2019), also may comprise Telles' *Continuum Naturale* beyond the limits of urban cores.

Urban agriculture, as it is practiced in the Lisbon Region, allows us to go across the most different types of settings, from the historical neighbourhoods of the Lisbon city, to its extensions and its metropolitan suburbs. Urban agriculture can be found at the margins of neighbourhoods, in industrial buffers, around derelict buildings and structures. It is one of the key activities in floodplains and valley bottoms, in many instances, it mediates between spaces that are more modern, more urban and rural places. Its presence, especially in the Lisbon city, is stronger the less maintained public spaces are. It has often been noted that urban agriculture may improve the food security of its inhabitants, prone to economic restraint. But the lack of municipal investment in the public spaces within social housing neighbourhoods or in the interstitial spaces between them and other neighbourhoods is a factor to be considered as well. In these cases, the residents themselves act as landscape maintainers, and do the work the council cannot or will not do.

On the other hand, and despite the lack of official recognition of urban agriculture in many municipalities—be it by excluding this land-use from the PDM or by not officialising existing spaces—our survey does demonstrate that this is an activity with a marked presence in the metropolitan landscape, that cannot simply be reduced to a leisure activity. Its relation to residential areas, but also to rural agriculture and to the spaces of other phases of the food system is necessary for their contribution to the local foodshed to be properly understood.

8.4 The Morphologies of Urban Agriculture —A Selection of Case-Studies

Portuguese legislation,[4] still active, includes a definition of typo-morphology, as being the characteristic of the urban fabric that results from the combination of urban morphology

[4] See Regulation 9/2009, published in Diário da República, 104/2009 – Série I, 29–5-2009.

and building typology. In the study of urban morphology, this concept has an uncontested centrality for understanding the structures, forms and transformations of urban fabrics. Urban morphology has until now mostly focused on the constructed elements of the landscape—namely buildings, streets and open spaces (Muratori 1959; Conzen 1960). The concept of urban morphology appears in this context as the identification of dominant typologies, especially of buildings, characteristic of specific time periods.

The study of these elements, of their relations and of their transformations across time and at several scales has been a key motivation of morphological studies, including those published in the Journal 'Urban Morphology' published by the International Seminar of Urban Form (ISUF) since 1997. Yet these studies have mostly dealt with space in its built aspect, studying the historical evolution of constructions and ultimately, the processes of expansion or compaction of settlements. More recent issues of the 'Urban Morphology' Journal have turned their attention towards urban green spaces, highlighting their importance for understanding urban growth historically (Whitehand 2019; Zhang 2019) or the relation of green spaces with fortifications in historical cities (Scitaroci and Maric 2019).

These works rightly acknowledge that unbuilt space also has a morphology, whose transformation is significant for understanding the processes of urban change as do the built areas. Urban agricultural spaces also have formal characteristics largely defined by their orientation towards food production and employ important physical elements such as fences and hedgerows, forms of water access and internal partitions that ensure a proper organization of produce. All of these elements help characterize a form of space usage that can be understood from a morphological perspective.

Many of the existing studies about urban agriculture in the Lisbon Region have focused on the location of practices, but less so with the formal characteristics of existing gardens. Here, we extract from our survey 18 examples to demonstrate the variation in spatial arrangements found on urban agricultural practices. In a sense, we try to adjust the legal definition of typo-morphology to this specific form of occupation. Our analysis thus assumed typo-morphology as the result from the combination of urban morphology and the organization of cultivations (Figs. 8.20, 8.21, 8.22, 8.23, 8.24, 8.25, 8.26, 8.27, 8.28, 8.29, 8.30, 8.31, 8.32, 8.33, 8.34, 8.35, 8.36, 8.37, 8.38 and 8.39).

Both allotment gardens and community gardens can be found across formal and informal examples. Most often, municipal regulations prohibit the use of trees—to avoid shading nearby gardens—but these are recurrent in informal examples, where the distribution of fruit trees is often in direct relation with the remainder of the space organization within the agricultural plot.

Informal examples tend to be more prone to irregularity. Plots often adapt to landscape elements (trees, water streams) and their dimensions vary, although in some cases, a generic rectangular geometry can be identified.

In the cases where municipal initiative or municipalization are verified, there is an overwhelming tendency towards the rectangular grid, albeit with varying degrees of regularity. While there is a certain ready-made aspect to the orthogonal composition, the rectangular grid of these urban vegetable gardens does recall a very ancient spatial model, that of the medieval *hortus catalogi*, whose geometric layout was to facilitate differentiation and classification, allying production with knowledge (Aben and DeWit 1999). Effectively, in these cases, beyond the grid formed by the different plots defined by the municipality with ridges and hedgerows, the internal logic of the plot often presents a further grid, whereby rectangles and squares, marked by footpaths, separate different types of species.

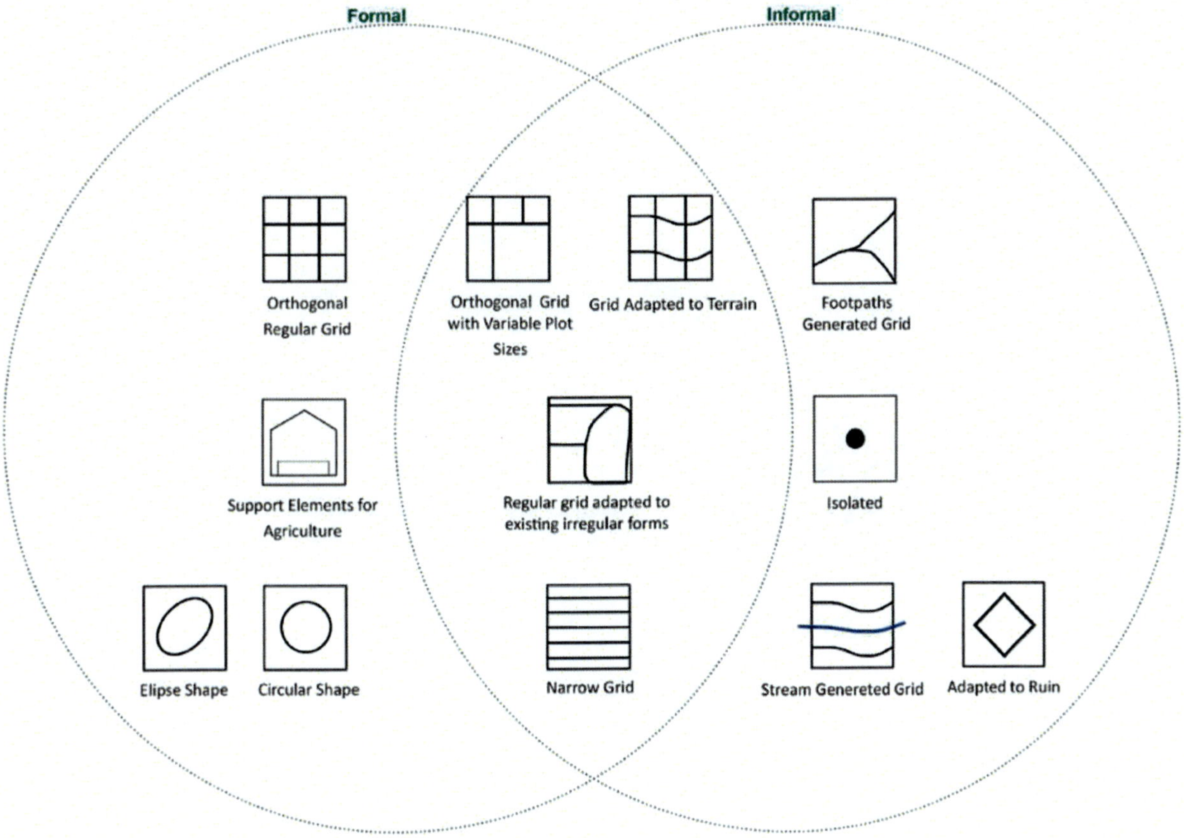

Fig. 8.20 Typo-morphologies of Vegetable Gardens. Source: Authors

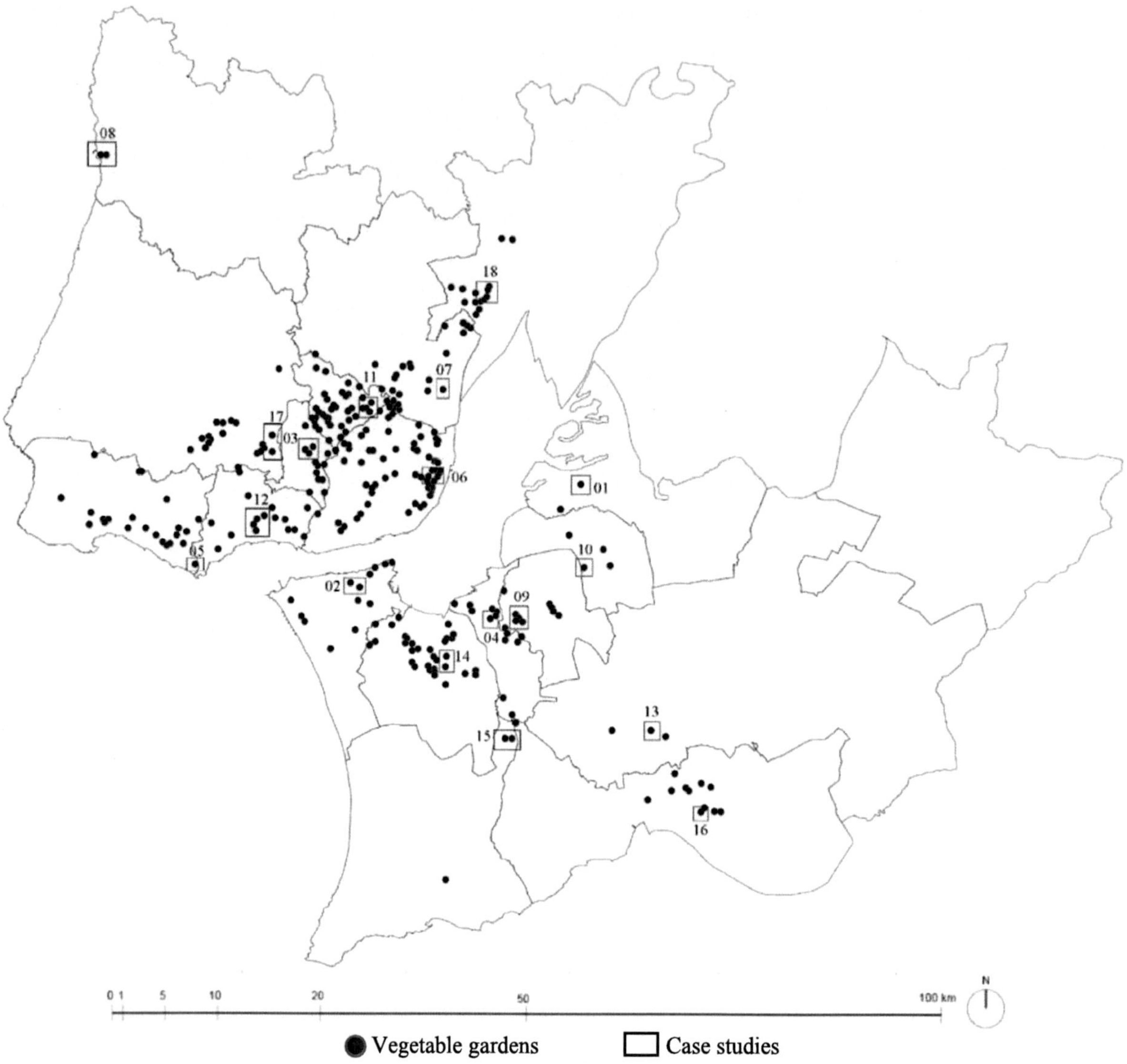

01. Alcochete
Social VG of Salinas do Samouco
02. Almada
Informal VG of the neighbourhood of Monte da Caparica
03. Amadora
Informal VG of Falagueira
04. Barreiro
Pedagogical orchard of the Principal city park
05. Cascais
Community VG of Quinta
06. Lisboa
Horticultural Park of Vale Fundão
07. Loures
Informal VG of the neighbourhood of Sacor (Bairro da Petrogal)
08. Mafra
Community VG of Ericeira
09. Moita
Horticultural Park of Vale da Amoreira
10. Montijo
VG of Afonsoeiro
11. Odivelas
VG of the neighbourhood of Codível
12. Oeiras
Informal VG of the neighbourhood of Laveiras
13. Palmela
Community VG of Palmela
14. Seixal
Agricultural Space of Quinta da Trindade
15. Sesimbra
Social VG on the Ecological Park of Várzea da Quinta do Conde
16. Setúbal
Informal VG of the neighbourhood of Bairro Azul da Bela Vista
17. Sintra
Social VG of Pego Longo
18. Vila Franca de Xira
VG of the Eco Park

Fig. 8.21 Vegetable Gardens (VG) case studies. Source: Authors

Orthogonal Regular Grid

01 | ALCOCHETE | SOCIAL VG OF SALINAS DO SAMOUCO

Fig. 8.22 Alcochete case study. Source: Authors

Orthogonal Regular Grid

04| BARREIRO | PEDAGOGICAL ORCHARD OF THE PRINCIPAL CITY PARK

Fig. 8.23 Barreiro case study. Source: Authors

Orthogonal Regular Grid

05| CASCAIS | COMMUNITY VG OF QUINTA DOS LOMBOS

Fig. 8.24 Cascais case study. Source: Authors

Orthogonal Regular Grid

13 | PALMELA | COMMUNITY VG OF PALMELA

Fig. 8.25 Palmela case study. Source: Authors

Orthogonal Grid with Variable Plot Sizes

08 | MAFRA | COMMUNITY VG OF ERICEIRA

Fig. 8.26 Mafra case study. Source: Authors

Orthogonal Grid with Variable Plot Sizes

11 | ODIVELAS | VG OF THE NEIGHBOURHOOD OF CODÍVEL

Fig. 8.27 Odivelas case study. Source: Authors

Orthogonal Grid with Variable Plot Sizes

17 | SINTRA | SOCIAL VG PEGO LONGO

Fig. 8.28 Sintra case study. Source: Authors

Fig. 8.29 Amadora case study. Source: Authors

Fig. 8.30 Moita case study. Source: Authors

Grid Adapted to Terrain

02| ALMADA | INFORMAL VG OF THE NEIGHBOURHOOD OF MONTE DA CAPARICA

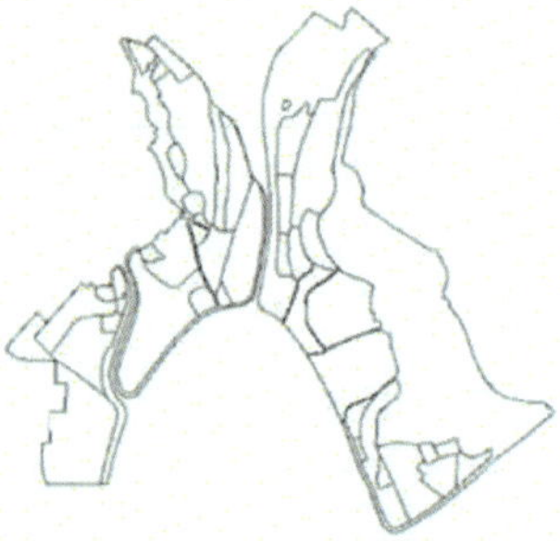

Fig. 8.31 Almada case study. Source: Authors

Grid Adapted to Terrain

16 | SETÚBAL | 1 INFORMAL VG OF THE NEIGHBOURHOOD OF BAIRRO AZUL DA BELA VISTA

Fig. 8.32 Setúbal case study. Source: Authors

Grid Adapted to Terrain

18 | VILA FRANCA DE XIRA | I INFORMAL VG OF THE ECO PARK

Fig. 8.33 Vila Franca de Xira case study. Source: Authors

Regular Grid adapted to Preexisting Elements

06 | LISBOA | HORTICULTURAL PARK OF VALE FUNDÃO

Fig. 8.34 Lisboa case study. Source: Authors

Fig. 8.35 Seixal case study. Source: Authors

Fig. 8.36 Oeiras case study. Source: Authors

Narrow Stripes

07 | LOURES | INFORMAL VG OF THE NEIGHBOURHOOD OF SACOR (BAIRRO DA PETROGAL)

Fig. 8.37 Loures case study. Note: In July 2021, when this Atlas was being finished, these vegetable gardens were cleared to allow the construction of individual housing. Source: Authors

Narrow Strips

10 | MONTIJO | VG OF AFONSOEIRO

Fig. 8.38 Montijo case study. Source: Authors

Circular Shape

15 | SESIMBRA | SOCIAL VG ON THE ECOLOGICAL PARK OF VÁRZEA DA QUINTA DO CONDE

Fig. 8.39 Sesimbra case study. Source: Authors

References

Aben R, de Wit (1999) The enclosed garden. History and development of the Hortus Conclusus and its reintroduction into the present-day urban landscape. Nai010 publishers, Amsterdam

Abrantes P, Rocha J, Marques da Costa E, Gomes E, Paulo M, Costa N (2017) Modelling urban form: a multidimensional typology of urban occupation for spatial analysis. Environ Plan B: Urban Anal City Sci 1–19

Cabannes Y, Raposo I (2013) Peri-urban agriculture, social inclusion of migrant population and Right to the City—Practices in Lisbon and London. City 17(29):235–250. https://doi.org/10.1080/13604813.2013.765652

Castel-Branco I, Saraiva M, Neto M (1985) As "hortas urbanas" em Lisboa. Sociedade e Território 3:100–107

Conzen M (1960) Alnwick, Northumberland—a study in town-planning analysis. George Philip, London

Costa S, Fox- R, GoodR SI, Treija S, Atanasovska JR, Bonnavaud H (2016) The position of urban allotment gardens within the urban fabric. In: Bell S, Fox- R, Keshavarz N, Benson M, Caputo S, Noori S, Voigt A (eds) Urban allotment gardens in Europe. Routledge, London, pp 201–228

Delgado C (2017) Mapping urban agriculture in Portugal: Lessons from practice and their relevance for European post-crisis contexts. Moravian Geographical Reports 25(3) 139–153. https://doi.org/10.1515/mgr-2017-0013

Delgado C (2018) Contrasting practices and perceptions of urban agriculture in Portugal. Int J Urban Sustain Dev 10:170–185

Dias AM, Marat-Mendes T (2020) The morphological impact of municipal planning instruments on urban agriculture. Cidades, Comunidades & Territórios 41 | 2020. http://journals.openedition.org/cidades/2991

Dias AM (2018) The shape of food: an analysis of urban agricultural shapes in Lisbon's greater area. MSc dissertation, Iscte—Instituto Universitário de Lisboa, Lisbon. http://hdl.handle.net/10071/19300

FAO (Food and agriculture organization) (2019) FAO framework for the urban food Agenda. FAO, Rome. https://doi.org/10.4060/ca3151en

Harper K, Afonso A (2016) Cultivating civic ecology: a photovoice study with urban gardeners in Lisbon. Portugal. Anthropol Action 23(1):6–13

Hespanhol R (2019) Agricultura urbana em Portugal: práticas espontâneas e institucionalizadas Confins [online] 43

Marat-Mendes T (2020) Designing for sustainability: retrieving a systemic role for urban form. Urban Morphol 24(2):235–238

Marat-Mendes T, Borges JC (2019b) The role of food in re-imagining the city—from the neighbourhood to the region. In: Juvara M, Ledwon S (eds) Proceedings of ISOCARP 2019 Congress, Jakarta, 9–13 September 2019

Marat-Mendes T, Lopes SS, Borges JC (forthcoming) Atlas da Agricultura Urbana da Região de Lisboa—Atlas of the Lisbon region urban agriculture. ISCTE-DINÂMIA'CET, IUL

Marat-Mendes T, Borges J, Dias A, Lopes R (2021a) Planning for a sustainable food system. The potential role of urban agriculture in Lisbon Metropolitan Area. J Urban: Int Res Placemaking Urban Sustain. ID: 1880960. https://doi.org/10.1080/17549175.2021.1880960

Marat-Mendes T, Lopes SS, Borges JC (2021b) Agricultura Urbana: Modelos de desenho urbano para um planeamento sustentável. In: Pinho (ed) Compêndio de Políticas Urbanas. FEUP-ISCTE-UA, Porto

Marat-Mendes T (2021) Urban form and urban metabolism. Recent research and academic trends conducted at the Lisbon Metropolitan Area. In: Llop C, Cervera M, Peremiquel F (eds) Forma Urbis Y territorios metropolitanos. Metrópolis en recomposición. Prospectivas proyectuales en el siglo XXI. Universitat Politècnica di Catalunya, Barcelona, pp 148–155

Messina T, Mourato JM (2018) Comunity Urban Food Gardens policy design: a study case with Almada (Portugal) and Paris (France). In Proceedings of the international scientific event, 26–27 April 2018, Lisbon, Portugal - Connections and missing links within urban agriculture, food and food systems, pp 57–61

Muratori S (1959) Studi per una operante storia urbana di Venezia. Instituto Poligrafico dello Stato; Roma

Paizinho C, Ferreiro M (2017) Práticas de economia solidária em iniciativas de agricultura urbana do concelho de Lisboa. Os casos do Vale de Chelas, da Alta de Lisboa e da Horta do Baldio. In: Associação Portuguesa de Horticultura (ed.) I Colóquio Nacional de Horticultura Social e Terapêutica. Associação Portuguesa de Horticultura, Estoril, pp 53–56

Pinto TC, Guerra I (2019) Housing policies, market and home ownership in Portugal: beyond the cultural model. Cidades Comunidades & Territórios 39:101–114

Šćitaroci MO, Marić M (2019) Morphological characteristics of green spaces in fortified towns and cities. Urban Morphol 23(1):27–44

Telles GR (1997) Plano Verde de Lisboa. Colibri, Lisboa

Weaver C (1984) Regional development and the local community. Wiley, New York

Whitehand JWR (2019) Green space in urban morphology: a historico-geographical approach. Urban Morphol 23(1):5–17

Zhang Y (2019) A spatio-temporal study of fringe belts and urban green spaces in Birmingham. UK. Urban Morphology 23(1):18–26

Part IV
The Food System of the Lisbon Region—From the Past to a Sustainable Future

In this fourth part, we present the results of the research that sustains this Atlas. Chapter 9 constitutes a case study. Being impossible to conduct an in-depth observation of the food system as it presents itself in all the Lisbon Region, we select a particular area, where we apply the analysis proposed in the Atlas, namely, in the chord that extends from the eastern end of the Lisbon city northwards to Vila Franca de Xira. Following the outline of the Atlas, we observe key planning instruments promoted here, followed by a description of the key food-related activities encountered, including urban agriculture. Finally, conditions for transforming the food system are assessed.

In Chap. 10, the key conclusions from the Atlas are briefly presented. The key reasons to promote a sustainable transition of the food system are pointed out. The key findings from each part of the Atlas are also broken down, from the shortcomings and potentials of planning policies towards the food system, to the results of socio-metabolic change in the Lisbon Region, and the role of grassroots action, especially in urban agriculture, to promote a sustainable transition. A set of main recommendations is presented at the end.

9 Putting the Food System into Perspective: A Case Study from Lisbon to Vila Franca de Xira

9.1 Introduction

The riverside area from the eastern end of the Lisbon city towards the north is today a repository of urban forms and territorial situations which largely exemplify important processes of modernization of the region in terms of its built structure, but also its social, economic and ecological fabrics. Farms, groves and vegetable gardens remain among factories, infrastructural corridors, between urban settlements and even within them. Apartment buildings, rural houses and small factories sometimes coexist within the same block. Social housing estates from several programs exist, alongside working class and middle-class neighbourhoods, illegal housing ranges from single-family homes to multi-storey buildings. Ridges and valleys intertwine in a complex mesh of forestry, farmland and fallow land. Deactivated factories face active industrial estates, some of which contain housing for workers and administrators. On the opposite bank of the Tagus River, are located the 'mouchões' (alluvium islets), and the immense flat extension of the Great Wetland. The rural and the urban, the planned and the informal, the old and the new, the natural and the artificial, all coexist in this area. The resulting spatial and functional diversity has problems—and potentials—of a very specific nature. This same diversity makes this territory particularly suggestive for reorganizing the regional food system.

This area was never poor in planning efforts, although, like in Almada (Baptista and Melâneo 2020), such efforts were often undercut by lack of political will to approve and implement the plans, while private construction often contradicted the ambitions of planners which municipalities tried to follow, even without the proper officialization.

The first attempt, which importantly considered the area as a single territorial unit, was designed by Étienne de Gröer with collaboration from his son Nikita de Gröer (1915–2001) and was finished in 1950. However, the industrial zoning they proposed needed amending, since it was too limited. But the touchstone of the failure of this plan—besides the lassitude of the administration—lied with the owners of industry and land, many of which used the industrializing wave as an opportunity for speculation, avoiding an organized concentration of industry in specific areas. The same inability to coordinate public interest with private property—an inability that was not accidental—determined the failure of the urban zoning as well. As he commonly did, Étienne de Gröer established rural buffers to separate different settlements while protecting agricultural activity, but these were doomed to be replaced by nearly continuous urbanization. More than any of its putative faults, this work failed to be implemented because it never went beyond the Foreplan phase which, as seen in Chap. 1, reduced it to a mere technical study without any binding power to regulate urban operations.

In 1964, a regional plan, the PDRL, clearly articulated this area with the Lisbon city, but even prior, an attempt at this articulation was advanced by the Chelas Urbanization Plan, for the Eastern end of the city, prepared and implemented by the Lisbon municipality, and proposing a mega-centre in Chelas whose scale was sub-regional, linking the centre of the city with the strong area of industrial development (GTH 1965; Folgado 2009).

While the territorial model proposed in by the Gröers would have been impracticable with the population rise that this chord faced throughout the twentieth century, landowners and construction industrials carried on several parcelment operations, often without proper authorization. The preferred housing form, the single-family home, was often put aside in favour of high-rise towers and slabs of apartment buildings. However, single-family houses with yards are the preferred typology in the illegal parcelments west of Highway A1,[1] an area that all planners had reserved as a rural buffer.

Yet most of available space was eventually filled with housing and industries too. Overtime, this territory has

[1] Many of these have already been properly legalized, while others are on the process of doing so.

T. Marat-Mendes et al., *Atlas of the Food System*,
https://doi.org/10.1007/978-3-030-94833-7_9

become a conurbation, in the sense given to this term by Patrick Geddes, a continuous urbanization where settlements tend to merge and form a visible continuity. This continuity is not uneventful, and its adaptation to the complex topography does impose occasional buffers between urbanizations, but built structures always seem to proliferate, either industrial complexes, apartment buildings or even low-density housing.

In this chapter, we observe this complex conurbation, seeking to highlight the history of its formation, putting its current forms and functions into perspective. Afterwards, we focus on the phases of the food system encountered in it, as well as on the spaces which, for different reasons, could be the starting points of positive changes towards a more sustainable—and local—food system.

9.2 The Regional Foreplan from Moscavide to Vila Franca de Xira (1946–55)

By the mid-1940s, Étienne de Gröer is officially responsible for a considerable part of the settlements in the Lisbon Region, which under Law 24802 required a PGU. The PUCS is approved in 1947, and in the following year, the Lisbon Masterplan is completed, although it will be rejected by the Central Government. In 1946, he is further tasked with the plan for Almada and for the sub-regional chord from Moscavide to Vila Franca de Xira, to which in 1948 is added the order for the Sintra plan. He experiments with all scales from the region to the building (Marat-Mendes and Sampayo 2010) and works in some capacity for nearly half municipalities of the Lisbon Region.[2] Overall, from the 16 municipalities that were officially included in the Lisbon Region in 1959, Gröer worked in seven, 6 of which in the Northern Bank of Tagus.

Historiography has naturally highlighted the PUCS (Pereira 2009; Marat-Mendes 2009), one of the few amongst his works with official approval and implementation (and the only one approved at the sub-regional scale), as well as the Lisbon Masterplan which, despite lacking official approval, had many of its proposals implemented by the city council (Silva 1994; Marat-Mendes and Oliveira 2013). His remaining proposals, as well as his metropolitan vision, are less studied and understood, probably because, regardless of the putative merits of his other plans, these remain as speculative exercises without practical confirmation. However, an observation of these works allows not only a peek into the reality of the Lisbon Region at the 1940s, as well as into the key dynamics of change which these plans attempted to encompass.

Perhaps the work Gröer developed with his son Nikita for the riverside chord of Loures and Vila Franca de Xira is one of the best examples of the suburban vision of this planner, as it dealt with a territory that was important for two different and competing reasons: on the one hand, the high agricultural value of this territory bathed by the Tagus waters, and on the other, the uniquely suited conditions it offered to industrial development: the presence of ports, the high-tension electrical wiring systems,[3] the water channels (Tejo and Alviela); the transportation infrastructures of national importance—including the existing National Road 1 (*Estrada Nacional* 1—EN1), later called EN10, and the highway already designed by the Portuguese Road Authority (*Junta Autónoma das Estradas*—JAE), as well as the trainway.

Another key motivation for development is that the rural terrains in this area were relatively cheap, prompting a spree of construction of industrial estates which, unaccompanied by housing construction, resulted in the overcrowding and subsequent degradation of the scattered and small villages that existed in this territory (Gröer and Gröer 1947).

Although this chord includes two different municipalities, it is developed by the Central State, namely by the General Directory of Urbanization Services (*Direcção-Geral dos Serviços de Urbanização*—DGSU), indirectly demonstrating the ways in which the New State handled the problem of industrial modernization. The efforts to structure and plan the industrial growth on the north-eastern extension of Lisbon happened when such process had already been started by the industrial proprietors themselves. The 'Regional[4] Foreplan from Moscavide to Vila Franca de Xira' (*Anteplano Regional de Moscavide a Vila Franca de Xira*) can thus be considered the first document to officially establish this northern Lisbon chord as a subregional unit. Two concerns guided this work: first, the stipulation of an adequate set of rules for the spatial development of industry, and second the location of areas for urban development destined to house industrial workers as well as other population (Gröer and Gröer 1947).

The development of industry and infrastructure in the Lisbon Region at the nineteenth and mostly the twentieth

[2] The limits of the Region were only formalized only in 1959, as seen in Chap. 2.

[3] Because the Carregado Power Plant was installed north of Vila Franca de Xira.

[4] Throughout the written documents that comprise this work, it is clear that the words 'region' and 'regional' are used to designate both the Metropolitan Region—i.e. the Lisbon City and its surrounding municipalities at both banks of Tagus—and the specific area of the riverside of Loures and Vila Franca de Xira. For clarification, this plan is, by today's standards, not a regional one, but rather an intermunicipal or a sub-regional one.

centuries was structured by a few conditions, namely the establishment of industrial port activities (in Lisbon and Barreiro), military roads allied with transportation and circulation (Alcântara—Xabregas in Lisbon and Algés—Sacavém around it), the structures of water and electricity provision and the key structure of collective transportation. These conditions made the north-eastern riverside extension of Lisbon a privileged territory for modernization (Folgado 2009).

While the territory from Moscavide to Vila Franca de Xira was very different from that of Costa do Sol, the planning methodology did not differ, at least in principle, from that of the PUCS. Thus, in a first moment, the area covered by the plan was surveyed in detail, noticing the variations on the topography, which includes a set of elevations between the margin of the Tagus River and the coastal areas in Vialonga and in the Trancão Floodplain. Notably, such a territory includes some of the most fertile land in the Lisbon Region, including the Odivelas–Vialonga Coastline (*Costeira de Odivelas–Vialonga*) and the Trancão Floodplain (Zêzere 2001). However, a set of other important rural territories were excluded from the Foreplan, namely the Great Wetland of Tagus (*Lezíria Grande do Tejo*) on the eastern margin of the Tagus, belonging to Vila Franca de Xira, as well as the three islets (*mouchões*) resulting from alluvial deposition on the Tagus riverbed, all of which were used for wetland farming (Rodrigues et al. 2016).

Industry was developing for about 15 years, but agriculture remained the predominant activity, and the planners highlight the 'excellent cultivations' (CSOP 1955, sheet 62), including cereals, orchards, vegetable gardens in the flat areas and olive groves and vineyards in the slopes. Furthermore, saltpans were noticed in the strip from the trainline to the river between Alverca and Sacavém, while between the trainline and the EN10 intensive vegetable gardens were found (CSOP 1955). Most of the territory's locations were small-scale villages, some of which were then becoming overcrowded from workers seeking shelter close to the factories they work on. Interestingly, the planners noticed that the overwhelming majority of housing in this chord was constituted by single-family homes with only some examples of collective buildings of more than two floors (Gröer and Gröer 1947).

The Foreplan established three functional zones: industrial, rural and residential. Industrial areas were placed on the strip between the Tagus riverside and the EN10, as well as in part of the Trancão floodplain and in Vialonga, designated as an industrial area despite its inland position. While this implied eliminating rural land, industrial facilities were to be separated by buffers of spaces of rural character, to uplift environmental conditions. With these industrial areas as reference, pre-existing settlements were selected for physical extension, to create urban units, while a few others were included in the rural area to be protected as villages. The official preference seemed to maintain lower densities which avoided great agglomerations of people, a tendency which was already in existence in the territory itself. This option was in accordance with the preferences of the time, since Salazar (1933: 200) publicly denounced 'the great Phalanstéres [or] the colossal constructions for the working-class' already at the early years of the New State.

The Foreplan predicted six types of settlement (CSOP 1955). First, locations with more industry than housing which needed to be improved (Sacavém and Alhandra); second, locations for expansion of industry and housing (Bobadela, São João da Talha, Pirescoxe, Santa Iria de Azóia, Via Rara, Póvoa de Santa Iria, Sobralinho, Catujal, Unhos,[5] Povos and Vala do Carregado[6]); third, settlements limited by geography (Alverca and Vila Franca de Xira); fourth—rural locations whose industrialization was to be slower (Vialonga and Castanheira do Ribatejo[7]); and fifth, the special case of Moscavide, which the Gröers considered a bad suburb of Lisbon, suggesting its annexation to the capital city and treatment as a fully urban area to refurbish. The sixth situation was that of rural villages, three in Loures (Camarate, Apelação and Frielas) and two in Vila Franca de Xira (Arsena and São João dos Montes).

A continuous rural buffer was created between towns, between industrial areas and on the hilltops, meant to maintain the existing farms and orchards. Unlike other cases where a rural buffer was merely seen as an area with restricted construction regime and subject to rules and activities that fall outside the scope of planning, in the Foreplan, Gröer was keenly aware of the importance of agriculture and rural industries in the local economy. According to the Foreplan regulation,[8] the rural buffer was reserved for agriculture and was compatible with small rural industries, which included slaughterhouses, dairies and cheese factories (as well as quarries and brick factories). Construction was limited to farmers' houses, houses for agricultural workers (permanent or temporary), sanatoriums or other institutional facilities, all of which were restricted in

[5] Catujal e Unhos are reserved for urban expansion at a later time, with priority given to the other cores (CSOP 1955, sheet 53).

[6] Povos and Vala do Carregado are not included in any of the drawings of the Foreplan which we could consult.

[7] Castanheira do Ribatejo is not included in any of the drawings of the Foreplan which we could consult.

[8] The definitive regulation of the Foreplan does not seem to be deposited in the Municipal Archives of Loures or Vila Franca de Xira, although most of its content is assessable through the technical report of the CSOP. A letter from the DGSU to the Loures Council dated 6 December 1956 includes as an attachment a copy of the full regulation for the Rural Zone.

terms of size, height, and relation with other built structures. A specific regulation was created for the rural villages, mandating the establishment of special preservation plans and borderlines, restricting industrial activities to those relying on no more than six employees and limiting the extension of existing buildings whether horizontally or vertically.

Considering that most of the industrial areas were obtained from eliminating rural land, the restrictions on the rural buffer may be seen as a compensation—a position that was reiterated by the CSOP, when it stated that harnessing a significative amount of land for agriculture was of strategic importance considering the proximity with a large consumption market—the Lisbon city (CSOP 1955, sheet 25).

The Foreplan reinstated the suburban vision Étienne de Gröer had kickstarted with the PUCS and the Lisbon Masterplan, but this vision was undercut by bureaucracy and by the rampant weakening of planning policies that followed the death of Duarte Pacheco in 1943.

According to the contract celebrated with the Gröers, their work previewed two phases. The first phase consisted of the foreplan, a general scheme of circulations, land-uses, delimitation of industrial areas and the regulations and legal frameworks for the spatial organization of the whole chord; while the second phase consisted of a further detailing of the territorial model proposed (Figs. 9.1, 9.2, 9.3, 9.4, 9.5, 9.6, 9.7, 9.8, 9.9, 9.10 and 9.11).

In 1950, the DGSU declared the first phase completed.[9] Yet only in 1955 was the Foreplan evaluated by the commission CSOP. Although this technical report praises the scrupulous and detailed survey at the basis of the Foreplan, and generally agreed with the territorial model proposed, it also stressed that the cartographical elements made available to the planners were outdated from the start, highlighting the need for revising it, especially with regards to industrial areas, which in the Foreplan were too limited (CSOP 1955). Both the CSOP and the planners highlighted the difficulties of planning this area and the expected industrial activities, since such works lack a structural vision that could only be provided by a regional spatial plan for Lisbon and by a national plan for industry and infrastructure (CSOP 1955).

The key motivation for the Foreplan was unmistakably the physical structuring of industrial modernization, as it was elaborated under the Law for Industrial Development[10] of 1945. This Law was a key example of how the New State established control over economic development, using corporative organizations to boost strategic industrial sectors through protectionism and other incentives—which included the survey of existing complexes and support for modernization of built structures. It is furthermore an antecedent to the Development Plans of 1950s, following the integration of Portugal on the Organisation for Economic Co-operation and Development (OECD). Indeed, the industrial estates in the Gröer Foreplan anticipate the formation of the conglomerates resulting from reorganization, as is the case with Covina and the Cimenteira Tejo, or the expansion of Sacor. However, the seriousness with which the implementation of the spatial plans was carried was very limited, and while until the 1960s the territory was to be more strongly infrastructured, there was virtually no control over urbanization.

The delay in obtaining the technical report, as well as other administrative problems, may have hampered the continuation of this process. The revision of the Foreplan was finished in 1956, but instead of Étienne and Nikita de Gröer, the author was Fernando António Lorenzini Borges Campos. The revision was limited to a remodelling of industrial zones, alongside a new regulation for industrial zoning, with the amount of land assigned for this activity significantly enlarged (in comparison with Gröer's version). Implicitly, the remainder of Gröer's territorial model was maintained.

Campos pointed out that although his predecessors left the municipalities with a balanced territorial model, there was no legal framework that allowed them to implement it, either actively or passively. Law 35.031, while safeguarding the legitimacy of the Foreplan, had set forward a general confusion in what was mandatory to follow and what was not, especially in a context such as the Portuguese, where the right to private property had always been understood to be in direct conflict with planning policy (Pereira 1997).

While some of the industrial areas of Gröer's Foreplan were already filled, there remained a chaotic organization. On the one hand, landowners on the riverside clinged to their land for agriculture or speculation, delaying the conversion to industrial activities (Campos 1956). On the other hand, industrial owners often bought the land surrounding their complexes, retaining them for future expansion. This resulted in new industries installing in rural land, while empty space remained in the area officially designated for such function (Campos 1956). Thus, the remodelling of the industrial zone and its regulation aimed to control this problem, while the author was clear that without a proper form of expropriation, it would be difficult to ensure implementation.

It is unclear why the Foreplan (and its industrial reformulation) were never carried on to the second phase, that of a properly designed plan. However, just 3 years after Campos' completion of the new industrial zoning, the State

[9] Letter of 20/09/1950 from the DGSU to the Loures council. Deposited at the Loures Municipal Archive, Process 2116—sheet without number.
[10] Law 2005, published in Diário da República 54/1945—Série I, 14-03-1945.

Fig. 9.1 Anteplano regional Moscavide—Vila Franca de Xira (1946–1956) by Étienne de Gröer and Nikita de Gröer (on left) and respective land-use scheme (on right). Source: Arquivo Municipal de Loures, AML_2116 and Authors

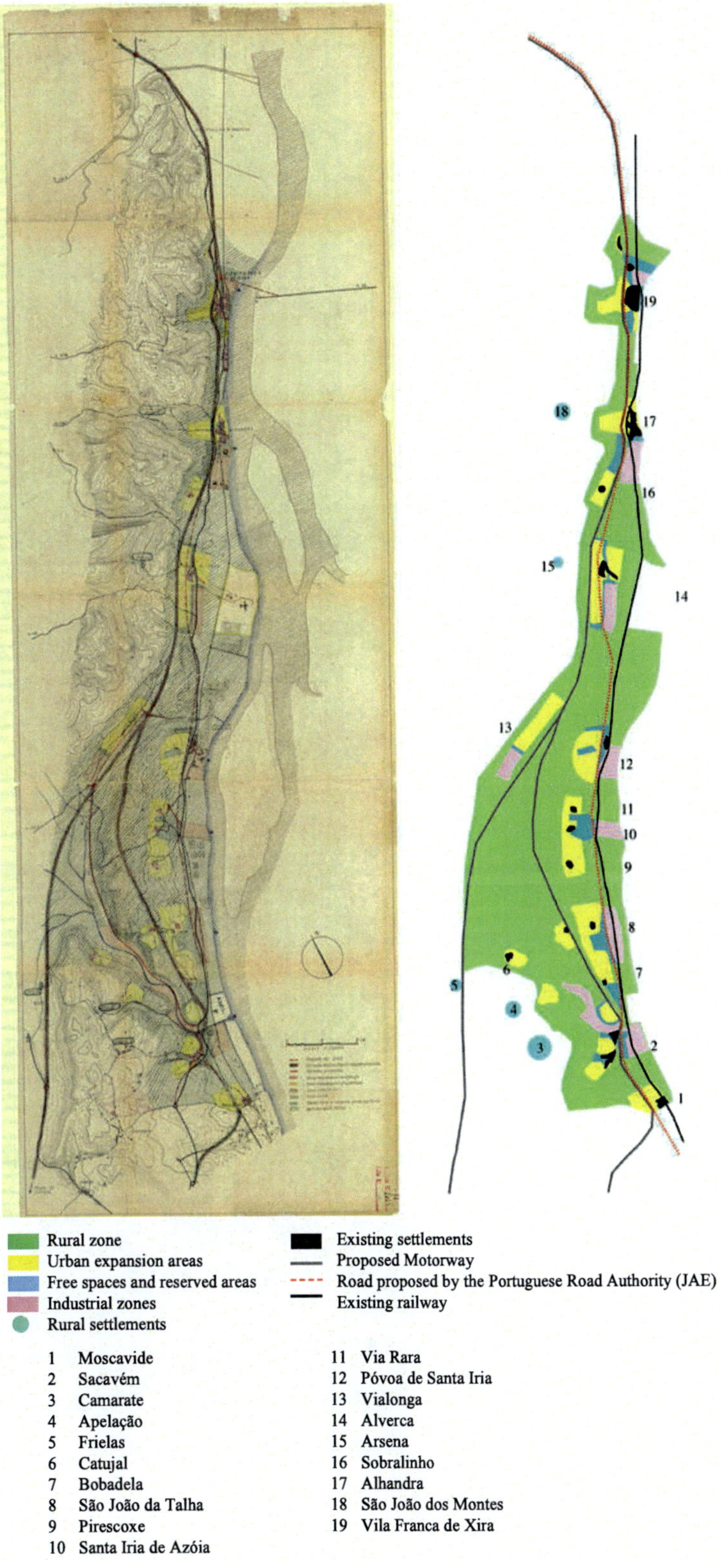

Fig. 9.2 Moscavide. Source: Authors

Fig. 9.3 Trancão Floodplan, with the Sacavém skyline on the background. Source: Authors

Fig. 9.4 Olaio Furniture Factory, Bobadela. Source: Authors

Fig. 9.5 Rural housing in São João da Talha. Source: Authors

agreed with the designing of the PDRL, with the corresponding legislation published in 1959. In this context, it would seem pointless to continue the works for a plan that threated to be always delayed in relation to reality.

Given the absence of a properly approved plan for the whole chord, the implementation of the territorial model achieved between the work of the Gröers and Campos is far from guaranteed. In practice, the implementation of the Foreplan depended upon the designing and approval of about 13 PGUs from two distinct municipalities,[11] both of which lack the funds, the technical skill and the de facto power to control the transformation of their territories.

For comparison, the PUCS predicted 11 urbanization plans, all of which were carried by Étienne de Gröer and

[11] Lisbon was not involved, considering that the Gröers' suggestion of annexing Moscavide to Lisbon was never followed through, with that town remaining, today, a part of the Loures municipality.

Fig. 9.6 Covina Glass Factory in Pirescoxe, and olive grove over the Tejo and Alviela Channels. Source: Authors

which were not subject to individual approval (Pereira 2009). In the territory of the Foreplan, the process was much more complicated, for while some PGUs were indeed developed, either in that form (Póvoa de Santa Iria—Vialonga, Alverca, Alhandra, Vila Franca de Xira[12]) or in the form of neighbourhood plans carried on by private industrial companies (Bobadela, Pirescoxe), other settlements (S. João da Talha, Catujal, Unhos) were never properly planned. Furthermore, not only the PGUs that were effectively designed were not implemented but also the rural buffers both between and around the assigned towns were extensively urbanized. While in some areas—notably Bobadela or Pirescoxe—the influence of the Garden City paradigm can still be felt, in others, urbanization followed widely different directions, especially from the 1960s onwards.

[12] The Vila Franca de Xira PGU, discussed in Chap. 1, seems to be in development at the same time as the Foreplan.

9.3 The Planning Situation After the 1950 Foreplan

The Foreplan designed by the Gröers and revised in its industrial dimension by Campos seems to have been quietly forgotten soon after it was finished. It never reached the definitive phase, always remaining as a Foreplan, and beyond sparse urbanization plans, mostly dealing with working-class neighbourhoods promoted by industrial owners, it was also never brought to a more detailed scale.

While Gröer had suggested that Moscavide be absorbed by the municipal borderline of Lisbon, such transfer never materialized. In 1960, the Lisbon Council is fully aware of the growth in industry that was taking place not only on the eastern end of the city but also on its northwards extension. By 1964, the Chelas Urbanization Plan, coordinated by architect Francisco Silva Dias, reaches its definitive version. Silva Dias, a former researcher at LNEC (d'Almeida et al. 2020) and a member of the Estramdura team in the 1961

Fig. 9.7 Working-class housing in Pirescoxe. Source: Authors

Fig. 9.8 Santa Iria de Azóia. Source: Authors

Survey on Regional Architecture, is a planner at the Lisbon GTH, which at the time is conducting the planning and design processes for carrying the housing programme set forward by Law 42454, which assigned the territories of Olivais and Chelas to the construction of large-scale and high-density housing, as explained in Chap. 2. The three areas—Olivais Norte, Olivais Sul and Chelas—add to the neighbourhoods of Madre de Deus and Encarnação, built in the 1940s as economic housing for public workers and a petty bourgeois population. This makes the eastern end of Lisbon a sector nearly all planned under municipal programmes, but Chelas stands in a somewhat different situation because it is the first work in the GTH to try and make sense of the implications of the strong industrial development extending towards Vila Franca de Xira (GTH 1965).

Fig. 9.9 Alverca. Source: Authors

Fig. 9.10 Póvoa de Santa Iria. Source: Authors

Fig. 9.11 Alhandra. Source: Authors

Thus, while the scale of the Urbanization Plan is obviously local, its scope aims to be sub-regional.

At the time, Chelas is the largest council housing project in the country. It had remained a rural suburb within the city for so long because its territory, a complex valley system with hefty ridges and hills, was very unappealing for urbanization by the private sector. The dimension of the urbanization plan, as well as the lack of financial means for the Council to ensure an integral operation determined that Chelas would be divided into seven sectors, each corresponding to a detailed plan and to a phase of implementation. While construction starts with the first re-housing estates in 1960, the most recent sector has been constructed as late as 2000. Because of financial restraint, the more the urbanization operations evolves, the more they concentrate solely on providing housing, dedicating less attention to public space and to equipment. Pictures from the urbanization process testify to the continued existence in the valley bottoms of horticultural gardens, sometimes occupying very large extensions and indirectly giving continuity to the original use of this territory as an agricultural area (Figs. 9.12 and 9.13).

The hilltops in Chelas—as well as those of Olivais—included several areas with olive trees and olive groves, indeed, 'Olivais' tanslates to 'Olive Groves', proving the continuity of this landscape, even in its rural state, with that of the northwards riverside chord. The urbanization of Olivais Norte and Sul was successful, with the GTH plans being fully or almost fully implemented, and including much of the predicted equipment, commerce and infrastructure. Chelas, while planned and implemented fragmentarily throughout the years, presented severe deficiencies in public spaces, infrastructure, equipment, services and commerce for decades (Heitor 2001; Borges et al. 2020).

However, at the early 1960s, the work of the GTH completed the development of the new city and the new suburbs on the eastern end and towards Vila Franca de Xira. Such an articulation is clear on the PDRL, which perfectly shows how, narrowed down by the airport, the strip from Chelas northwards forms a clear axis—although not a conurbation.

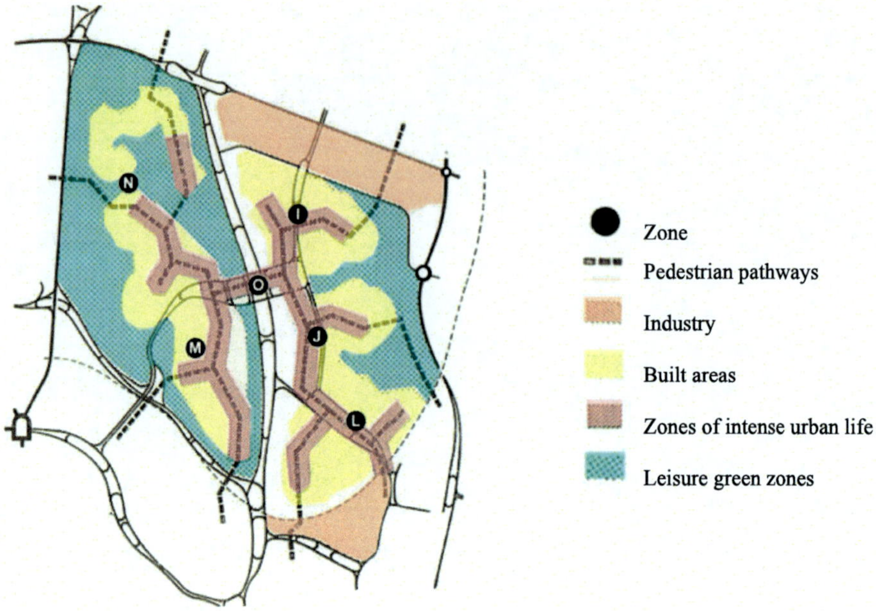

Fig. 9.12 Chelas Urbanization Plan (1964) coordinated by Francisco Silva Dias (1965). Source: GTH, 1965

With the PDRL finished in 1964, the chord of settlements separated by rural buffers was confirmed as the territorial model for this area. Perhaps because Campos is one of the authors of the PDRL, the territorial model in the northern extension was not especially altered in relation to the 1956 version. Even within Lisbon, the large area of Chelas was separated from Olivais Sul and from the existing city westwards by a large green buffer.

But while the PDRL clearly maintains the treatment of this riverside area as a single territorial unit, for over 40 years, such a consideration would no longer be sought. Indeed, with the democratic planning policy and the rise of the PDM, the organization of this territory was conceived municipally. However, in 2002, when the PROT-AML was finished, in its map of territorial dynamics, the northern extension chord was again revived as a unit. With a notable change, that it is now linked with the Amadora–Sintra chord, through the borderline of Lisbon with Loures and Odivelas. The classification is of 'critical urban area', socially and urbanistically degraded and lacking in infrastructure and equipment (CCDRLVT 2002). This corridor covers some of the most challenging areas of the Lisbon Region, including many housing areas for the urban poor, from social housing estates to illegal neighbourhoods, shanty towns and urban areas built over coastline (*costeira*) areas. The old centre of Lisbon is classified in the same way, testifying to its population decline and physical frailty at the time. While the Moscavide–Vila Franca de Xira is now separated from the eastern end of Lisbon which had been its continuation in the 1960s, this latter area is classified as a 'problematic space', pointing out the fragmented and unstructured physical settings, environmental degradation or losing population and activities, reaching urban decline (CCDRLVT 2002). Thus, while this north-eastern chord is now classified differently, in both classifications attributed, it is clear that the area needs investment and intervention, whether in its physical or social dimensions, or both.

The PROT-AML seems to favour the use of the Sacavém–Vila Franca de Xira chord as an area with privileged conditions for industry and logistics, in continuity with the dominant function of this area since the mid-twentieth century. A particularly unique situation here is the creation of the Lisbon Region Supply Market (Mercado Abastecedor da Região de Lisboa—MARL) in 2000. Until then, this function was fulfilled in Entrecampos, right in the Lisbon city, between its business area (Marquês de Pombal–Saldanha) and its university area, in a market which overtime became unsuited for its function.

The MARL was placed in S. Julião do Tojal, Loures, given its proximity to Lisbon and also to the productive regional areas. However, the PROT-AML does not particularly highlight this function, nor does it create any specific conditions to promote the installation of food industries in the vicinity—probably since such concern is automatically attributed to the private sector.

At the same time, the area is excluded from the more relevant rural zone, with the exception of the Great Wetland of Tagus, which is naturally included to bridge the areas of Mafra to those of Benavente (outside the Lisbon Region) and Alcochete and Montijo. As discussed on Chap. 3, this inclusion in the rural zone of the PROT-AML also means that it lacks any specific spatial arrangements or programming, while stating that this land is particularly prone for agricultural funding opportunities.

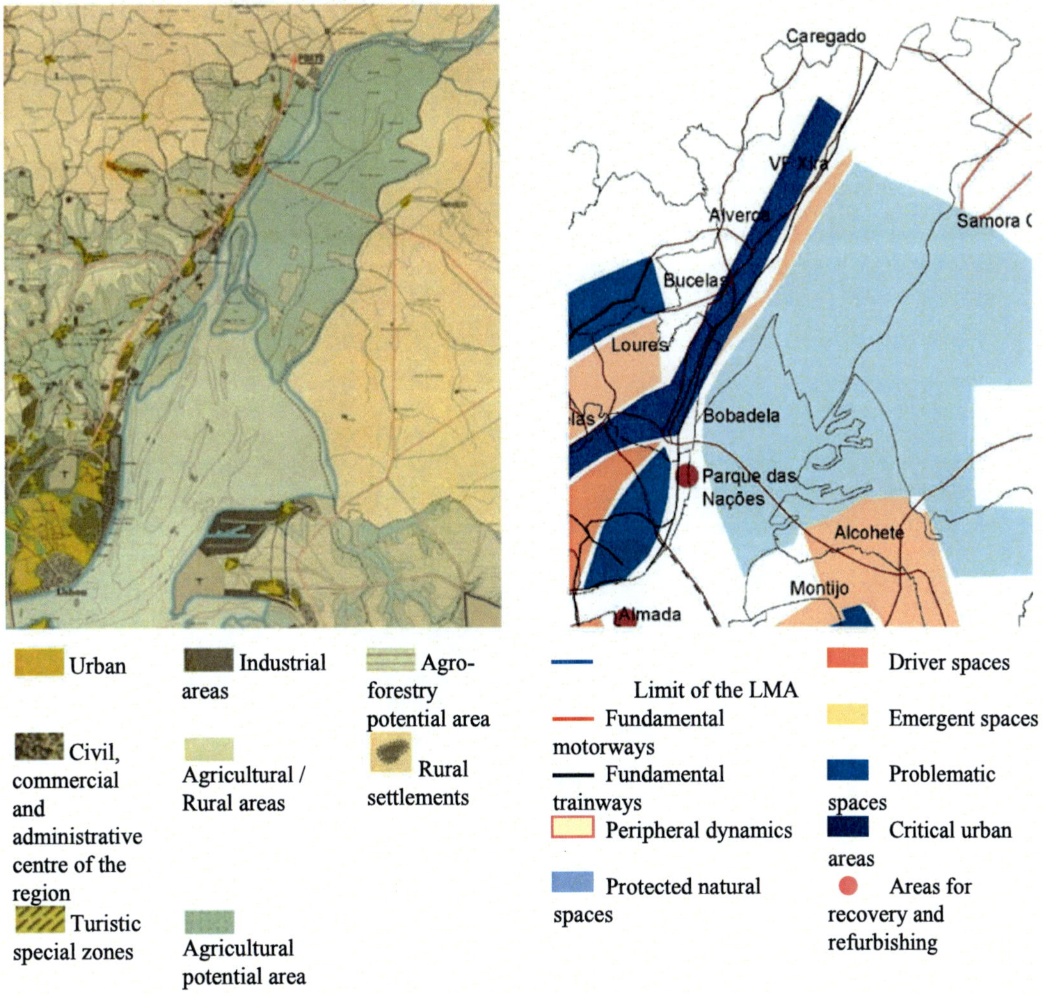

Fig. 9.13 Lisbon–Vila Franca de Xira axis in regional spatial planning—the 1964 PDRL (left) and the 2002 PROT-AML (right). Source: Arquivo Municipal de Loures, AML_2116 and Authors (Adapted from CCDRLVT, 2002, 32)

As happened earlier with the Foreplan of the Gröers and Campos, as well as with the PDRL, the PROT-AML was not implemented—indeed, its legal status is unclear, as we specified in Chap. 3. It received approval, but was withdrawn in 2007 for revision, a process which did end in approval. Thus, many of its proposals were never fulfilled. On the other hand, both municipalities had approved PDMs before the PROT-AML, and prepared new updated versions ever since—Vila Franca de Xira had its first in 1992 and the second in 2009; while Loures had the first in 1993 and the new one in 2015.

The current Vila Franca de Xira PDM (CMVFX 2009) was prepared after the 1998 Planning Act. Its land-use scheme, while subjected to the fundamental division between urban and rural, is very detailed. It presents specific dispositions for both rural and the urban soils, often leaving unclear whether the municipality would prefer to preserve and refurbish its past structures or the creation of new ones. This potentially makes activities and constructions more flexible.

Agricultural space comprises three subcategories, from soils with greater agricultural potential, included in RAN to less productive soils from RAN, and others not included in RAN but still productive. Notably, the most productive soils are not regulated by the PDM, but rather by RAN. The PDM does regulate directly the remaining rural solid, preferentially maintained for agriculture, although some forms of housing, agroindustry, tourism, leisure or equipment are not precluded. In this subcategory of Additional Agricultural Space, outside the RAN, allowed uses include agroindustry, transformation and selling of the produce of agriculture, livestock or forestry. The corresponding construction regime is small scale and low density.

In general, the urban areas of the PDM correspond to the urban settlements predicted in the Gröer Foreplan, although

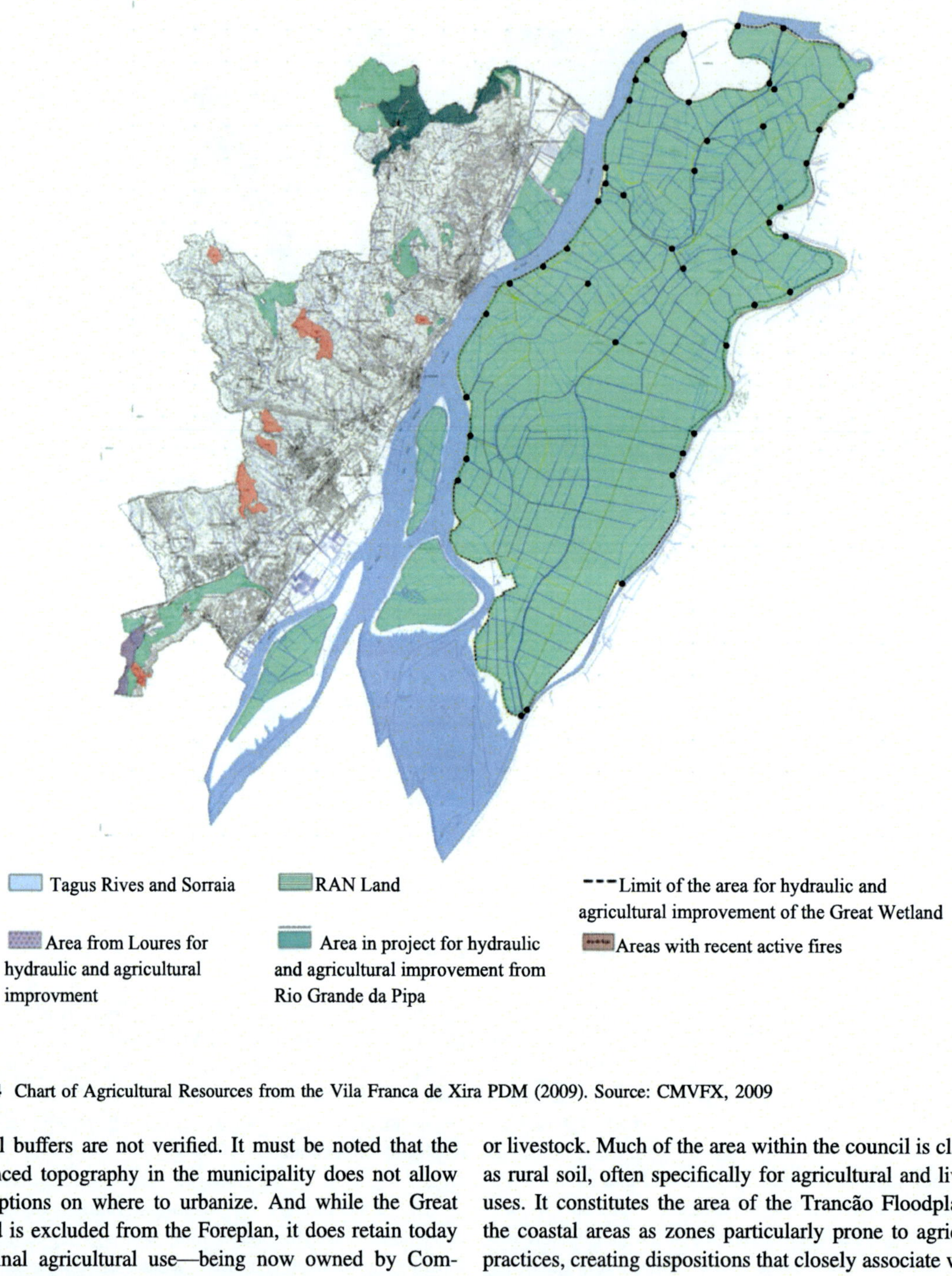

Fig. 9.14 Chart of Agricultural Resources from the Vila Franca de Xira PDM (2009). Source: CMVFX, 2009

the rural buffers are not verified. It must be noted that the pronounced topography in the municipality does not allow many options on where to urbanize. And while the Great Wetland is excluded from the Foreplan, it does retain today its original agricultural use—being now owned by Companhia das Lezírias, a state-owned company.

The PDM for Loures was approved in 2015 (CMLoures 2015). It highlights the need to preserve agricultural land from unplanned urbanization, loss of productive soil and landscape degradation, problems which had a strong impact in Loures in the past. Agri-industrial activities are constituted as a specific activity, for transformation of agriculture or livestock. Much of the area within the council is classified as rural soil, often specifically for agricultural and livestock uses. It constitutes the area of the Trancão Floodplain and the coastal areas as zones particularly prone to agricultural practices, creating dispositions that closely associate with the food system in more than its productive phase, specifically through the creation of a support centre for agriculture, including several sorts of agricultural practices and biological markets (Fig. 9.14).

Moreover, in the same areas, the PDM proposes a reconversion of abandoned farms and structures associated with traditional agriculture, simultaneously expanding agricultural land and improving architectural heritage. Unlike in Vila Franca de Xira, the territory of Loures extends well beyond the area affected by the Gröer Foreplan. Because of its topography, Loures has been naturally polynucleated, with a complex network of villages, towns and cities separated by fertile land and forestry areas. This polynuclear structure has proved difficult to manage overtime, with the municipality of Odivelas forming from the southwestern area of Loures in 1998. On the riverside area, most of the land is classified as urban (of different types), with a few industrial estates and free spaces. Areas for equipment and buildings of public interest are also a relevant land-use. The only relevant exception if the floodplain of Trancão, with an extensive area classified as a highly productive rural area and the remainder as an area for conservation or as a natural space.

In Vila Franca de Xira, a special chart of agricultural resources surveys the land included in the RAN, as well as the complex infrastructuring of the Great Wetland, with its system of irrigation ditches, trenches and roads, as well as water gates. On the western margin of the Tagus River, urbanization and industrial complexes dominate, and there are only a few rural areas scattered in among topographical variation. In the riverside area of Loures where urbanization continues towards Lisbon, settlements are placed between the Tagus and the Trancão. The floodplain of the latter constitutes one of the most extensive agricultural areas of the whole municipality of Loures, while between settlements, there are almost no areas classified as rural.

Today, this is a conurbation, extending from the high-density suburban neighbourhoods within the Lisbon city (Olivais, Chelas) until Vila Franca de Xira in a nearly uninterrupted urban chord. The most contradictory forms and uses coexist without coherent articulations or separations (See Fig. 9.22), alternating and mingling in every landscape unit. It took over half a century of interventions from the municipalities to ensure that at least some urban amenities were introduced in many of these spaces.

While the space for food production was present in the Foreplan and while Gröer rightly acknowledged its importance and its potential, it was the key sacrifice to make way for the suburban extension that supported the growth of the capital city and its metropolis.

The separation of urbanization policies from rural activities, contributed to suggest that rural buffers lacked a proper function, and while this can be unsurprising in the territory of the Foreplan, it also happened with the PUCS, despite its full approval (Marat-Mendes 2009).

Perhaps more than in any other of Étienne de Gröer's works for the Lisbon Region, the partition between urbanism and agroforestry policies unquestionably undercuts the Foreplan interpretation of Howard, ultimately favouring speculative processes that lead to the emergence of illegal neighbourhoods, many of which extending over most of the planned rural buffer.

9.4 Opportunities for a Better Food System

Our survey on the food system of the Lisbon Region, presented in Chap. 5, includes many examples from the territory from eastern Lisbon to Vila Franca de Xira. Given the centrality of some examples—notably the MARL and the Great Wetland of Tagus—we opted for keeping the key examples of our survey all on the same chapter. With this option, we confirm that in this territorial unit, there are indeed many important spatial and built elements of central importance for the regional food system.

We may recall that this chord is present with agriculture of several scales and types (see Figs. 5.16, 5.27, 5.33, 5.38 and urban agriculture case studies 06, 07 and 18), factories and storage (see Figs. 5.50, 5.51, 5.53), several kinds of food venues (see Figs. 5.74, 5.79, 5.80, 5.85, 5.86, 5.87) and restaurants (see Figs. 5.100 and 5.104). A particularly important element is the presence of the MARL (Figs. 5.62, 5.63, 5.64, 5.65, 5.66), a uniquely representative structure of the contemporary food system made of long-distance flows of food produce which arrive in immense quantities to the surroundings of the largest agglomeration—that of the Lisbon city, to be distributed from thereon out to markets, supermarkets and stores of closer proximity to the population of consumers.

Thus, from territories of the highest productivity (the Great Wetland, the Trancão Floodplain) to the regional supply market to several industrial units, the existing conditions for reorganization of the food system within a local, sub-regional and even regional perspective are already favourable. Obviously, since the Lisbon Region concentrates nearly a third of the continental population of Portugal, it would be impossible for all the food needs to be met strictly from within the region itself. At the same time, transformation is necessary, and opportunities to articulate different phases of the food system, by promoting and supporting a more systematic organization, must be highlighted. Thus, in addition to the structures and spaces already noticed in Chaps. 5 and 8, here we will highlight spaces of opportunity which, under a specific programme for renewing the food system, could become the key spaces of a new foodscape, simultaneously improving the food system and the physical structure of the metropolitan region—a step that is particularly important since many of the main reasons for these areas to be classified as 'problematic spaces' or 'critical

Fig. 9.15 'Saloia' and rural housing in different states of conservation: Sacavém, Loures (top, left), Bobadela, Loures (top, right), Camarate, Loures (bottom, left) and Alverca, Vila Franca de Xira (bottom, right). Source: Authors

urban areas' in the 2002 PROT-AML have not changed so drastically since then.

A first area that would justify closer attention is the physical condition of 'saloia' housing, as well as of other remanents of rural architecture which remain in many of the settlements of this conurbation, even in cases such as Sacavém, Alverca or Alhandra. Such rural remains are intertwined with modern buildings, due to an urbanization process that favoured the demolition of individual plots and their fulfilment with taller apartment buildings. In many cases, these remanents are vacant, abandoned and often even derelict. It is not difficult to imagine that, lacking protection, their destiny is demolition (Fig. 9.15).

Yet rural architecture is one of the physical elements in this sub-regional chord that more strongly recalls the bygone past of this territory as a hinterland for food production. The protection of 'saloia' architecture and other forms of rural construction can thus be directed specifically at the physical characteristics of buildings or it can be more comprehensive, and prompt connections to the land, to cultivation spaces or even to consumption and trade, depending on its context. Regardless, to use old rural structures to revive the role of food in urban conviviality is an opportunity to bring back the rural into the urban realm without the nostalgic—and borderline reactionary—overtones of ruralist discourse, but instead using food as a key activity and function of the human habitat. While many cities are promoting solutions like gastronomic neighbourhoods or blocks (Parham 2015; Manning 2016), the position of 'saloia' housing in the settlements of this chord may suggest a different approach, where these gastronomic spaces are disseminated in the urban fabrics. This way, instead of monofunctional concentration, convivial food spaces would be in proximity to different neighbourhoods and areas within each settlement, while their rehabilitation, if articulated with optimization of vacant land in the vicinities, could contribute to improve both the built and the natural landscape. Indeed, the general situation of 'saloio' buildings in all the Lisbon Region demonstrates the mutism of spatial policies in Portugal in face of private property as well as in face of the need to create quality public spaces, both in urban settlements and outside them. Such a programme for a network of food spaces in 'saloia' architectural examples could be negotiated with the owners of derelict properties, to distribute benefits and responsibilities, for example, promoting a publicly funded reconstruction of derelict structures in exchange for periods of concession of that structure for food-related activities, from small artisanal factories to canteens and restaurants.

The problem of derelict or abandoned rural buildings is added by the problem of untreated spaces, which is extremely common, ranging from the intersticial spaces between different buildings or neighbourhoods, to the 'green' spaces in their precincts, which often are merely constituted by grass. Such spaces provide little to no landscape quality and their effect—if they have any—is vaguely aesthetical, adding nothing to the environmental quality of the settlements in question. This situation can be verified in areas without construction, but also in areas planned by the private and the public sectors, and even in urbanizations of a very recent date. Notably, there are many examples in this chord of privately led urbanizations where all care and all design are placed in the buildings, with the open areas merely responding to legal requirements, lacking any treatment or landscape arrangement (Figs. 9.16 and 9.17).

Such spaces, often of a considerable scale, would have an obvious use as spaces for agriculture or, should this be less appropriate, as agroforestry. If no interest is demonstrated by residents in having vegetable gardens—and often, there really seems to be none—then the use of these green buffers as spaces for planting fruit trees is a viable solution. Agroforestry withstands urban activities better since, unlike vegetable gardens that do demand a certain enclosing to protect the fragile vegetable shoots during their growth, fruit trees can perfectly coexist with the typical equipment of urban green spaces, including children playgrounds, kiosks, benches and tables or even work-out circuits.

On the other hand, it is very common in Lisbon and its suburbs to find derelict olive groves, while fig trees and loquats are also common to find. These trees, which seem to survive even without special care, must be understood as wasted resources, and their rehabilitation, as well as a proper harnessing of what they produce, must be promoted. The creation of teams for treating these abandoned fruit trees and for stowing their produce is yet another example of how food activities can simultaneously improve the landscape, optimize resources and create jobs. Should vacant or underused 'green' spaces in and around urban neighbourhoods be mobilized for agroforestry, such teams will have plenty to work, and plenty of food can be retrieved.

Another point of opportunity has to deal with industrial estates. As discussed in Chap. 4, Portugal practically transitioned in the second half of the twentieth century from a rural to a service economy, with industry fundamentally underdeveloped. The Lisbon Region was one of the key territories for the attempts at industrial development, notably in this chord, prompting the Gröer Foreplan—and the Campos revision—as well as the Chelas Urbanization Plan. However, the industrial fabric in the northern chord did not resist for a long time. Many estates were and are abandoned and derelict, leaving large areas without any use and

Fig. 9.16 Unused land: Cristo-Rei Urbanization in Moscavide, Loures (left) and in Santa Iria de Azóia, Loures (right). Source: Authors

Fig. 9.17 Unused land: derelict olive grove in Camarate, Loures (left) and vacant space in Vialonga, Vila Franca de Xira (right). Source: Authors

Fig. 9.18 Unused industrial spaces in Bobadela, Loures (left) and Póvoa de Santa Iria, Vila Franca de Xira (right). Source: Authors

generating a landscape of decay. Both buildings and outside spaces exist in these poor conditions. Quite often, the solution has been the refurbishing of such buildings for services and for the cultural and creative industries. An alternative solution, taking heed of the proximity of these industrial areas with rural fields and of the availability of large vacant spaces, would be to promote or support innovative food businesses, cooperative productions, seeking to bring transformation of local produce as close to production spaces as possible (Fig. 9.18).

From the survey we conducted on the food system phases in the Lisbon Region, it is clear that there is an abundance of practices and spaces, although most of them are limited to the more conventional practices of recent modernity. Since the less conventional practices, including periurban and urban agriculture, are not properly acknowledged in planning instruments and are often conducted at the margin of the commercial food system, there are few options for the resulting produce other than self-consumption. A scaling-up of local cultivations, making the best use of unused, underused, vacant and derelict open spaces would increase the volume of production (Crowley et al. 2021). In such initiatives for recovery, an opportunity is generated to advance sustainable practices like organic and biological agriculture, agroecology and integrated production (Ferreiro et al. 2020; Salavisa et al. 2021a, b; Henriques et al. 2021). To keep this increase from being absorbed by the international food system and instead direct it to consumption within the region, intermediary spaces, where produce can be stored, transformed (if necessary) and traded are needed. Given the large dimensions of these industrial estates, it would be possible to create multifunctional spaces towards those goals—and giving the recent popularity of such food quarters and food districts (Manning 2016) there is a probability that these activities would attract both people and businesses to areas which, as they stand, are voids without any proper function.

9.5 A Integrating Urban Agriculture

One hundred years ago, all the eastern territory of Lisbon towards Vila Franca de Xira was rural. Be it with vegetable gardens, vineyards, olive groves, orchards or ploughed land, this was home to different forms of agriculture. And of agricultural labour, as that described in the early books of Redol (1971), who described the brutality of seasonal farmwork, which attracted outsiders from the rest of the country to the flat plantations of the Great Wetland. This rural tessellate was ravaged by modernization, by the uncoordinated construction of industrial estates and housing neighbourhoods of all forms and densities. But perhaps it was precisely because of this savage transformation that, on the recoils and slopes of the territory, some forms of agricultural life managed to survive, or to adapt. The complex diversity and almost site-specific adaptability (Parham 2015; Napawan 2016) characteristic of periurban agriculture has a privileged space in this chord. But the relative isolation of the Great Wetland of Tagus, crossed by the bridge road without really opening to it has maintained it quite untouched by this spree of modern change. And the flat 'mouchões', always visible and yet always reclusive, cut off from the land, also remain practically untouched as well.

In both Loures and Vila Franca de Xira, there are numerous examples of agricultural and horticultural practices, although these are not absorbed by the PDMs. Notably, all the spaces where we found agricultural practices of a non-commercial scale are classified either as urban land or as leisure spaces for landscape framing. In general, this demonstrates that agricultural practices are usually considered from the perspective of economic viability (Parham 2015), which means that some practices, especially those which are more characteristic of periurban spaces within metropolitan regions are still hard to be included in planning instruments. Small and larger vegetable gardens, agricultural explorations of several dimensions, olive groves whose status is unclear, walled farms—all of these are present in this chord, in both municipalities.

In 2009, the Lisbon Council officialized the first urban horticultural park, precisely in Chelas. This remains the largest municipal urban agricultural space, although it covers only a part of the valley bottom on the northern area of Chelas. On the southern area, where urban agriculture also had a very expressive presence, vegetable gardens have been recently cleared to make way for tree areas, while some specific spaces are permitted for urbanization (Fig. 9.19).

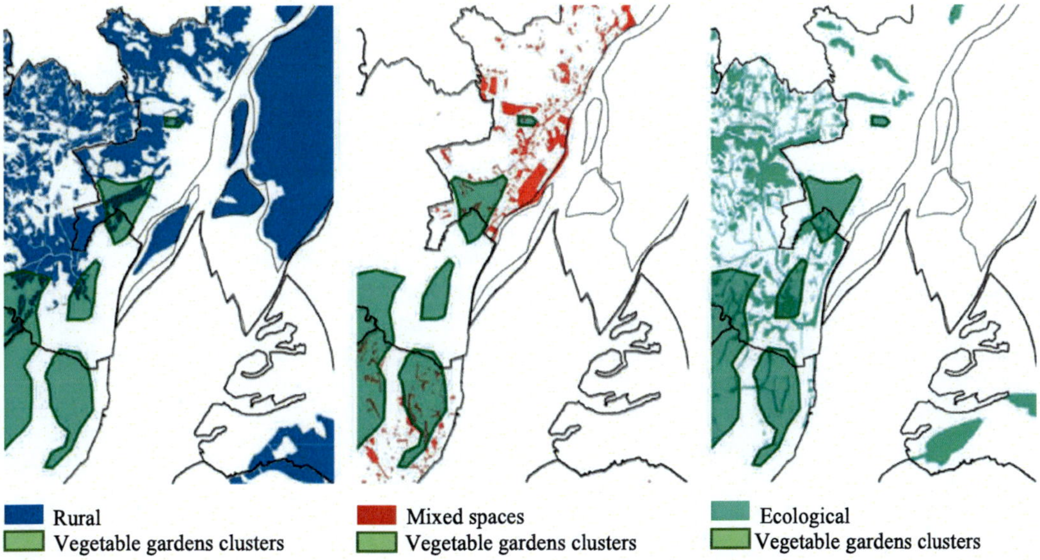

Fig. 9.19 Main clusters of urban agriculture over municipal land-uses—rural, mixed and ecological. Source: Authors

Fig. 9.20 Examples of urban agriculture in the neighbourhoods of Chelas, Lisbon. Source: Authors

In the past decade, urban agriculture has become a much more common practice, not only in Lisbon but also outside it. In the three municipalities of the chord from eastern Lisbon to Vila Franca de Xira, there are many examples of vegetable gardens which have undergone municipalization, while numerous informal examples remain also. These occupy a variety of situations, from the borderlines of housing neighbourhoods to more classic fringe belt situations, where topography impends construction and the local residents eventually start to use the leftover space for food cultivation. A particularly good example is precisely the Chelas Valley, which includes different types, morphologies and scales of urban agriculture, in both municipal and informal framework (Borges et al. 2022). The horticultural park is located in the steepest hill of the valley, while in some of neighbourhoods, this practice has been verified in urban plots left without construction.

Through the Loures riverside, there are several examples, both in fringe belts and in leftover urban space. To these, one may add the numerous olive trees and olive groves that remain either on farms which survived the urbanization spree of the 1960s and 1970s or on the corridor spaces above the water channels of Tejo and Alviela pass, where construction is obviously precluded. These spaces were especially favourable to the creation of 'courelas', which were strips of kitchen gardens linked to houses in the vicinities. These 'courelas' remain to the present day—and have even provided the name for an illegal neighbourhood[13] in Pirescoxe—and often exist as the backyards for the ground floors of collective buildings.

Another generator of cultivation spaces are the corridors of the EN10 and the highway A1, while in the area of Póvoa de Santa Iria there is a very large strip of vegetable gardens on the margins of the trainway. Póvoa de Santa Iria indeed includes one of the most significative clusters of vegetable gardens in this chord, with several municipal allotment gardens and large extensions of informal gardens on the northwards fringe belt on the way to the town of Forte de Casa. Vegetable gardens become less numerous and smaller the closer one gets to Vila Franca de Xira, where only a few examples can be encountered.

[13] This neighbourhood, constituted by detached houses with surrounding yards, has recently been legalized.

Fig. 9.21 Examples of informal urban agriculture in Sacavém, Loures (top, left), Bobadela, Loures (top, right) and Póvoa de Santa Iria, Vila Franca de Xira around the trainway (bottom, left) and on the northwards fringe belt (bottom, right). Source: Authors

This has to do with the increase in rural space—since settlements in the Vila Franca de Xira municipality are larger but more contained by topography—as well as with the large number of individual houses which include a private yard, often used for agriculture.

The key challenge in this area is probably not the scaling-up of urban agriculture, but its links with rural agriculture as well as with the municipal ecological structures of both municipalities. Because of topographical constraint, in this chord, it is possible to articulate the large agricultural areas around cities and towns with smaller vegetable gardens, which penetrate into the urban settlements themselves. Problems of spatial fragmentation are pointed out in the classification advanced by the 2002 PROT-AML. The presence of these different types of productive activities could be used to respond to the challenges of greater connectivity. The proposal of a CPUL (Viljoen et al. 2005) would encounter here a literally fertile ground, uniting three important points: rehabilitation of derelict rural explorations, creation of soft mobility circulation through continuous paths linking urban allotment gardens—as already done in the Eco Parque vegetable gardens in Póvoa de Santa Iria—and an intensification of the links between urban and rural spaces (Figs. 9.20, 9.21 and 9.22).

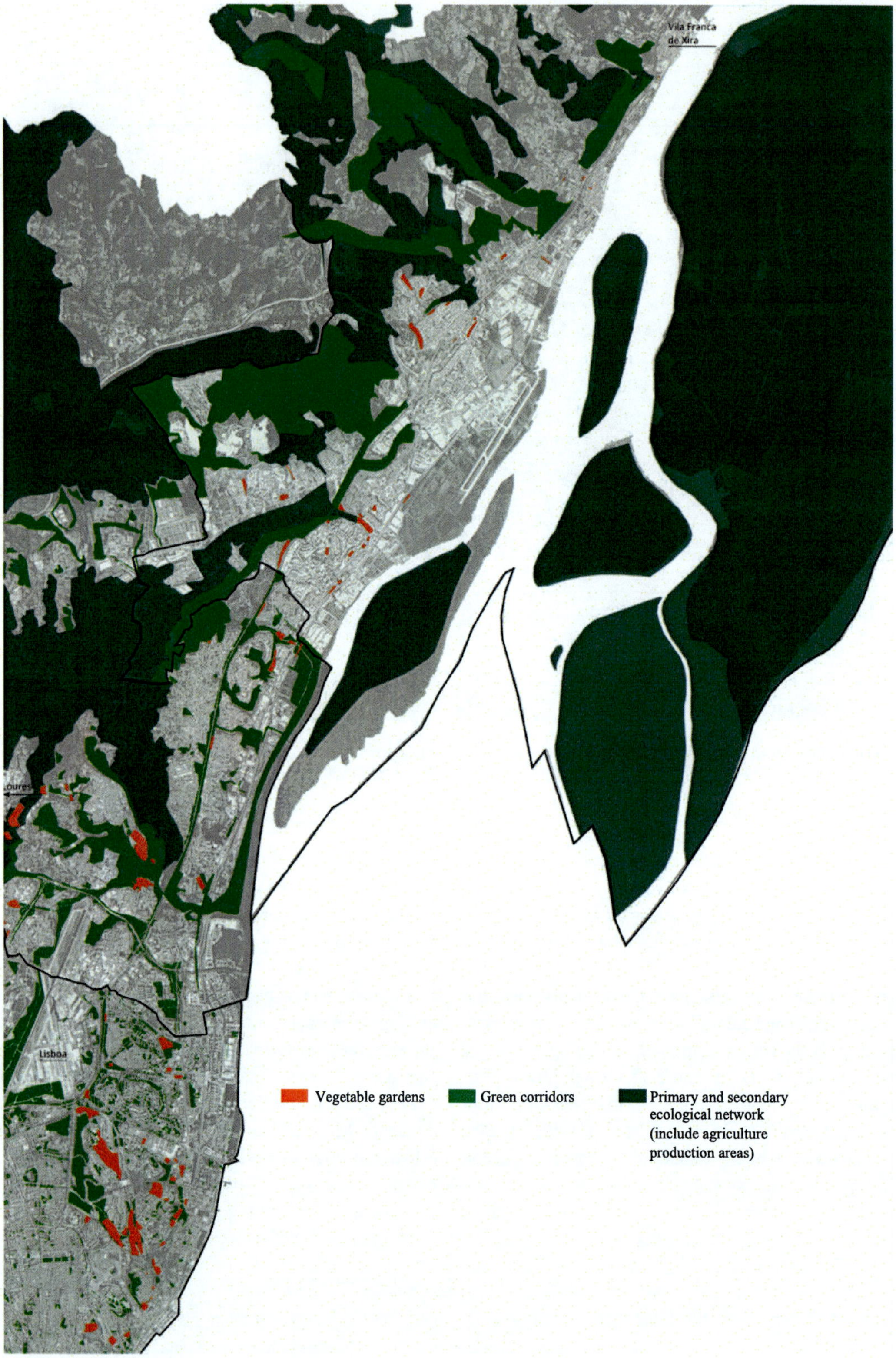

Fig. 9.22 Vegetable gardens, green corridors and ecological network in the Lisbon–Vila Franca de Xira axis. Source: Authors

9.6 A Continuous Foodscape?

As discussed above, all the food system phases are present in the chord from eastern Lisbon to Vila Franca de Xira, and several types and scales are included, from the familiar self-consumption vegetable garden to the metropolitan-scale supply market. Yet in the shelves of the supermarkets and hypermarkets of the Lisbon Region, there is no significative presence of produce from within the region.

The challenge thus is how to articulate the phases of the food system so that they do function in a systemic manner within the region, and to locate and design the spaces where such articulation is studied, promoted and practiced.

If the planning opposition between urban and rural land often responds to the most convenient organization of space towards the needs of the (capitalist) production system, in this chord, such organization has visible impacts, since it was a heavily industrialized suburb of the capital city, where factories and other industrial estates emerged in ways that they did not in other places of the country. Although the Gröer Foreplan seeks to harmonize industrial development with the maintenance of agricultural supplies for the capital city, this was not to be the future development of this area which, locked between the Tagus and the A1, became intensely urbanized, almost without interruption, and often with promiscuous coexistence of industrial and residential activities.

Moreover, the topographical conditions the this chord also favours the distinction between the urban areas and the areas where rural land-uses remain, with the ramifications of the Trancão Floodplain and its coastlines westwards, and the 'mouchões' and the Great Wetland eastwards. Although, as we saw on Chap. 1, it was speculated that in the future a new settlement could be created on the opposite side of the Vila Franca de Xira bridge, such urbanization has never come to pass. At the same time, the settlements on the urban areas have gained tremendous density, with a proliferation of apartment buildings in all towns and cities on the chord.

This coexistence of the urban and the rural in such close proximity can be used to generate new fluxes of food produce to make the food system more regional. The large production areas are located around the conurbation where transformation and consumption spaces are located. To these, one may add the numerous examples of spaces of opportunity and for landscape recovery, on which additional areas for cultivation, transformation, trade or consumption may be placed. From the survey and the interpretation of the existing activities, a plan can be programmed and designed, seeking to enlist which other activities and spaces may be more helpful to compensate for absent activities, to advance innovative practices and to increase the production of fruits or vegetables which imply a dependency from outside sources.

Such a reorganization of existing activities and addition of new ones can facilitate the urban–rural links and enlarge the input of local producers and business owners on the regional food system from a city-region perspective. Moreover, such a strategy for transitioning the food system towards sustainability has the potential to gather local actors around a coordinated effort, while the role of the public sector may be ensured through new spaces to create through the rehabilitation of vacate and derelict areas. Such a role can be regulated through the creation of Food Councils and the design of food policies, similar to examples already in existence in several countries (Marat-Mendes et al. 2021). The regional industrial fabric may also be enriched by such a strategy, creating a reasonable and realistic assessment of the needs of the region and of which produces must be encouraged.

A programme aimed at increasing the circularity of the regional food system implies shorter distances between the different phases—however, the ambiguities of food miles as a sustainability indicator has been noticed in the literature on the subject (Sonnino 2013; Hardman and Larkham 2016). Thus, the programme must also prevent forms of transportation, seeking to promote the decrease in food miles as well as the use of sustainable forms of mobility.

The CPUL proposal (Viljoen et al. 2005; Viljoen and Bohn 2012) has been acclaimed for promoting a more sustainable urbanism and a better balance between urban and rural. It is one of the first examples of an urban model that aims at dissolving the limits between urban and rural without reducing rurality to a generic green backdrop, precisely highlighting that forms of cultivation are as complex and variegated as urban forms. However—and following a suggestion often hinted by the authors of the CPUL proposal—a continuous structure of food production will be much more sustainable if it opens up to other phases of the food system, using the produce from agriculture immediately in the surrounding areas. Furthermore, the creation not of a continuous productive urban landscape but of a continuous foodscape, where productivity and other activities intertwine, opens up new opportunities for architects and urban designers to participate in the territorial renewal that such a strategy can kickstart.

The notion of 'continuum naturale' by Telles (1997), which he applied in his 'Green Plan' for Lisbon, serves as the best reference to understand the role that urban agriculture can have in sewing the urban and the rural, as well as in bringing biodiversity and rurality into cities. This 'continuum naturale', if it integrates agriculture, may effectively function as a way to reconnect urbanites with the land and with nature, reminding those who live surrounded by artifice that nature too has its joys and its forms of sociability and conviviality, where the 'continuum naturale' encounters the 'continuum culturale'. Indeed, vegetable gardens at the peripheries of many European cities were once precisely spaces for taking walks, eating and making friends (Parham 2015), a form of leisure that, unlike those we seem to prefer now, coexisted harmoniously with productivity and with the environment (Figs. 9.23, 9.24, 9.25, 9.26, 9.27, 9.28 and 9.29).

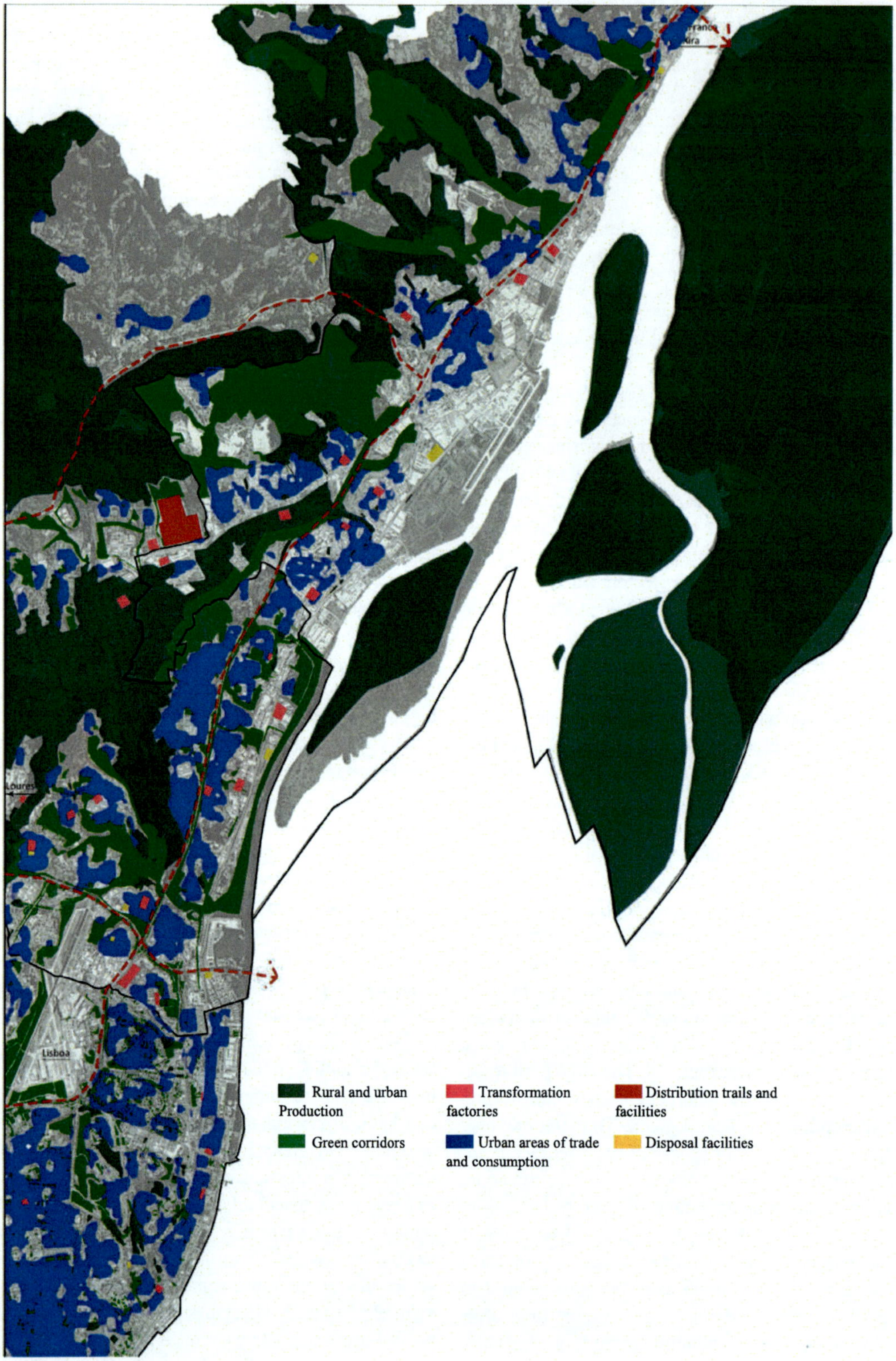

Fig. 9.23 Ecological network and food system related spaces in the Lisbon–Vila Franca de Xira axis. Source: Authors

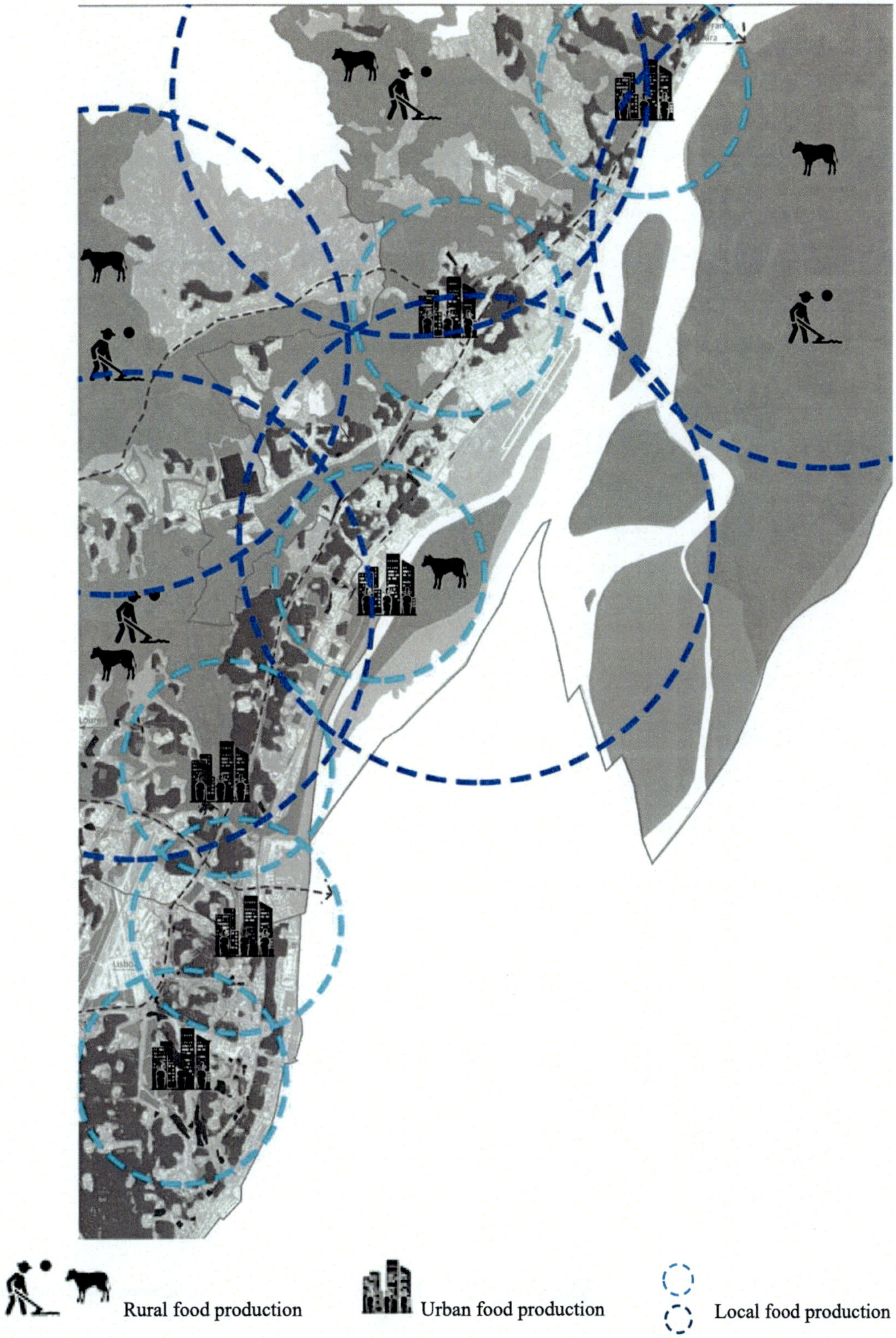

Fig. 9.24 Valuing and integrating the surrounding (peri)urban and rural fields, strengthening rural–urban linkages. Source: Authors

Fig. 9.25 Linking local food production and processing, regenerating the regional industry. Source: Authors

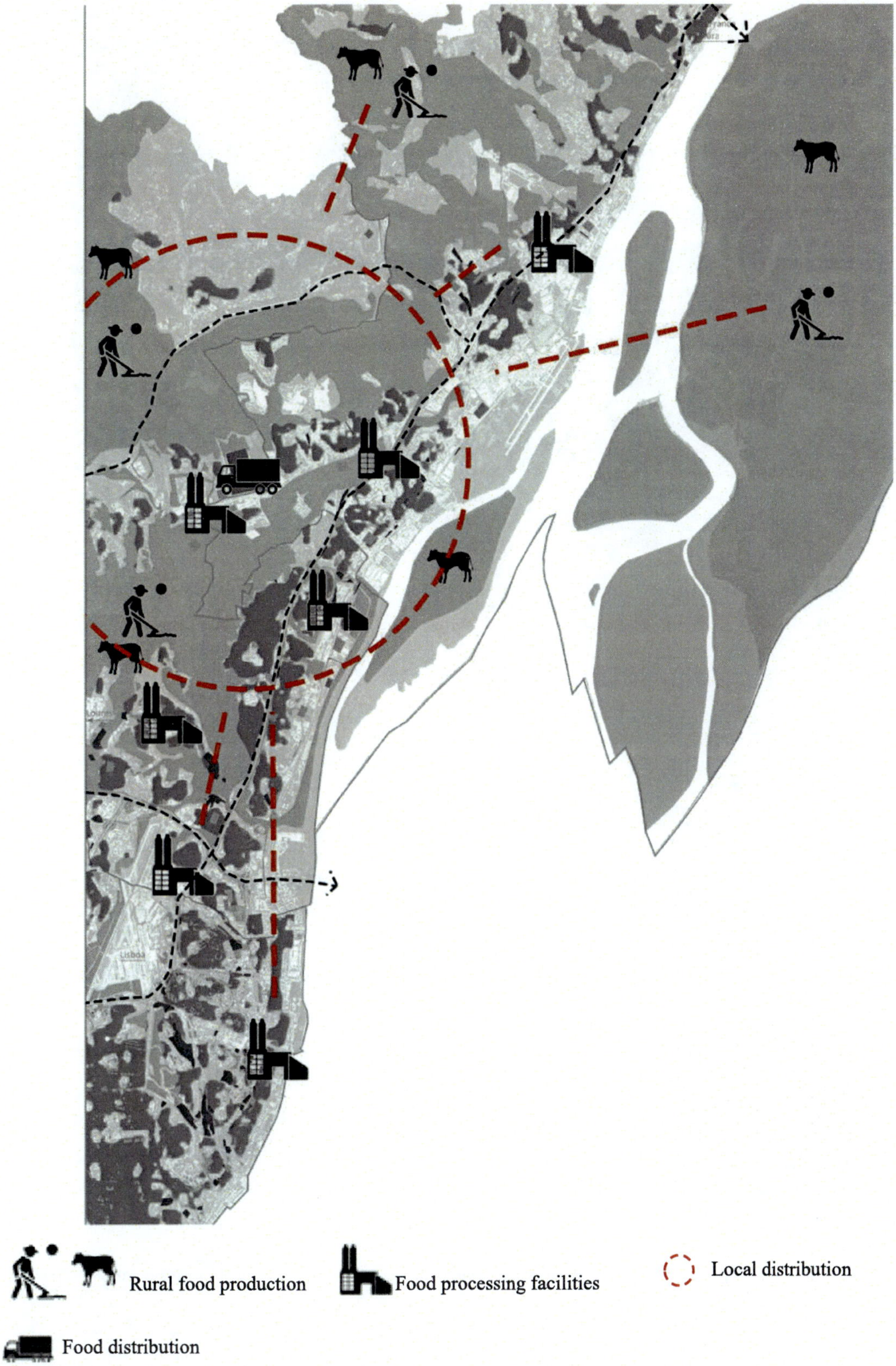

Fig. 9.26 Prioritising local distribution organizations and more sustainable means of transportation. Source: Authors

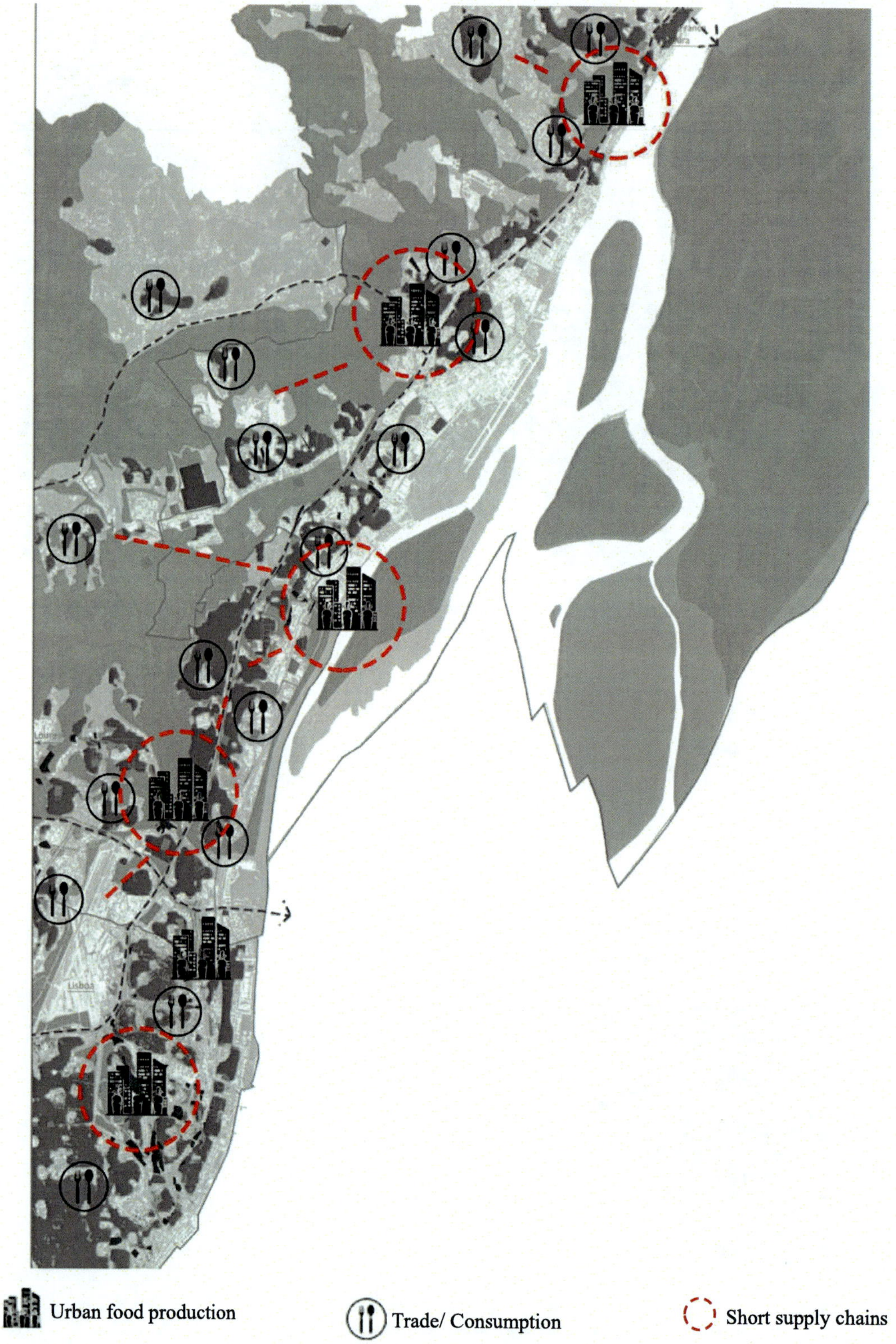

Fig. 9.27 Promoting urban alternative distribution circuits, such as proximity distribution in soft mobility. Source: Authors

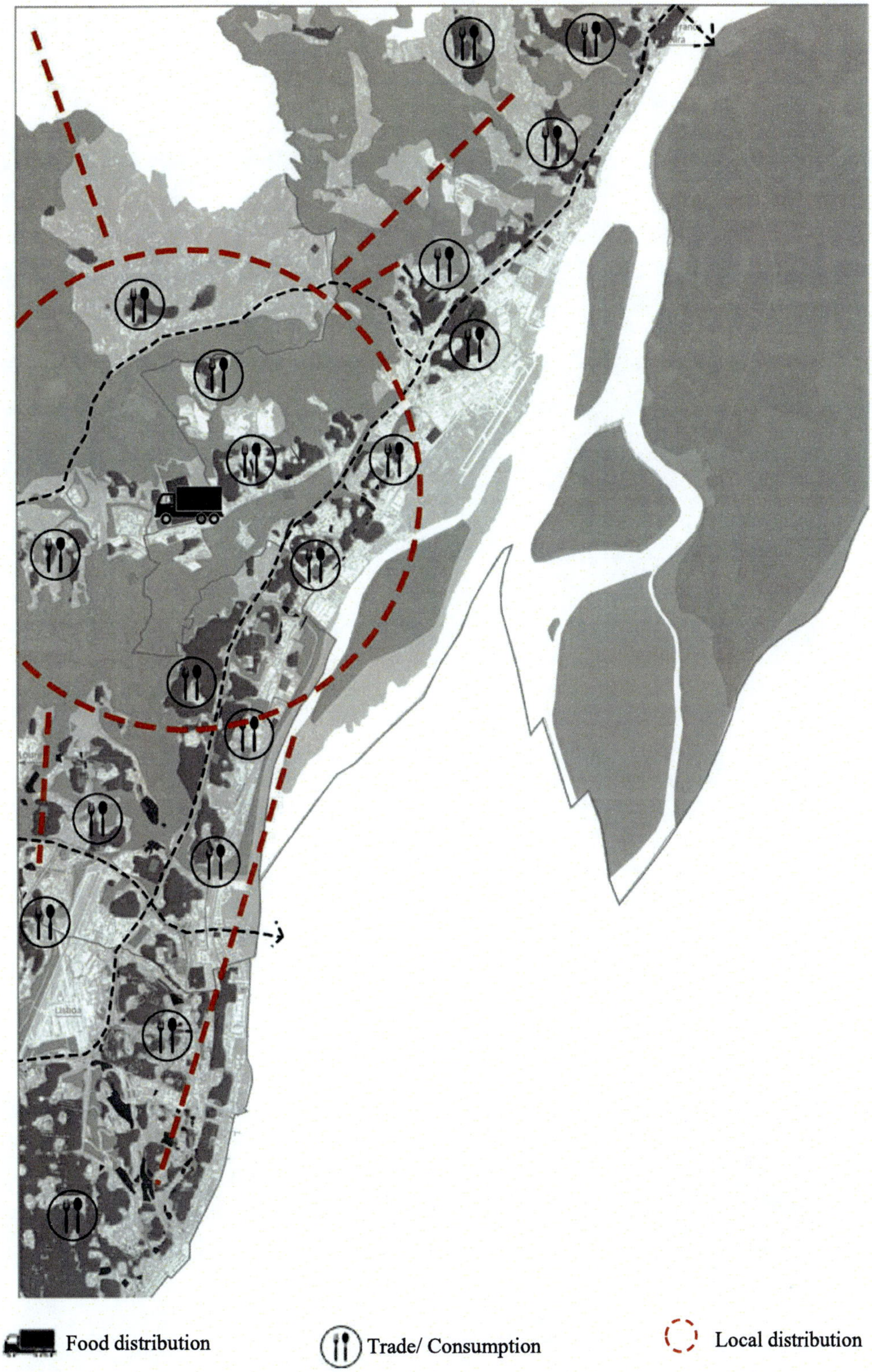

Fig. 9.28 Promoting access to healthy food and enhance urban and regional localized consumption. Source: Authors

Fig. 9.29 Increasing organic waste recycling or reducing food waste and emission reductions, through the functional articulation of the various phases of the food system. Source: Authors

References

Baptista LS, Melâneo P (2020) Cidade consolidada. In: Baptista LS, Melâneo P (eds) Almada: um território em seis ecologias. Câmara Municipal de Almada, Almada, pp 11–16

Borges JC, Marat-Mendes T, Lopes SS (2020) Chelas Zone J revisited: urban morphology and change in a recovering neighbourhood. In: Strappa G, Carolitti P, Leva M (eds) Urban Substrata and City regeneration. Proceedings of the 5th ISUFitaly International Conference, Rome, 19–22 Februray 2020, pp 751–760

Borges JC, Lopes SS, Fernandes RDP, Marat-Mendes T (2022) Planning at the edge: urbanism and socio-political transition in Chelas, Lisbon. Plan Perspect. https://doi.org/10.1080/02665433.2021.2001364

Campos, FALB (1956) Anteplano Moscavide – Vila Franca de Xira – remodelação. Arquivo Municipal de Loures, Processo 2116 – folhas 122–109

CCDRLVT (Comissão de Coordenação e Desenvolvimento Regional de Lisboa e Vale do Tejo) (2002) Plano Regional de Ordenamento do Território – Área Metropolitana de Lisboa PPROT-AML. Published in Resolução do Conselho de Ministros n.o 68/2002. Diário da República, Série I, Nº 82, 8 Abril 2002, 3287-3328. http://www.ccdr-lvt.pt/pt/plano-regional-de-ordenamento-do-territorio-da-area-metropolitana-de-lisboa/54.htm

CMLoures – Câmara Municipal de Loures (2015) Em Loures o Passado tem Futuro – Santo António dos Cavaleiros. Loures, Câmara Municipal de Loures

CMVFX - Câmara Municipal de Vila Franca de Xira (2009) Plano Director Municipal De Vila Franca De Xira. Warning 20905/2009. Diário da República, Series 2 — N.224 — November 18th

Crowley D, Marat-Mendes T, Falanga R, Henfrey T, Penha-Lopes G (2021) Towards a necessary regenerative urban planning. Insights from community-led initiatives for ecocity transformation. Cidades, Comunidades Territórios [Online] Sp21 | 2021. https://doi.org/10.15847/cct.20505. UIDB/03127/2020. 83.

CSOP – Conselho Superior de Obras Públicas (1955) Parecer dirigido ao Ministro das Obras Públicas. Arquivo Municipal de Loures, Processo 2116, pp 75–22

d'Almeida P, Marat-Mendes T, Toussaint M (2020) Portugal's rising research in architecture and urbanism: the influence of international research centers and authors. J Urban Hist 1–28. https://doi.org/10.1177/0096144220968078

Ferreiro MF, Salavisa I, Bizarro S, Soares M (2020) O sistema alimentar em Portuga. Cidades, Comunidades & Territórios, p 41. http://journals.openedition.org/cidades/3026

Folgado D (2009) A nova ordem industrial. Da fábrica ao território de Lisboa: 1933–1968. PhD thesis, Universidade de Lisboa - Faculdade de Letras, Lisbon

Gröer É, Gröer N (1947) Plano Regional de Moscavide – Vila Franca de Xira – esboço. Arquivo Municipal de Loures, Processo 2116 – folhas 9-2

GTH – Gabinete Técnico da Habitação (1965) Plano de urbanização de Chelas. Câmara Municipal de Lisboa, Lisbon

Hardman M, Larkham PJ (2016) Informal Urban Agriculture. The Secret Lives of Guerrilla Gardeners. Springer, London

Heitor T (2001) A Vulnerabilidade do espaço em Chelas: uma abordagem sintáctica. Fundação Calouste Gulbenkian, Lisbon

Henriques CN, Pinto TC, Costa P, Marat-Mendes T (2021) Tracing Lisbon metropolitan area's foodscape. Cidades, Comunidades Territórios 42. http://journals.openedition.org/cidades/4065

Manning J (ed) (2016) Food and the city. Urban Des 140:14–39

Marat-Mendes T (2009) O PUCS e os vazios planeados. Novas oportunidades para o ordenamento sustentado da Costa do Sol. In: Pereira M (ed) O Plano de Urbanização da Costa do Sol - Uma visão inovadora para o território. Câmara Municipal de Oeiras, Oeiras, pp 92–122

Marat-Mendes T, Oliveira V (2013) Urban planners in Portugal in the middle of the 20th century: Étienne de Gröer and Antão Almeida Garrett. Plan Perspect 28(1):91–111

Marat-Mendes T, Sampayo M (2010) Étienne de Groer: the scales of urban intervention in the Lisbon Territory. In: Proceedings of the 1st European architectural historians network conference, Guimarães, Portugal, pp 32–39

Marat-Mendes T, Isidoro I, Catela J, Pereira J, Borges J, Lopes SS, Henriques CN (2021) Drivers of change: how the food system of the Lisbon Metropolitan Area is being shaped by activities, initiatives and citizens needs towards a sustainable transition. Cidades Comunidades Territórios Special Issue 41–62

Napawan NC (2016) Complexity in urban agriculture: the role of landscape typologies in promoting urban agriculture's growth. J Urban: Int Res Placemak Urban Sustain 9(1):19–38. https://doi.org/10.1080/17549175.2014.950317

Parham S (2015) Food and urbanism—the convivial city and a sustainable future. Bloomsbury, London

Pereira MIBR (1997) P.M.O.T. – A regulação quantitativa e a qualidade do ambiente urbano. MSc Dissertation. Universidade Técnica de Lisboa, Lisbon

Pereira M (2009) O Plano de Urbanização da Costa do Sol – o pioneirismo de um plano sub-regional. In: Pereira M (ed) O Plano de Urbanização da Costa do Sol - Uma visão inovadora para o território. Câmara Municipal de Oeiras, Oeiras, pp 24–41

Redol A (1971) Gaibéus. Europa-América, Amadora

Rodrigues M, Freire P, Fortunato A, Alves E (2016) Characterization of the hydro-agricultural development of the Lezíria Grande de Vila Franca de Xira and of the Mouchão de Alhandra. Laboratório Nacional de Engenharia Civil, Lisboa

Salavisa I, Ferreiro MF, Soares M, Bizarro S (2021a) Sistema alimentar e sustentabilidade: o papel das políticas públicas e dos atores locais. In: Pinho (ed) Compêndio de Políticas Urbanas. FEUP-ISCTE-UA, Porto

Salavisa I, Ferreiro MF, Bizarro S (2021b) The transition of the agro-food system: lessons from organic farming in the Lisbon metropolitan area. Sustainability 13:9495. https://doi.org/10.3390/su13179495

Salazar AO (1933). Conceitos económicos da nova Constituição — Discurso radiodifundido da U. N., em 16 de Março. In: Discursos, Vol. I, Coimbra Editora, Coimbra, pp 199–204

Silva CN (1994) Políticas urbanas em Lisboa 1926–1974. Livros Horizonte, Lisboa

Sonnino R (2013) Local foodscapes: place and power in the agri-food system. Acta Agric Scand Sect B Soil Plant Sci 63(1):2–7. https://doi.org/10.1080/09064710.2013.800130

Telles GR (1997) Plano Verde de Lisboa. Colibri, Lisboa

Viljoen A, Bohn K (2012) Planning and designing food systems, moving to the physical. In: Viljoen A, Wiskerke JCS (eds) Sustainable food planning: evolving theory and practice. Wageningen Academics, Gelderland

Viljoen A, Bohn K, Howe J (2005) CPUL—continuous productive urban landscapes. Routledge, London

Zêzere JL (2001) Distribuição e ritmo dos movimentos de vertente na região a norte de Lisboa [report]. Centro De Estudos Geográficos Área de Geografia Física e Ambiente. https://core.ac.uk/download/pdf/237195915.pdf

Food: A Necessary Agenda Towards Our Common Future

10

10.1 Introduction

We have defended so far that food and food systems are of importance for architects, urban designers and urban planners for three fundamental reasons: first, they allow us to critically observe our conceptions of space and the way our habitats are structured; second, they allow us to observe and imagine socio-metabolic transitions, insofar as those include the transformation of the territory by human labour; and finally, because they allow us to identify popular initiatives regarding not only food access itself, but also the relationship with the landscape and with space.

In this chapter, we revisit the process of conception of this Atlas, as well as the key motivations and goals established for it. We then present the key conclusions, emphasizing the main aspects of the Lisbon Region food system and its evolution across history until the present. Moreover, the background, motivations and aims behind the Atlas and its relation to the work developed in projects MEMO and SPLACH is discussed, highlighting the role attributed to tools of visual characterization. Afterwards, we briefly overview the key results of each chapter, and, based on the resulting perspective, provide suggestions for a sustainable transition of the metropolitan food system and the spatial planning of its activities.

10.2 Resources in Metropolitan Space: A Journey in Research and Teaching

Urban planning and urban design disciplines have long been called to contribute for building more sustainable communities. Recognizing, describing and measuring urban metabolism is a fundamental step towards a sustainable transition and it calls therefore for appropriate tools to derive from the examination of the city better and suitable habitat solutions (Rogers 1998; Marat-Mendes et al. 2015; Marat-Mendes 2021). Such efforts have the utmost importance to provide society with effective tools to improve the built and natural environments and to make cities attractive and functional habitats without jeopardizing the environmental context which the city integrates. Yet although great advances have been made by the engineering disciplines, largely in the development of methods to quantify and measure the material flows that operate in the urban system, within the urban design context the study of the metabolism remains somehow obscure, obfuscating the necessary advances to direct the designing of the urban realm and of its necessary infrastructures towards a sustainable transition (Marat-Mendes 2020). While cities have often been studied from the standpoint of their physical structures, the morphology of their fabrics and the dynamics of urbanization activity, there is an increasing recognition of the role of nature in cities, the role of cities in the environment, and the critical role that flows and stocks of resources play in ensuring the urban life we often take for granted. Valuable transdisciplinary contributions from the domains urban design, engineering, urban history and environmental ethics have advanced important concepts from which to construct our perceptions on the problem of the metabolism of cities and its history (Fischer-Kowalski 1997; Oswald and Baccini 2003; Kennedy et al. 2011; Gandy 2014; Purdy 2015).

As recalled by Marina Fischer-Kowalski (b. 1946), the Austrian sociologist and social ecologist, a sustainability agenda is not one of urban designers or engineers, but of the overall society (Fischer-Kowalski 1998). Therefore, further efforts should be made on translating sustainability perspectives into urban design, calling a wider audience to participate, by demonstrating the intrinsic and everyday impact of urban systems in social and individual life. Resources such as food and water are case-in-point: no individual living in a city, no matter how alienated from the rural world, can deny the irreplaceable character of such resources, brought through hidden or discrete flows from outside to the city.

One way to clearly disseminate this message is through visual representation. That is the strategy employed in this Atlas, which seeks to provide a visualization of the historical and contemporary forms that operate on the territory of the

T. Marat-Mendes et al., *Atlas of the Food System*,
https://doi.org/10.1007/978-3-030-94833-7_10

Region of Lisbon to manage the access to resources and ensure proper sanitary and nutritional conditions for urbanites. From a better acknowledgement of the importance of resources, but also of the complicated physical and non-physical structures and practices that make them available, society can better understand the central importance of the territory. And from such recognition, decision-making—and support for decisions to make—can be more informed, more self-aware.

This Atlas integrates a survey of the forms assumed by the food system operating in the contemporary Lisbon Region, and is built over contributions gathered from a continued research process, coordinated by the first author of this Atlas at Iscte—University Institute of Lisbon, approaching the study of the urban metabolism from the perspective of visual characterization. Visual elements such as cartography and photography provide an obviously significative support to the study of urban form, and they also allow the mapping of the patterns of territorial use and transformation, especially if considered from an historical perspective. Moreover, if a more sustainable use and administration of resources implies better strategies and solutions from urban design and urban planning, visual elements are particularly suited to the ways in which architects and designers tend to think and work.

This Atlas retrieves the articulation between the perspective of socio-metabolism and that of design (urban and architectural), framing the history and the future possibilities of design within a multidisciplinary perspective, aimed at promoting more sustainable solutions. To do that, it builds upon research contributions provided by two research projects, MEMO and SPLACH, both financed by the Portuguese Science Foundation and other European Community funds. While the first and fourth authors of this Atlas were research members of Project MEMO, for Project SPLACH, all authors were part of the research team.

Project SPLACH provided an opportunity to further develop the metabolic visualization for Lisbon Region already attempted at MEMO, trough the development of a first Atlas dedicated to the analysis of Water and Agriculture sources at the Lisbon Metropolitan Area, in two specific periods of twentieth century, 1900 and 1940 (Marat-Mendes et al. 2015). Key water-access elements, such as wells, aqueducts, channels, fountains, water mills were then identified. The work furthermore observed the relation of water access with economic activities, specifically agriculture. Indeed, by the early twentieth century, water was used for agricultural, industrial or domestic uses.

The present Atlas, dedicated to the food system of Lisbon Region, continues the promotion of the metabolic visualization of territorial change. This exercise is considered by the authors as having the utmost importance, as it can clarify the ways in which architectural and urban forms are encapsulated in socio-metabolic processes. By focusing on the example of food and all the activities that ensure it, from production to consumption and waste, the Atlas aims to provide a different outlook into the historical process of formation of the metropolitan region around the Portuguese capital, while seeking to approach both the food system and the role of design in shaping it, from a sustainability perspective.

The Atlas is meant to portray the dynamic processes of physical, environmental and social transformation which have occurred in this particular territory, focusing on the twentieth and twenty-first centuries, precisely those where transformations were more radical and impactful. By focusing on the problem of resources in the metropolitan region, we aim to understand its environmental and metabolic performance:

> Yet such performance depends on how society recognizes urban form and controls it or the occurrence of flows and other dynamic roles throughout the natural and built environments. The key issue is exactly this. Society has, through time, perceived urban form and its inherent role in relation to the environment differently. Thus, the occurrence of specific environmental dynamics and flows have also varied through time. (…) Therefore, the flows and the dynamics occurring in such distinct cultural landscapes do also operate differently in relation to urban form (Marat-Mendes 2020).

Water, agriculture and food demonstrate this problem perfectly. They force us to carefully re-examine the functions of space around us, to question the provenience of the basic elements of our sustenance and to search for the 'hidden' elements which allow the city to thrive but are systematically kept from human sight by modern and contemporary values regarding sanitation and health (Gandy 2014). When we focus our attention on the resources we need and the distance they travel to get to us, and the way this distance in ensured, we see a vastly different city from the one typically assessed by architecture and urban morphology, the city of buildings, of streets, squares and leisure gardens. Sewers, water treatment stations, supply markets, distribution routes for heavy transportation, waste management facilities—these structures and activities, a common mark of the metropolitan periphery, are seldom within our sight, and seldom recognized for innovation in spatial or architectural arrangement. And yet, without them, urban life as we now conceive of it would be impossible. Such spaces and activities have effect been studied and understood by those concerned with urban systems, but their presence among the concerns of architects, urban designers and scholars of both these areas is still scant. Urban morphologists have recently directed their attention at the fine grain and wide diversity of green spaces within and around cities, but deeper studies of the morphological relations between the city and the systems—including resources—that sustain it are still necessary.

In agrarian societies of the past, water was a determinant factor for the maintenance of the economy, and its flow traversed to the urban landscape, making use of conventional urban forms, such as streets, buildings and squares. Agricultural use and urban domestic or public use were often both guaranteed by the same flow in a combined manner (Marat-Mendes et al. 2016). In what regards agriculture and food, it is also possible to perceive several differences with implications on the urban form, between the dominant agrarian economies from the contemporary one. The central importance of water-access structures, irrigation structures, agricultural support buildings and the dependency of certain food transformation activities with their surroundings were self-evident in the territory, as it was transformed (however slowly) by those living and working in it. But the relations between structures and infrastructures for water and food is much less evident in contemporary cities, and planning guidelines provide scant solutions to such issues. Nevertheless, contemporary society also depends on food to survive and on agriculture to sustain such production of food. New processes of food production and transformation therefore emerged with deep impacts on the environment (Steel 2020). The urban and architectural forms mobilized to physically realize these new processes were not always the most creative nor the most sustainable. Often, they applied industrial models of spatial organization to agriculture and to open space. Food commerce was increasingly integrated in large shopping areas, on public space but also in interior spaces such as shopping malls where all different kinds of commerce come together (Cachinho 2007). At the margins of this model where the key force is trade and profit, other forms of food production developed—not always in direct opposition to the commercial system (Parham 2015; Steel 2020)—which maintain a certain amount of local produce and often of self-consumption. Retrieving the spatial and physical implications of both these food-related activities is the ultimate goal of this Atlas.

Beyond the materials and findings retrieved from the above-mentioned projects, this Atlas also benefits from the research background experience of all its authors. Their experiences and research background also has a presence in this Atlas, while its key conclusions will continue to inform the research to be developed by the authors in the near future.

The first author is an architect (Faculty of Architecture of the University of Lisbon, 1994) with an MSc in Land-Use Planning and Environmental Planning (Faculty of Science and Technology from the New University of Lisbon 1999) and a Ph.D. in Architecture (The University of Nottingham 2002). In 2003, Teresa Marat-Mendes was appointed as an Assistant Professor at Iscte—University Institute of Lisbon, for the Department of Architecture and Urbanism. She became the responsible lecturer for the fourth-year Urban Project studio, connecting the study of urban form with sustainability, through urban design proposals for the contemporary city. That course offered an opportunity to test methodologies to engage architectural students with the study (analysis and proposal) of urban form with a more critical interest towards the processes of change associated to its inherent historical, socio-economic and environmental context throughout time, which ultimately determine the sustainability of urban form (Marat-Mendes 2002, 2004). The identification of the ground rules, or the Plan Principles, that regulate the original forms and guarantee their evolution over time, proved to be useful tools for students to determine which physical conditions or characteristics are maintained across time, and which are subject to alteration, transformation or destruction.

In this process, 'change' and 'time' emerged as the central features to evaluate urban form performance, according to several contexts under analysis (Marat-Mendes 2015a, b). Adaptability, flexibility, resilience and continuity emerged as the possible capacities of urban form to evaluate its performance throughout time and change, and consequently determining levels of sustainability to be ascertained. The application of this methodological framework at the Urban Project Studio was promoted in regard to the city students usually work on, Lisbon, but also to its Region or Metropolitan Area. Planning proposals for the city and the region were considered for analysis, including the 1930s and 1940s Urban Proposals for Lisbon by architect-urbanists Groër and Faria da Costa namely the residential neighbourhoods of Alvalade, Almada, Areeiro, Costa da Caparica and Restelo currently considered as remarkable examples of sustainable urban design, but also the PUCS, the first approved urban extension of Lisbon (Marat-Mendes and Oliveira 2012, 2013). Comparison of planning ambitions with territorial reality also proved effective tools for assessing the transitions between urban–rural realities, including the management water, land and crops (Marat-Mendes and Cuchí 2008; Marat-Mendes 2016). The specific cases of water and green spaces, studied alongside urban form, were effective tools of analysis of the results and the changes occurred in the territory of Costa do Sol (Marat-Mendes 2008).

Since 2011, the academic activity involved also the launching of a new course, Ecological Urbanism, offered to architecture students but also of other disciplinary areas. This new course provided a rich opportunity to enhance the sustainability debate from the metabolic perspective. Since 2017, also the thematic of food systems was introduced at the Urban Project studios and more recently, at the final-year project of architecture (Marat-Mendes 2017), while rural space has been also the theme of dissertations and thesis (Cabrita 2008).

The academic work developed at the Iscte-Lisbon University Institute Urban Project studios was naturally connected to the research conducted by the first author in the same timeframe, including the approach to the spatial dimensions and the visual characterization of urban metabolism (Marat-Mendes 2009), in particular in concerning water and green spaces. Such approach to urban metabolism, was initially explored in collaboration with colleagues from the Universitat Politécnica de Catalunya, notably with the research team led by the architect Albert Cuchí (b. 1957) and his students (Cuchí et al. 2008, 2010). This provided important inputs to advance the study of urban metabolism from the perspective of urban material analysis. The results would then be applied in a study for Santiago de Compostela (Cuchí et al. 2008) but also for the Lisbon Metropolitan Area (Marat-Mendes and Cuchí 2008), through the analysis of several pilot cases located at Costa dol Sol (Marat-Mendes and Cuchí 2007).

These works helped to identify a generalized fragility in spatial planning policies and in urban planning studies, consistent with researchers who have already argued that to improve urban design, greater acknowledgment of the metabolic functioning of the urban realm is required (Oswald and Baccini 2003; Kennedy et al. 2011). Moreover, effective assimilation of social history and environmental change are necessary to inform the implications of different urban and architectural forms. However, the challenge of making the complexity of urban metabolism understandable and useful to architectural students—and to a general audience—demanded the use of means which clearly identify with the materials architects and designers use to work. Visual elements proved effective in this regard, since they allowed architecture students, naturally inclined to consider problems from a spatial perspective and in relation to a specific site. Through mapping and comparison of photographic evidence, the more 'abstract' aspects of metabolism, as well as the inevitable effect of urban transformation in the environment and in the overall socio-ecological context, were all better understood by students, ultimately helping them clarify strategies towards sustainable design. This Atlas represents the attempt to compile and systematize the tools and data collected throughout the research and teaching journey where they were originally experimented.

The second and third authors of this Atlas, Sara Silva Lopes and João Cunha Borges, are both architects and former Architecture students from the Department of Architecture and Urbanism of Iscte—University Institute of Lisbon. At DINÂMIA'CET-IUL, they were research members for Project SPLACH and provided key contributions to the further improvement of the visualization process initialized at the MEMO project. Their strong involvement on the survey of the Lisbon Region forms focusing on its food system and urban agriculture contributed to a more substantiated and more realistic portrait of the ordinary forms exhibited by contemporary socio-economic dynamics and their spatial outcomes in the context of the Lisbon Region, but also in the lives of its inhabitants, and the features its economic fabric (Marat-Mendes et al. 2021d). Sara and João were also particularly engaged in problems related to housing neighbourhoods. In their individual researches, but also in their participation on the SPLACH Project, they emphasized the relations between the architecture of residential neighbourhoods and their public space—which often implied analysing the importance of food production in the everyday life of residents and of cultivation as an integral part of the landscape and the public space of neighbourhoods both in the Lisbon city and in its metropolitan region. These aspects, as well as the social history of key examples of modern housing development in the Lisbon Region, are concerns which integrate their respective PhD research works, supervised by the first author of this Atlas (Borges and Marat-Mendes 2019; Borges et al. 2020, Lopes et al. 2021; Borges et al. 2022).

Finally, the fourth author, Patrícia Bento d'Almeida, a postdoc researcher at DINÂMIA'CET, initiated her research activities at Iscte—University Institute of Lisbon, precisely through her integration at the MEMO research team, contributing to the first attempts of visual characterization of the metabolic functioning of the Lisbon Region. She was also a contributor for the development of the Atlas of Water and Agriculture of Lisbon Region 1900–1940 (Marat-Mendes et al. 2015). At present, focused on the history of architecture research, she provided valuable contributions for the identification and gathering of specific archived sources of information regarding Lisbon Region architecture and urban planning. Her former participation at the IAPXX, provided important methodological insights into the present Atlas, although guided by different aims. Moreover, her knowledge on the research conducted by architects at LNEC helped identify the conceptual and intellectual roots of contemporary planning instruments, as well as in establishing a critical perspective on the historical evolution of planning policy in Portugal (d'Almeida et al. 2020; Marat-Mendes et al. 2021a; d'Almeida and Marat-Mendes 2021).

10.3 Better Food Systems for Better Habitats

Food shapes our lives through values, laws, economies, homes, cities, landscapes, attitudes towards life and death, and perhaps because its impacts and implications are so overwhelming, it is almost too big to see (Steel 2020). Like nearly all aspects of contemporary life, food is organized, managed and ensured mostly within a capitalist system—and one of the key promises of capitalism was indeed that it ended scarcity. While social, political and economic crisis in the agrarian regime were caused by a crisis in productivity due to climatic shock, crisis in the capitalist world were never of scarcity. Yet throughout the already long history of capitalism and of the increasingly liberalized market, hunger has not been eradicated (Castro 1946; Steel 2008; Morgan 2013). The low prices achieved with the current global food system has not alleviated hunger and often, food relief policies encourage hunger and obesity by 'pushing unhealthy food onto the poorest' (Brinkley 2013: 245). It is hardly surprising, then, that ending poverty is still one of the chief goals of UN international agendas, including the 2030 Agenda for Sustainable Development (UN 2015), which makes it obvious that this is still a problem to be solved. Developing nations still struggle to feed their people, while in the developed countries there is a problem of overfeeding and waste (Steel 2020).

More than failing to end hunger, the latest development of capitalism, austerity capitalism, has even prompted new forms of poverty including the 'alarming spectacle of *nutritional* poverty, one of the most insidious forms of poverty' (Morgan 2013: 3–4, author's italics). Farmers produce daily nearly 2800 cal per person, enough to feed the entire world population under a properly organized food system—yet 850 million people go hungry in the world, and double that number have nutritional problems, including overweight and obesity (Steel 2020). Even in Europe, nearly 5% of the population risks malnutrition, with high numbers or vulnerable groups including the poor, the elderly and the sick (Reisch et al. 2013). How can food be still out of reach for so many in a world that produces such a massive, almost impossible to grasp, amount of wealth?

The plight of hunger in the modern world has been difficult to discuss as a public problem, and people have been deaf to it for reasons that are moral, economic and political. Moral, because hunger forces our rationalistic society to acknowledge the prevalence of animal instinct, a reluctance which can also be observed with regards to sexual taboos (Castro 1946). But while sexuality had psychoanalyst Sigmund Freud (1856–1939) as its avid student which affirmed the necessity of debating it, the problems of food and hunger have not yet had such luck (Castro 1946). In the realm of economics and politics, the dominant minorities who control wealth and production seek to construe a vision of the food system whereby it is a purely economic phenomenon, downplaying the fact that wealthy nations buy tons of food from countries where people still die from hunger or malnutrition—all with the consent of political systems (Castro 1946; Moorcroft 1972).

Hidden very far away from view, the places where our food originates are sometimes stricken with poverty, hunger, devastation and degradation. However, when food reaches us, it comes perfectly packed, seemingly trivial and cheap. In this industrialized modern world, the importance and value we attribute to food is diminished because we got used to pay very little for it and to access it easily from a vast multitude of outlets (Steel 2008, 2020). The polarization between urban and rural is a key reason for this unrealistic vision of food, and it is one area where the problems of the food system and the problems of the city and the urban systems overlap perfectly. Indeed, the urban–rural schism prompts us to unconsciously consider the city as separate from the natural environment, while failing to acknowledge the massive productive structure that is necessary to feed over 3 million people—the current population of the Lisbon Region, relatively small in the European context. We need to become more aware that (urban) artifice does not replace nature but is rather placed upon it—the preservation of environmental quality does not imply countering the pitfalls of city life by compensating elsewhere, but rather to promote environmental balance within cities themselves.

At the end of this chain of unequal and unbalanced organization of global food flows, agrifood business has witnessed a rise in large-scale corporations which often control significant proportions of retail sales, with some retailers achieving international proportions (Reisch et al. 2013), a situation which may even intensify in the aftermath of the COVID-19 pandemic, since lockdown measures may result in the collapse of food commerce of smaller scales. These food economies of scale, due to their globalized structure, often manage to escape regulations. Furthermore, this situation is helped by the fact that policies dealing with food, nutrition, environment, health and social cohesion are seldom integrated, while policies for sustainable food consumption are scant (Reisch et al. 2013). The endemic weakness of Portuguese planning policies, which include rural space but are generally ineffective in promoting its sustainable use, are case-in-point: they fail to articulate with other policies, and despite the increasing calls for sustainability, are powerless to promote local production and consumption. Perhaps a key challenge for future policies for sustainable transitions is to acknowledge the need for integration—in a sense, there are wise and well-constructed policies already created, but they often remain sectorial and without a proper territorial scope. The case of food is particularly important here, because its role in territorial organization is nearly absent from planning instruments, at least in Portugal—a situation which may no longer be acceptable or recommendable (Fig. 10.1).

Environmental problems (Fig. 10.1) like climate change, (mass extinction, deforestation, soil erosion, water depletion, declining fish stocks, pollutions, antibiotic resistance and diet-related diseases all result or are deeply impacted by our misguided attitudes towards food (Steel 2020). Indeed, no phenomenon is more convenient to the study of relations between human groups and their regional context than that of food (Castro 1946).

Hence, solutions to create better food systems are also a matter of social justice at a global scale. It is impressive that

Fig. 10.1 Mural painting on pollution and environmental disaster by LAP in Moscavide, Loures. Source: Authors

more than nearly 50 years after the publication of Moorcroft's (1972) anthology on architecture and ecology, where food has such a strong role, we are still striding to generalize the acknowledgement of how much problems of inequality between Global North and South depend upon the global organization of food systems. Indeed, Brazilian physician Josué de Castro (1908–1973), who was elected chairman of the FAO (for the mandate of 1952–1956), has long called attention not only to how hunger is much more pervasive than we were (and still are) ready to admit. However, already in the 1940s, Castro rejected the Malthusian assumption that overpopulation was the cause of hunger, as agricultural land became insufficient to feed everyone. In a bold book that might have been slightly ahead of its time, Castro (1946) pointed out the dependency of underdeveloped nations on developed ones—often a survival of colonial socio-economic structures—and specifically the alienation of native agricultural land to make way for monoproductive latifundia as the real problem. Only this mobilization of massive territorial extensions to produce food for the West could explain that the natives of these productive countries could starve while produce tons of foodstuffs.

Thus, countering the current organization of global food systems is a priority area to achieve better environmental conditions and more justice between different nations. While one must avoid romanticizing the abilities of local territories to produce their own food, it is also important to notice that the severity of the social and environmental harm being done in the current situation is so pronounced that all initiatives, including those of a small familiar scale, can help. Systemic international change, based on firm policies that dare to counter installed practices are the only way global food systems and their injustice will be improved. However, this must not stop us from acknowledging the importance of individual and community action, and more importantly, it must not stop us from seeing in these actions the seeds to more sustainable and fair alternatives.

10.3.1 Portuguese Urban Planning—Learning from the Past?

The first theme discussed is that of spatial planning, whose history in Portugal we overview on Part I. Chapter 1 focuses on the roots of planning policy in nineteenth-century planning policies, developed at the rise of liberalism and its dispute with the monarchy. This legislation introduced a schism between urbanism policies, on the one hand, and agroforestry planning on the other alienating (the few) cities from the majority of the territory, still rural and managed by widely different rules, a schism which would last until our own time. Yet back then, urbanism sought to achieve hygiene and infrastructuring conditions, which were severely in lacking in many settlements throughout the Portuguese territory (Lôbo 1995; Campos and Ferrão 2015). While planning was no longer prompted by emergency situations, as it had been in the eighteenth century, its key goal was to tackle underequipped and unsanitary settlements lacking in basic resources, which are reported in most planning instruments until the 1960s.

The dictatorial regime sets the basis for a truly modern and consequential planning policy, based on the political influence of Duarte Pacheco, but also on the intellectual contribution brought by a handful of competent urbanists, both foreigners and Portuguese who specialized abroad, notably Étienne de Gröer, Faria da Costa, Paulo Carvalho e

Cunha and João António Aguiar, to mention those more relevant within the Lisbon Region.

The world we 'see' in the descriptive documents of planning instruments of the first half of the century is one where modern life—and modern sanitation with its hydraulic systems and its cleanliness ideals (Gandy 2004)—has not yet arrived. If the villages and small towns around Lisbon were in these conditions, for instance, lacking in proper water supply and sewage systems (Marat-Mendes et al. 2015), there is little reason to believe the remainder of villages existing in the country would not share these problems. Granted that some of the problems pressuring the Lisbon surroundings were particular to the demographic growth they experienced with the increasing urbanization and modernization of the capital. But seen the absence of such infrastructures in Palmela or Vila Franca de Xira, on the outer borderline of the Lisbon Region, there is hardly any reason to believe that in the remote villages of the Portuguese countryside, some of which the 1961 Survey on Regional Architecture depicts so perfectly, conditions would be better. Indeed, popular indignation among village dwellers was felt when the New State developed the Agrarian Colonies, which offered to outsiders better conditions than those of the locals.

Agricultural policies like the Wheat Campaign lacked a detailed territorial basis, general urban improvements did not apply to the villages and thorps of the Portuguese interior, and agricultural policies largely benefited landowners. Thus, and while there were hydraulic and infrastructural works ordered by the State, the rural world and local-based foodsheds were excluded from modernization. Some sensible solutions were proposed to maintain a functioning hinterland amid the development of the industrial belts of Lisbon—on which modernization also depended. But these resulted more from the sensibility of planners and architects than to an appropriate planning framework that allowed a proper modernization of agricultural infrastructure or a foresight for villages and other settlements. The possibility of such a course of action is already demonstrated in the projects for the Agrarian Colonies of the JCI, supported by land aptitude surveys, hydraulic infrastructure and models for a 'casal' where housing is designed alongside stables, poultry houses, storage units and other agricultural support elements. While some Agrarian Colonies proved unsustainable throughout time, Santo Isidro de Pegões has somewhat stood the test of time—many descendants of the original settlers still live in the village, its agricultural and agroforestry structure is still functional, its Wine Cooperative is still running and aside from the occasional abandoned 'casal', most are inhabited and have been personalized and adapted without losing the original visual structure that allows us to recognize them as parts of a larger plan.

However, the technical and design process of Santo Isidro de Pegões is unparalleled in most urban plans of this era. Urbanization Plans, the key instruments for territorial transformation during the New State, conceive of green space as a generic space meant as a reserve for fresh air and for urban decoration. Often, among the advantages of single-family housing, the availability of outside space for a kitchen garden is mentioned, but there is little evidence that such goal may have had any impact in the specific design or sizing of such outside space.

While the 1960s mark the definitive rise of the construction industry and the real estate sector as key elements of the Portuguese economy, the already faint role of food-related activities in planning structures withered even more considerably. We saw in Chap. 2 how Portuguese architectural and urbanism culture transformed from the 1950s onwards. Important steps are taken towards a proper ordinance of the territory of the Lisbon Region, notably the 1964 PDRL, the first instrument to conceive of the region as a unit, and to promote an integral model which, while spatially hierarchized, did include a complete set of activities, from residential areas to industrial, agricultural, forestry and services. An important precedent had been set by architect José Rafael Botelho during his time as planner for the municipality Almada. He advanced a proposal for a national park, where he anticipates the importance of landscape preservation, without ignoring the existence of important fishing areas in Sesimbra and Setúbal. Throughout his work as a municipal chief-architect, he promoted not only a proper planning of urban areas but also of the rural metropolitan structure, surpassing his assignments and projecting the Lisbon Region into the future. Yet these plans do not get approved, and the vision they propose is far from what came to be the reality of this territory. In spite of planning instruments, food supply started a process of reconfiguration, from the elimination of farms and groves by suburban development or the weakening role of municipal markets (despite their proliferation from the 1940s to 1990s) in a metropolis increasingly supplied through supermarkets (Salvador 2019). In this sense, Portugal is another example where mobility infrastructure changed the morphology of periurban settlement and the land-use scheme around the city (Frayne et al. 2009; Parham 2015; Battersby 2017).

In Chap. 3, we observed the democratic planning policies, which for the first time instate a planning structure where, at least in principle, all territory is included and prone to spatial rules. In practice, however, emphasis still laid on urban territories, and when environmental concerns come to be recognized, in the 1998 Planning Act and then even more directly in its 2014 replacement, it is mostly for countering the urbanization pattern of the country under a liberal market.

While PDMs classify land as rural, the key planning aspect does not regard rurality—in the sense of cultivation or forestry at all—but really the regime of construction restrictions. It would almost be more rigorous to speak of these spaces as spaces of prohibited construction. Rurality implies dealing with a plethora of issues related to land-use, hydraulics and productivity which have nothing to do with construction and which fall completely outside the scope of the PDM. Especially after the 2014 Planning Act, but already prior to it, the other key concern of rural spatial planning is the control over the conditions for land-use conversion. The role of municipal planning in the rural landscape seems to be that of containing urbanization. Necessary as such contention is, it also means that the properly rural aspects of the territory are somewhat unbound by spatial planning regulations and are even immune to the sustainability concerns that are so correctly defined in the 2014 Planning Act. At the limit, one may wonder why environmental concerns preclude suburban sprawl but allow intensive or super-intensive olive groves to occupy large latifundia, despite their abusive and highly prejudicial impacts in water-use and in soil degradation.

Beyond the understated planning culture applied to rural territories, the current planning system, conceived largely to respond to specifically urban problems, also neglects the specific needs and characteristics of periurban space. Indeed, this is a natural consequence of the urban–rural opposition, since understanding the city as opposite to the countryside reduces reality to two poles of territorial arrangement that are simply not in accordance with reality, as those who study the territory have long been defending (Banzo 2007; Portas et al. 2007; Domingues and Travasso 2015).

Beyond the properly productive aspect, rural space is not accurately encompassed by Portuguese planning structures, also with regards to villages themselves. Having always been outside the scope of (urban) planning policy, villages remained outside of the scope where architects and urban planners usually worked. Thus, little solutions were advanced to propose or imagine new, 'modern' villages with their cultivation spaces, their forms of transforming the landscape through labour and thus produce the food on which all, urbanite or rural, depend.

10.3.2 History and Metabolism—Land and Labour

In Chap. 4, we reviewed two surveys conducted by Portuguese architects at different times of the twentieth century—the one on Regional Architecture conducted in the late 1950s and the IAPXX on the early years of the twenty-first century. Although each has its own focus and its own research goals, from the comparison of both it was possible to extract a portrait, a visual representation of the food system of each epoch (Marat-Mendes and Borges 2019a; Marat-Mendes et al. 2021b). That food-related structures and buildings are present in two different surveys which did not seek to portray them specifically, proves how all-pervasive food is, being omnipresent in rural as well as urban territorial contexts. Both surveys also demonstrate key changes in land-use and labour patterns (Marat-Mendes and Borges 2019a). They testify to the late industrialization of Portugal, and to a belated process of socio-metabolic transition whereby Portugal passed from an agrarian to a service economy. An industrial phase, always shy, is specific to metropolitan territories, although in these, the physical marks on the landscape are undeniable, as testified by the IAPXX.

After the late 1950s, when the steps towards an industrial wave—even if controlled by the State—start being taken, the transformation rhythm of the Lisbon Region landscape accelerates. The study of urbanism and architecture from the 1960s and 1970s has rightly emphasized the construction of large high-density social housing neighbourhoods, mostly in Lisbon (Nunes 2007; d'Almeida 2015; Borges and Marat-Mendes 2019; Borges et al. 2022), to privately developed suburbia (Ferreira 2010) but a considerable importance has also been rightly given to the proliferation of illegal neighbourhoods (Lages 2017; Pinto and Guerra 2019). A less noticed aspect in such studies has been the loss of rural land that was implied in most of these processes, both officially planned and not. The pressure to solve the ongoing housing crisis which was bound to worsen with industrialization was understandably higher than the need to preserve the regional foodshed, which itself threatened to become insufficient to feed this growing population. Indeed, if the villages near the industrial areas—as we saw on Gröer's report for the Moscavide–Vila Franca de Xira Foreplan—rapidly started to show signs of collapse, there is little reason to believe that regional supply, which had never been self-sufficient (Niza 2017), would stand the test any better.

Food critic Jane Grigson (1928–1990) criticized in 1978 the urbanization of rural peripheries, which alienated local food production, thus taking away local resources (Parham 2015). By this time, a similar process was abundantly visible in the Lisbon Region: after the proletarianization of peasants came the urbanization of farms and horticulture, a displacement of people followed by a displacement of activities.

A renewed socio-metabolism was effectively created in Portugal and boosted by the follow-up of the 1975 Constitution. Conditions of modernity, closer to what was verified in the rest of Europe became attainable for the Portuguese as they seldom had been before. It must be reminded that even at the peak of economic power, at the time of colonial expansion, the Portuguese people remained exemplarily poor and culturally isolate, even backward, with wealth being distributed among royalty and the aristocracy, as noted by several of our visitors (Mónica 2020). Thus, the instauration of a democracy, the entrance in the EEC and the move towards a service economy all seemed like changes too positive to allow an afterthought about the rural—and indigent—world that was being erased. But the problem we have come to increasingly notice with this new socio-metabolism, where something so fundamental as our food is mediated by powerful corporations and brought to us from all points of the globe, is not only that it eliminated the productive belt around large cities but also that urbanites did not, on the long run, necessarily benefit from it. For, as Grigson importantly hinted, the loss of the regional foodshed is also a loss of resources, of autonomy and, at least on some levels, of environmental quality (Parham 2015), while promoting starvation and malnutrition in the poor productive countries (Castro 1946).

In Chap. 5, we present the locations of key food-related activities within the Lisbon Region, as well as a selection of photographs of built and unbuilt structures to which these correspond, selected from more than 8000 photographs taken between 2019 and 2021 for the SPLACH Project. This selection aims to testify to the diversity of forms, scales and types of practice within each phase of the food system, although a conditioning factor has been the obtention of permits for photographic reproduction which kept certain images from being included. Our travels through the Lisbon Region were done by public transportation, walking and also by car, to reach to more isolate points where no other mobility option is available. Our survey encountered the remains of the rural past of artisanal agriculture, but also of industrialized activities, including mass production of crops. However common these spatial forms and economic activities may be, they still demonstrate the character of food-related activities in the region.

The signs of socio-metabolic transition may be observed with special intensity not only on the changes occurred in the food production but also in consumption, the food activity that is perhaps more strongly identified with the urban realm. The rise of privately owned 'public' spaces such as shopping malls are a clear sign of the new affluent world of the second half of the twentieth century, whose interiors replicate, within their own architectural aesthetics, the structures of the traditional city (Cachinho 2007), namely the commercial street, the plaza which becomes the food court, even transportation is somewhat readapted in the form of rolling stairs and lifts. In Portugal, there were only a handful of relatively small commercial galleries until the late 1970s (Figueira 2016), but a key sign of the new democratic society ready to integrate Europe was the inauguration of the Amoreiras Shopping in Lisbon, whose megastructural design (a shopping mall topped by a set of three office towers and luxury housing) gave a definitive breadth to the commercial post-modern architectural language of Tomás Taveira (n.1938) (Broadbent 1990). In the following decades, from urban supermarkets to large hypermarkets, sometimes located on stand-alone pavilions at the outskirts of urban settlements, proliferated in the country and in the Lisbon Region. Many examples of markets and commercial spaces we saw in our survey on the Lisbon Region food system did not have the elaborate architectural design of Amoreiras, but it is undeniable that this new large-scale food outlet became a fundamental aspect of the Portuguese landscape and of Portuguese food access.

While in villages, towns and cities of smaller dimension it may be easier to combine local production with global-scale trade, in a metropolitan context, the dependence upon food brought from outside and the resulting privileging of consumption intensifies a dependence upon outside spaces (where food is produced) and from the system linking the city with those places (who organized, distributes and trades). When a metropolitan region, such as that of Lisbon, is home to nearly a third of the national population, the level of dependency is high and the opportunity for profit is also high. Such dependency and opportunism must be accepted as factors of pressure for a transition of the food system, which can no longer be ignored or casually promoted through sporadic policies.

The shrinking hinterland of the Lisbon Region witnessed between the two surveys is deeply related to the transition from a circular to a linear metabolism, largely due to the management of waste. The proximity of agricultural and livestock activities in the periphery gave an obvious destination for urban and domestic waste, while in suburbanized metropolitan space, waste reused for agriculture is residual, while the change in lifestyle implied also means that waste is much higher individually than it was in the world of the 1961 Survey, prior to the introduction of plastic and of industrially produced goods. This change alone makes it challenging to recycle the tremendous amount of waste produced in a metropolitan region that concentrates nearly a third of the national population, but the proximity to food cultivation and livestock constitute a privileged opportunity to revive metabolic circularity within the region.

Indeed, waste disposal, one of the areas where food system research is less numerous (Brinkley 2013), is a global problem, but its causes are not uniform: on the Global South, food waste is caused by lack of proper infrastructure, while in the North the cause is one of oversupply, with people consuming more food than they need and wasting a very considerable amount (Steel 2020).

A key challenge to establish the Lisbon Region from a proximity foodshed perspective is to ensure its proper recognition as a region, which is in contradiction with the political interests of the past decades. Delays and undercutting in regionalization hamper also the gathering and comparison of specific data, from the past and also from the present. Throughout the Atlas, we saw the country divided into regions in several examples: Orlando Ribeiro's (1987) geographical history, the Survey on Regional Architecture (AAVV 1961), the cookbook of Maria de Lourdes Modesto (1982), the IAPXX (OA et al. 2006), the Agricultural Statistics (INE 2018) and the Agricultural Census (INE 2021). Yet the regional division they present is extremely uneven—the same division is not encountered twice. The Lisbon Region, in its relations with the larger region of Estremadura or with Ribatejo and West still stands out for its instability, including occasionally in the relationship between the two banks of the Tagus River, since the Northern Bank is one administrative district (Greater Lisbon) and the Southern Bank another one (Setúbal).

10.3.3 Urban Agriculture and Spatial Planning —An Overview

In Chap. 7, we briefly overviewed the definitions of urban agriculture and its key advantages and challenges as identified in the literature. While there is a vast literature on this subject, being a particular popular subtheme within the food system, there is unanimity in considering it a positive practice towards more sustainable cities, as well as very common a phenomenon in cities of the Global North as well as the South. However, this is not an interest merely among scholars, but also an opportunity acknowledged by decision-markers and planners.

At the international level, the 'Urban Food Agenda' (FAO 2019) recommends an increase in links between rural producers and urban markets through urban and periurban agriculture. Intensification of urban–rural links, sustainable rural development and diversified food systems which encapsulate informal and small-scale production are also defended. All these suggestions are supported by the idea that in metropolitan regions, externally produced food is to be articulated with an orientation of production and commerce towards a city-region framework, through short supply chains (FAO 2019). To fulfil such an agenda, most metropolitan territories cannot go without a general transformation.

The literature has noticed that urban agriculture, as a practice, has tended to intensify in periods of crisis. Less attention has been given to urbanism projects that acknowledge the centrality of food, which have been advanced at moments of transition. For example, Fourier understood that industrialism meant a new world order, which he tried to prevent, while Howard and his Garden City followers were trying to tame its effects. The second postwar period, a planned economy are paramount for recovering from war destruction, also produces several examples, from Hilberseimer's to Kurokawa's to Constant's. The CPUL model and the CitAgra also gained traction in the context of the post-2007 financial crisis. Only 'Agronica', the utopian project by Andrea Brazi seems to emerge at a time of relative economic stability, in the early 1990s.

In Chap. 8, we discuss a survey conducted at the SPLACH Project on urban agriculture in the 18 municipalities of the Lisbon Region, from which a total of 315 examples were collected. The diversity of typologies surveyed in the Lisbon Region is considerable, presenting several forms of occupation of vacant land with different scales of food production. Being verified in the less central areas within the Lisbon city, but also on the metropolitan periphery, we can indeed confirm that peripheral spaces, home to an important part of the population (including the commuters who work in Lisbon), also shelter urban forms in constant renovation, spaces where the contemporary city is made (Banzo 2007). These urban forms, we must again underscore, are not only limited to constructed elements but also in forms of occupation and transformation of the land.

However, confronting the existing practices with their legal framework in spatial planning and with the current management structures, there is a lack of correlation between planning dispositions and the de facto practice of urban agriculture. Reinstating the assumed fundamental opposition of rural and urban land, planning instruments in the Lisbon Region are suppressing a specific forms of land-use whose consistency is testified by their presence in all municipalities, even in those—like Palmela, Mafra or Alcochete—where (rural) agriculture is a dominant activity. However, in many municipalities, an official recognition and the assimilation into land-use schemes is still to come. Urban agriculture is a common practice among migrants and the urban poor, proving its value for food security and political representation for these social groups (Cabannes and Raposo 2013; Wekerle and Classens 2015; Purcell and Tyman 2015).

The unpreparedness of existing spatial planning instruments to respond to the challenge of cultivating food in cities —a social aspirations and an environmentally positive activity—stems from a planning culture which relies heavily on the partition between urban and rural soils as a generalizable rule guiding all spatial planning instruments.[1] The vacant urban spaces, valley bottoms and infrastructural corridors where it is typically found may be currently

[1] Let us recall that in the democratic planning scheme, while municipalities have relative flexibility to establish functional land-use categories, the key partition between urban and rural is imposed by the Planning Acts and applies to all municipalities—except for those whose classification is wholly urban, as is Lisbon.

acknowledged for ecological importance, but seldom for having a productive importance. This maintains modern tendencies to disconnect people from the land, a process meant to promote proletarianization, but which had the collateral effect of disconnecting them also from the origin of the food they consume (Steel 2008; Faraud 2017). The current planning approach presents other pitfalls. First, it assumes urban land must be necessarily urbanized, continuing the 1980s–2000s growth which slayed countless rural soils and largely contributed to a 'housing bubble' still unresolved (Mourão and Marat-Mendes 2016). Second, while the 2014 Planning Act represents an important step towards the formalization of green and ecological structures, these are removed from a far-reaching sustainability strategy in which a productive green structure cuts across urban space towards the countryside, functioning as a symbolic and functional link between these two realms (Viljoen et al. 2005, 2012). Meanwhile, the environmental benefits of urban agriculture and its increase of circular metabolism, biodiversity and waste recycling (Faraud 2017) are also unacknowledged as areas where planning could make a decisive contribution, but where it also can gather data for constructing well-established visions for the future transformation of the territory.

Urban and periurban agriculture demonstrate the valid critiques raised against planning structures risk becoming outdated, notably by the authors of 'Non-Plan'. Promoted by Reyner Banham (architectural critic, 1922–1988), Paul Barker (writer and journalist, 1935–2019), Peter Hall (planner, 1932–2014) and Cedric Price (architect, 1934–2003), 'Non-Plan' was a critique of postwar planning that proposed a 'plunge into heterogeneity' with self-organization and social aspirations as the keys to spatial organization (Banham et al. 1969: 436). While there is a provocative extremism in the proposal to have spatial management without planning, this essay rightly calls our attention to the danger of becoming crystalized, hampering innovation and transformation.

Greater attention is needed to the spaces daily transformed by local residents, whose use stands in ambiguity between urban and rural uses. These 'mixed land-uses' occupy relevant areas within several municipalities in the Lisbon Region. Instead of interpreting them as opportunities for future urbanization, they should be used as privileged points of contact between the city and the countryside, the type of space that was once defined by food—and which may be again.

Despite the wide array of specific forms of urban agriculture, strategies regarding community food production have sometimes been criticized for problems such as fragmentation, decentralization and chaotic organization, environmental communities engaged in 'thinking globally, acting locally' underestimating their links with the central state in the landscape of policies and regulations (Hardman and Larkham 2016). A similar point is brought forward by Delgado (2017) regarding the specific case of Portuguese initiatives—both private and public—which involve urban agriculture at several capacities. Specifically, this author notes that in Portugal, initiatives related to urban agriculture are still raw, rooted in unprocessed food products which lack integration in the complete food chain and with extremely weak processing of non-food products such as compost. Furthermore, this sector is more about self-consumption than business, that it is found predominantly on metropolitan cities, and even there it is a relatively young phenomenon and that it lacks a food system approach that considers new resources exploration (Delgado 2017; Henriques et al. 2021).

At the European level, urban agriculture is also overlooked by the current Common Agricultural Policy (CAP) and its reformulation in the European Funds programming 2021–2027 (EU 2018). The CAP focus on family farming, sustainability, innovation and bio-economy could easily encapsulate urban agriculture, where forms of artisanal cultivation and family-based labour still persist, but all this is excluded. The FOOD 2030 strategy (EU 2017) focuses on implications of food production for health and the environment but lacks a properly established spatial dimension. For the Portuguese context, where planning policies have historically been detached from other sectoral policies, such lack of spatial orientation has all probability of rendering such strategy irrelevant in any urban context.

At the national scale, the Family Agriculture Statute[2] was created in 2018 to help small farmers solidify their work through public financial support and optimized legal requirements. Among its goals is the promotion of locally produced food and the improvement of distribution chains. Nevertheless, the underlying territorial scope is rural—among tis goals is the halting of rural desertion—without any concession to the particularities of urban and periurban agriculture which, in many cases, could contribute significantly to achieve some of those goals in a metropolitan context. Yet, urban agriculture forces us to acknowledge the plurality of contemporary agrarian realities (Napawan 2016), a 'complexity' which implies also looking beyond rural territories, a conceptual leap the Portuguese State still resists.

Thus, the omission or understatement of urban agriculture in spatial planning instruments is combined with another omission in agricultural policy (Marat-Mendes et al. 2021c). In a sense, precisely because urban agriculture is itself

[2] Decree 64/2018, published in Diário da República 151/2018, Série I, 7-8-2018.

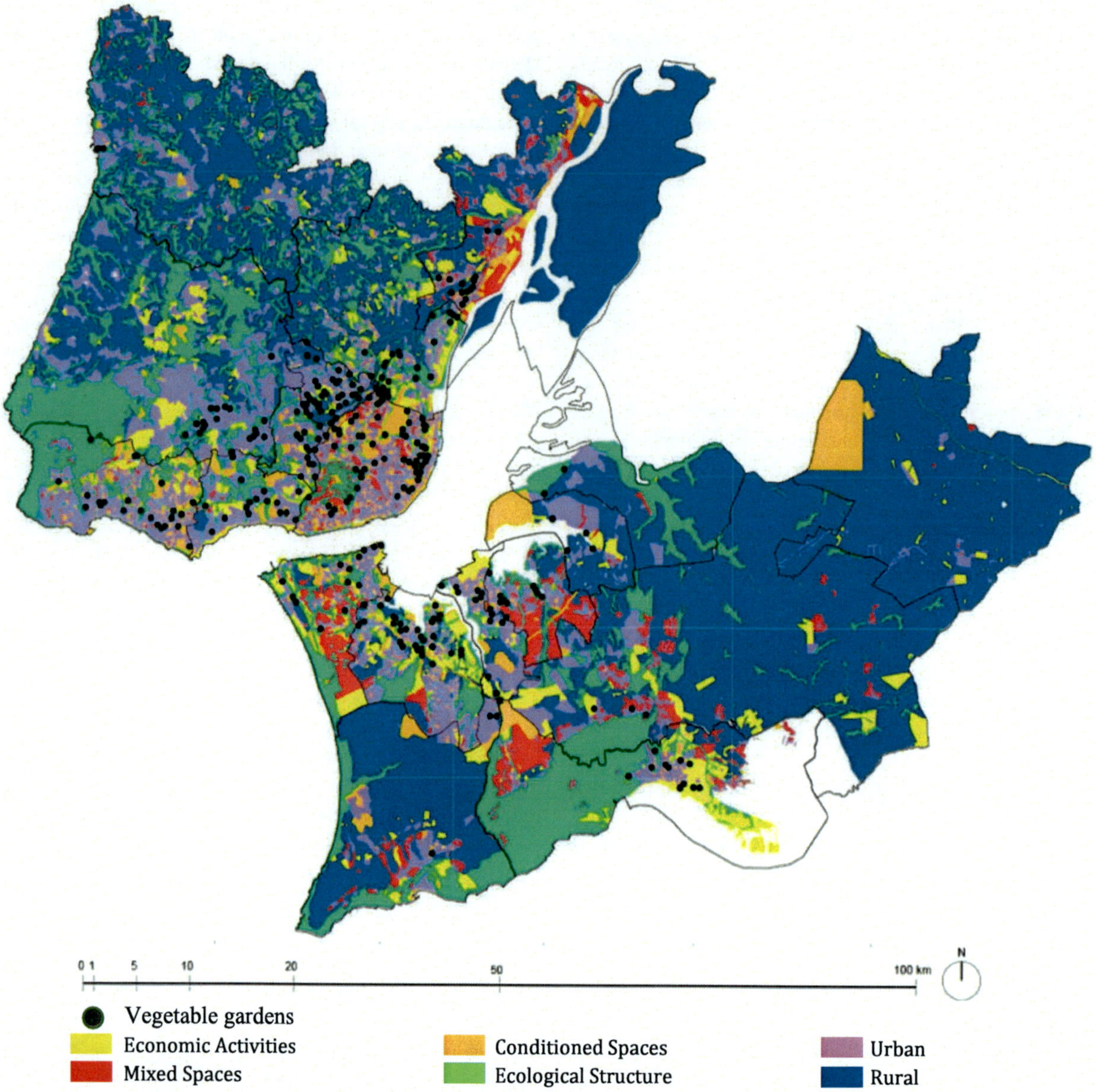

Fig. 10.2 Vegetable gardens by classes of land uses. Source: Authors

ambiguously placed between urban and rural, it easily escapes policies strongly directed at either of these realms. The democratic PDM has been critiqued for assuming the problems of large cities as generalizable for all territories (Baptista and Henriques 1985). Urban agriculture, which is in Portugal a fundamentally metropolitan phenomenon (Delgado 2017), demonstrates that planning policy may need to evolve, in the near future, towards a model that allows instruments more closely adapted to specific realities, instead of appealing to an ideal type of territory (Fig. 10.2).

10.4 A New Metropolitan Food System?

Considering the key themes observed in this Atlas and the key conclusions we derived from them, the food system of the Lisbon Region has a potential for sustainable transformation, as much as it needs to undergo such transformation to counter the many important negative aspects of its current social, environmental and physical problems. Although food will not be able to resolve every issue that needs to be acknowledged and tackled as an agenda for the Region, there is no doubt that it can assist and inspire many solutions for that sustainable transition. What follows is a set of ideas suggested by the research process that fed the Atlas.

Urban design operates as a geographic agent marking 'the physical configuration of human occupation on the ground', and going much beyond our 'conceptual and territorial binaries—society-nature, town-country, city-hinterland—[which become] inadequate to describe the contemporary condition of urbanization' (Ibañez and Katsikis 2014: 8). Urban planning, at its several scales, must also assume a similar role at several scales, fulfilling its prospective functions by presenting an integral vision for the future of the territory as a complex of resources (Ferrão 1999).

Current urban food systems are a delicate area where changes, from small to drastic, are necessary, not only for social reasons—resilience, elimination of hunger and poor nutrition, access to resources and sources of labour—but also for ecological reasons—countering of environmental degradation and resource depletion and better use of land. All these problems have obviously a spatial aspect and thus, we may 'design our way out of the problem' (Manning 2016: 39). In Portugal, such a design problem is met with a policy problem, since the key challenge of the Lisbon Region food system is not spatial. All the phases of the food system exist in the regional territory, and a considerable—though perhaps not outstanding—diversity also exists in all phases. Agriculture and livestock present diversity, as do transformation industries, and spaces for trade and consumption. Yet these phases are sparse around the region and, mobilized by a market dynamic, do not function systematically. Functional articulation is the critical aspect of the 'regionalization' of the foodshed.

Environmental protection has assumed an increasingly important role in spatial planning policies after the 1970s and the 1980s, but much remains to be done with respect to strategies which, following the suggestion of the FAO, effectively create urban–rural synergies and the integration of small-scale local and regional supply as a valid element of the food system. The reorganization of the regional foodshed and of the territory is powerful enough to prompt a larger socio-metabolic change, in accordance with scholars who have long been suggesting that 'the circulatory dynamics of (…) socio-ecological processes offer an opportunity to reshuffle old binaries into new categories of analysis and intervention for design' (Ibañez and Katsikis 2014: 8). In a sense, planning structures may be revised according to the reality of the territory and its resources, instead of territory and resources being subjected to presupposed planning structures which may hamper an adequate management of the existing reality.

Within the legal framework of Portuguese spatial planning, sustainable urban design has been often considered from the perspective of CO_2 emissions and energy supply, but territorial productivity and circular metabolism have been less emphasized in the transition debate. Municipalities do not seem equipped or even willing to manage the food system, which raises the question of who its agents might be.

The fact that the Lisbon city, following the example of Porto (Oliveira 2006) has created and approved a PDM whose land-use scheme is based on the morphology of its urban fabrics is already a good sign of a planning instrument that takes the existing spatial conditions as its basis. However, if such efforts for greater specificity and adequacy stop at urban fabrics and no akin solutions are promoted with regards to unbuilt cultivated space, the conceptual framework that has limited the scope of planning solutions will not be resolved. A morphological survey on agricultural practices—from city allotment gardens to latifundia—is therefore necessary to sustain a proper spatial model for the future of the regional foodshed.

The current agendas of intervention, especially those promoted by the public sector, can also be deeply enriched and diversified by accounting for the fundamental role of food in metropolitan organization. While in other countries there has been an increased role in new technologies and innovative ways to produce or transform food, in the Lisbon Region, the overwhelming majority of food-related activities either belong to traditionally artisanal or industrial structures, or else are niche local venues, too recent to have expanded (Marat-Mendes et al. 2021e). However, the increase in agricultural activity and on the UAA—in this region as in the country (INE 2018)—calls for some optimism with regards to the future. At the same time, the persistence of certain forms of food production or transformation does not necessarily mean a persistence in the practices they convey. As is often the case with buildings, so with open spaces and derelict built structures new uses can be fostered by old forms—thus simultaneously promoting local food and the improvement in spatial structure.

Moreover, programmes for social housing (now euphemistically called 'affordable rents') are being revived in the municipal agendas of the metropolitan Portuguese cities, but we must question whether this should only be conceived in urban settings. The growth of agricultural activities and the entrance of younger people—often with superior education—on the area of food production (INE 2018) may signify a reorientation of social values and aspirations, whereby rural space is no longer as unattractive as it once seemed.

The emergence of 'cottage core' aesthetics, briefly discussed in Chap. 4, demonstrates that dismay with urban life and its excesses is no longer a question for the elderly or the wealthy. Derelict and abandoned spaces on the rural land of metropolitan regions must be seriously taken as an option for affordable housing. The Lisbon Region certainly has the physical conditions to do so, although the relative isolation and lack of mobility alternatives (except the car) constitute challenges which such a programme would have to counter in order to be attractive and feasible as a contemporary habitat.

The optimization of rural spaces without proper cultivation is often hampered by their statute as private property—which calls for serious consideration of the way we treat rural spaces as well as our attitudes towards waste of space and resources. To avoid questioning the limit of private property rights, we consent to have unused land in our metropolitan spaces, since we are wealthy enough to purchase food produce from countries where people die from hunger and malnutrition. But even without considering the matter of private property, waste of space with productive potential is largely caused by the fact that farming viability is usually measured by the size of the operation, thus excluding many practices that have proved proliferous in urban and periurban habitats (Parham 2015). Urban agriculture—including in its particular expressions all over the Lisbon Region—provides plenty of examples of how adaptive food production may be. A regime of temporary use, of 'allowed informality', similar to what is verified with agriculture in cities, could be reasonably studied as a way to maintain rural landscape productive and well maintained.

The Portuguese planning history already provides examples of an effective planning methodology for rural space. Although the moralistic rhetoric of the New State about the supposed purity of rural life is even less acceptable nowadays, it must be noted that the JCI counted on the work of competent technicians and professionals—from architects to foresters—some of which even proposed solutions that the State, in its defence of the interests of the landowning bourgeoisie, chose not to pursue (Barreto 2017). However, the technical studies of the JCI on agricultural potentials, of effective hydraulic systems, of markedly rural architectural design with the 'casal' including the house but also agricultural support equipment, poultry houses, stables and kitchen gardens, remain among the most comprehensive methodologies for rural urbanism developed in Portugal. While revisiting the notable work that was done in Santo Isidro de Pegões must not make us oblivious to the oppressive nature of the dictatorship, we must also acknowledge that unsustainable urban growth must be countered by well-designed proposals for other forms of habitat, including on the countryside.

Indeed, similar programmes can be developed within the perspective of the new forms of regional organization verified in many world cities, characterized for their progressive (instead of regressive) localism which prompts democratic capacity-building, sustainable practices and cross-scale solidarity (Sonnino 2016). In the Peterchurch study presented in 'Designing for survival' (Nick Roberts quoted in Moorcroft 1972), there is already a proposition for a set of building and territorial arrangements to increase local food and energy productions and aiming towards regional self-sufficiency. His study concludes by defending a process where small compact urban cores are incepted into carefully planned and managed agrarian land, where physical growth happens through slow and organic infill, responding to the emergence of new community groups and their initiatives to fulfil their lifestyle. Such an approach is then recommended for the revitalization of village communities and for kickstarting neighbourhood-based alternative lifestyles (Moorcroft 1972).

The creation of housing programmes that articulate housing-access with local food cultivation are an option for a sustainable metropolitan transition. The Statute of Small Familiar Agriculture, valid only for rural territories, can be articulated with programmes for repopulation of rural settlements, offering more diversified housing (and lifestyle) options and countering the current tendency for abandonment of the countryside which has rendered entire villages, even in the Lisbon Region, completely deserted and derelict (Filipe 2014).

The role of peripheries is also a pivotal aspect of a sustainable transition of food systems. Peripheries are often perceived—in Portugal also—as spaces of anonymity, violence, monofunctionality, fragmentation, where social bonds and public space are threatened or at least in process, but they have also been recognized for a long time, at least among scholars, as places of innovation and singularity (Banzo et al. 2007). This process of reconfiguration has a very determining spatial dimension. In the Lisbon Region, peripheries are complex tapestries of different types of neighbourhoods—from social housing to private developments to illegal construction to old villages—of industrial estates—some functional, others abandoned—rural remnants—some well preserved, others ravaged by recent development or by neglect. All of this is mostly sown by motorways of several scales, and the level of equipment and amenities available varies greatly.

A sensible planning structure that can accurately encapsulate the specificities of such a territory and its morphologies, while maintaining feasible sustainability goals is necessary, but it cannot stand on ready-made oppositions. The dependence of the Portuguese planning policy on Roman Law is largely at the basis of the current reliance on abstract principles as the guiding force of planning instruments, but such a model cannot avoid—and has not avoided in Portugal—to create its standards having a specific reality in mind. In Portugal, planning has tended, at least originally, to respond to the challenges of large cities (Baptista and Henriques 1985), even though the country only has two. More emphasis needs to be placed on territorial surveys as the basis of a planning instrument—following the methodological tradition of Patrick Geddes—and on the integration of goals at several interlinked scales. Municipal autonomy is an important achievement of the democracy—during the dictatorship municipalities were controlled by the State not only through financial means but also by the fact that the president was selected by the government. It is one of the scales of power that lays closer to the people, and thus it must be protected. However, an adequate planning structure, and an adequate organization of resources in a logic of proximity cannot be achieved without going beyond municipal borderlines. Regional planning—undercut in the same way that regionalization has been—is to be a cornerstone of socio-metabolic reconfiguration, and it can indeed function as a stimulus for cooperation and sharing between different municipalities.

Peripheries will most likely be understood in a new light under a regional plan, not as satellites of the large city, but as places where urban and rural interplay, with different possible combinations of urbanity and rurality (Parham 2015). In a sense, while in the city agriculture represents a disruption with common assumptions about the urban, in the periphery, agriculture may instead function as pathway for recognition and identity. By not acknowledging—or misunderstanding—peripheral metropolitan territories, planning schemes risk smashing this identity.

Historically, in countries like France and the UK, rural spaces at the outskirts of cities were often spaces of leisure for the working-class, offering space for promenading, conviviality and trade in street fairs, but also in taverns (Parham 2015). Such a tradition is not unknown in the Lisbon Region, if we remember the nineteenth-century description of Chelas and its horticultural gardens as the walking space preferred by the popular masses by Romantic writer Almeida Garrett (1899–1854) (Borges et al. 2022), or the more generic 'bohemian' horticultural gardens mentioned by Couto Viana (quoted in Modesto 1982).

A similar social atmosphere may be still encountered today in Lisbon, for instance, in the Relógio Street Fair, which not only gathers people from all social classes, from Chelas but also from the rest of the city. Yet street fairs have an important presence in the Lisbon city, and examples can be found, such as the Biological Street Fair of Príncipe Real, which caters to an upper class clientele which, given its centrality, also appeals greatly to tourists. Some of the vegetable gardens we visited also present themselves as spaces of conviviality—even if reserved to the social circles of horticultors. This includes improvised sheds with tables and chairs in informal or associative examples, and wooden tables with equipment for barbecues in some, albeit a minority, of municipal gardens. Such social gatherings around food are indeed common in the Lisbon Region, and they range from more exclusive forms of conviviality to more organized and socially selective forms. Such street fairs and spaces for meals can easily be articulated with production from the vegetable gardens themselves—and they often are. Promoting such activities encourages local produce but may also help reconsider food habits. As discussed, in Modesto's (1982) survey, recipes with meats are limited, although currently they are a part of the eating habits of the majority of the population. As people move to cities, traditional rural diets based on grain and vegetables become replaced by Western urban meals where meats and processed foods play a key role (Steel 2020), a transformation which can be sensed in Portugal. The incentive of urban agriculture, but also for small street fairs and community meals may be an effective way to promote a more balanced diet, centred on the produce of horticultors. Periurban agriculture, and urban agriculture to a sense, may operate on the modern food system or maintain a more place-based character (Parham 2015), however, especially when municipal authorities are involved, rules imply that produce must be for household consumption, a rule which could be revised to allow trade under specific circumstances.

Yet, in terms of planning instruments, urban agriculture seems to be frequently interpreted or perceived as a secondary activity for entertaining the poor and the retired, a reductive vision demonstrated by the narrow and unimaginative morphological solutions used in municipal allotments, and by the reluctance to definitively preclude construction in their area (Dias 2018; Marat-Mendes et al. 2021c). A larger set of design typologies could constitute a useful tool for promoting the productive and visual potential of urban vacant land, while adapting to different aspirations and productive capacities. More varied solutions could also account for different social groups, physical capacities, knowledge, culture and interests, encouraging creative use of public and vacant spaces. The difference between regular municipal plots and irregular grassroots plots in many municipalities suggests that involving farmers with the design process would eventually generate more diverse and adaptive typologies (Dias 2018; Dias and Marat-Mendes 2020; Marat-Mendes et al. 2021c).

The food system breaks down into several scales, ranging from the global scale of international trade but also on the cellular or individual level (Brinkley 2013). Understanding these several scales is key to design policies and spaces that seek to resolve food supply at all these scales and at every phase of the food system. Agriculture can be practiced from mass production to a kitchen garden improvised on a slope. Similarly, street fairs may have different scales, from a municipal street fair to a yard sale. And consumption spaces are often key elements of city public space, in the form of kiosks or esplanades, but the physical elements they employ (chairs, tables, preparation stations) can be used as equipment for public space. Only through a minute and site-specific network of food-related activities can we take significative steps towards urban food governance, which in many cities is creating an important platform to build social and cultural capacities to face sustainability challenges and promote inclusive food policies (Sonnino 2016).

Among contemporary proposals for a stronger role of food in the city, the CPUL model (Viljoen et al. 2005, 2014) is perhaps the most notable, prompting two books and discussed at many anthological scholarly editions and specialized press (Manning 2016). Indeed, the CPUL touches the nerve of the problem by stressing the need to reclaim space within cities to cultivate food. Not only does urban soil exclude the possibility of agriculture as an activity, production is arguably the phase of the food system that, by its character and the space it requires, is more difficult to incept in city space. Yet food can support a reshaping of the territory—urban and beyond—that exceeds the production phase.

In Chap. 9, in the chord from Chelas to Vila Franca de Xira we demonstrated the coexistence of all phases of the food system—including the critical presence of the MARL—as well as the existence of several urban problems of industrial decay, rural abandonment and urban disorganization, all of which could be transformed in facilities for increasing local food. Obviously, the Lisbon Region does not have the conditions to feed itself entirely—but its territory may be more and better used, its socio-metabolism may become more circular. Policies thus must focus on encouraging regional consumption and mobilizing to that end the available territorial resources and creating conditions to counter rural abandonment and decay.

The increase in agricultural business—despite its already mentioned traps—is certainly a positive change for the countryside, but as a cultural phenomenon (which it undoubtedly is), it could be an interesting opportunity for cities too. Urban and periurban agriculture teach us to be less being less short-sighted with regards to the adaptability and variety of forms, scales of production and spatial morphologies that food production may assume. This is a form of grassroots knowledge on landscape and land-use which can be recognized and integrated into planning structures and into agricultural incentives, in turn helping these to promote overall sustainability. Indeed, the practice of urban agriculture is not sustainable per se (Hardman and Larkham 2016; Parham 2015), and planning can indeed predict and guarantee sustainability standards.

There are obvious advantages in promoting the municipalization or formalization of informal cases, but such a process must stem from the recognition that this activity is legitimate, that it plays a part in the ecological structure of the city, the municipality and the region, and that as such it deserves and even demands specific protection. Moreover, it must stem from a correct assessment of the aspirations of existing practitioners, with the priority of creating the best possible correspondence between the horticultors (individually or collectively) and their potential contribution to the local and regional food system. In some municipalities, 'solidarity' vegetable gardens and eased access for the unemployed already exist—but instead of seeking to help ensure part of domestic consumption, urban agriculture can be seen as a solution, or at least a temporary solution, for the indigence that grants eased access to land for food cultivation. Food can provide the basis for enlarged sets of public goods and create new spaces of solidarity (Sonnino 2016) but is can also provide labour, from which results not only food but a properly managed landscape which is available to the pleasure and quality of life of all inhabitants.

That urban agriculture, a much more common activity than most of us would think, naturally favours the consumption of locally produced food is obvious, but its contribution can even go further than that. A critique raised to industrial agriculture is that by selecting the species more adequate to mass production, it standardizes production and thus has often contributed to the extinction of certain varieties of vegetables, fruits and even animals—for example, Cornish Cauliflowers and Ansault Pears are both extinct, as are 86% of apple varieties once cultivated in the US (Nuwer 2014). Agriculture on the scale that it can be practiced within cities and even on their peripheries, is particularly suited to cultivation of species that would be impossible—or unprofitable—to mass produce (Parham 2015). Although variables such as weather and soil conditions obviously have to be considered as well, urban and periurban agriculture could become nurseries of endangered vegetables and fruits, saving them from obliteration by a culture where only market-oriented food production can thrive (Figs. 10.3 and 10.4).

Agroforestry is another important area to highlight in the Lisbon Region. Always remembering the impossibility of rendering the metropolitan region fully self-sufficient, it does however make sense to try and ensure as much produce as possible of foodstuffs we know to be less significant in the

Fig. 10.3 Plum trees in the Murganheira Estate, Oeiras. Source: Authors

Fig. 10.4 Loquats and fig trees in the Murganheira Estate, Oeiras. Source: Authors

regional production system. The Agricultural Census demonstrates that fruit production is among the least relevant produces of the Lisbon Region (INE 2018). In the Agricultural Statistics of 2017 we can see that production of all fruits except apples and pears are very residual in comparison with other—obviously larger and more fertile—regions. While we found only one example of an urban agricultural space largely filled with fruits, in the Murganheira Estate in Oeiras, it demonstrates that the amount of space necessary for a couple of common urban agricultural gardens suffices to plant some trees—in this case, plumtrees of different varieties, peach trees, vineyards, loquats, olive trees and fig trees. Moreover, in most cases of illegal or associative vegetable gardens, fruit trees proved very common. Such a practice can be promoted in public spaces in ways that the cultivation of vegetables and grains cannot, and the supply —however small—such an option could bring must be seen against the fact that in many fruits, production is low not only in the Lisbon Region but in Portugal as a whole.

A similar point can be made about livestock within cities, a rare phenomenon in the Lisbon Region, but which is often considered critical to organic waste management, since animals can process leftover food and turn it to manure for agriculture (Brinkley 2013).

If we know that mobility infrastructures were a key to territorial development of the Lisbon Region territory, generating or transforming the pattern of settlements and of land-use, then maybe such transformation can be turned into a positive dynamic, by articulating soft mobility, especially bicycle tracks (Ferreira et al. 2020) as generators of new land-uses or maybe, on the long run, as generators of new forms of settlement.

Recognition of grassroots activities unforeseen by planning instruments is not only a form of recognition of the importance of grassroots urban intervention. It would also constitute a way from which planning could critique itself and change towards to the demonstrated aspirations and interests of the people whose home territory plans regulate. In many spaces, both in cities and their peripheries, landscapes are untreated, ill-used and barren. Urban agriculture has been a form of popular landscaping, responding more closely to the characteristics of existing spaces and the absence of official landscaping.

In a sense, cooperation between decision-makers—including planners—and the civil society present themselves as the key for a sustainable transition of the Lisbon Region food system. This implies not only the recognition and validation of existing practices, but also a serious consideration of their impacts and an imaginative strategy that takes the best practices to a larger scale, using them to revitalize the territory. A territory which displays diversified activities, a large productive potential and a wide range of settlement types, which presuppose a wide range of social and ecological organizations and of lifestyles.

While the Agricultural Census (INE 2021) gives us signs that agricultural business is, instead of declining, actually progressing, this must not be necessarily taken as a positive situation. Before observing the growth or lack of growth in the food production sector, we must clarify our aims, and conceive what food systems we want for the future of our communities, our land and our world.

For Josué de Castro (1946) the knowledge on the feeding situation of peoples and of the resources mobilized for it are fundamental for a social revolution whereby money is put on the service of humankind, not humankind on the service of money. On the wake of WW2, Castro believed that this revolution was under way, but at the dawn of the twenty-first century, this is less obvious. To reconceive the role of food in our lives, we may really need to question of the foundational aspects of our social and political lives—namely, the notion of progress.

Progress has been an important factor in the lives and the politics of the Portuguese, partly because the country was comparatively backward and stagnant until the twentieth century, and its modernity was hampered by a dictatorship which despised all progress that it could not control. The democratic era seemed to promise to the Portuguese the modernity, the progress, that had been historically withheld by politicians and elites. But is progress inherently good? We know the notion of growth—on which the notion of progress naturally rests—has been deeply criticized for its inherent unsustainability (Meadows et al. 1972; D'Alisa et al. 2015).

Perhaps a helpful metaphor for the desirable future of food systems can be retrieved from a classical—although polemic—aphorism by Philosopher Walter Benjamin (1892–1940):

> Marx says that revolutions are the locomotive of world history. But perhaps it is quite otherwise. Perhaps revolutions are an attempt by the passengers on this train – namely, the human race – to activate the emergency brake (Benjamin 1940: 402).

Our ideas of food are currently embedded in our ideas of markets and business—in other words, in the capitalist organization. Capitalism stands on the idea of growth and progress as the moral principles which justify its form of exploration and organization. Socio-metabolism, environmental protection and the food system all defy us to look back onto the very morality of progress and wonder up to what point is progress really desirable. Obviously, progress in labour rights, in the recognition of the rights of minorities and in scientific knowledge are desirable. But if climatic conditions continue to deteriorate and resources keep being depleted at the current rate, in a few decades there will be no livable world on which to enjoy those desirable progresses, even we achieve them. So the progress of capitalism, especially when it implies the destruction and overexploitation of resources and a threat to the environment must be understood differently—and it may not be as desirable. It is uncomfortable to accept that urbanization cannot go unlimited, that economic growth cannot be the chief societal goal and that the accumulation of capital has to have restrictions—because all those ideas sound like they come from a conservative past, which for many nations, including Portugal, even resonate with the patronizing rhetoric of dictatorships. Benjamin's idea of revolution as the brake of progress, as a way to save ourselves from its harms, may give us hope in the years to come. Being a primary source of life, an essential element of everyone's lives, a change for the better—a change towards sustainability—may be prompted by the restructuring of the food system. And, as many spaces on the cities, the peripheries and the rural settlements of the Lisbon Region testify to, if we do activate the emergency brake on the progress of food business, plenty of solutions can be achieved, with creativity, knowledge and cooperation. If we aim to move in such a direction, maybe in the future we can live in a world where, to adapt the idea by Josué de Castro (1946), we can have food business at the service of people, and not people at the service of food business.

Such a change implies new ways of thinking and conceiving the way we eat. But it is most definetly a spatial change too, towards which adequate arrangements, facilities and strategies will need to be designed—quite literally. If the history of architecture and urbanism teach us anything is that creative or radical visions for the future were seldom inspired by merely following whatever trends defined a given moment, but rather by architects—and other authors—being keen observers of their contexts, sensing underlying problems and anticipating solutions and opportunities, on which they based their often utopian proposals. Some of these—think of Howard's Garden City or Wright's Usonia—have remained influential until today, even if somewhat adulterated by their interpreters.

Marat-Mendes' (2002) four 'ground-rules' of sustainable urban form are adaptability, continuity, flexibility, and resilience. Each of these can be confirmed in whatever remains of the Lisbon regional foodshed. Agricultural and horticultural practices, olive groves and orchards—all of these seemed to have been displaced by the fury of belated urbanization. And yet, over half a century later, those food production spaces that seemed to have disappeared have only acquired different forms, different meanings and different purposes. Perhaps this is the underlying principle of metropolitan food activities, especially production: since they are conceptually precluded from the city, they can only thrive with an interplay of fertile soil and a fertile imagination. Indeed, the vegetable gardens lying at the fringes of social housing estates of around infrastructural corridors may not be the first image that comes to one's mind when imagining a city. And yet, taking a walk in any of these spaces, they have nothing to do with the most pessimistic predictions of our oversaturated cities.

The challenges opened by the need to transition the current metropolitan food systems have been made clear by over two decades of scholarly research and international political agendas. There are endless possibilities for designing new ways of using space, of using it better and more sustainably, of creating better landscapes, growing better food and engage urban dwellers not only with the ecosystems they inhabit, but also with the particular city or neighbourhood they inhabit—and from which it is all too easy to become disconnected, as more and more of our lives take place online. Now it is time to take it to the drawing board! (Fig. 10.5).

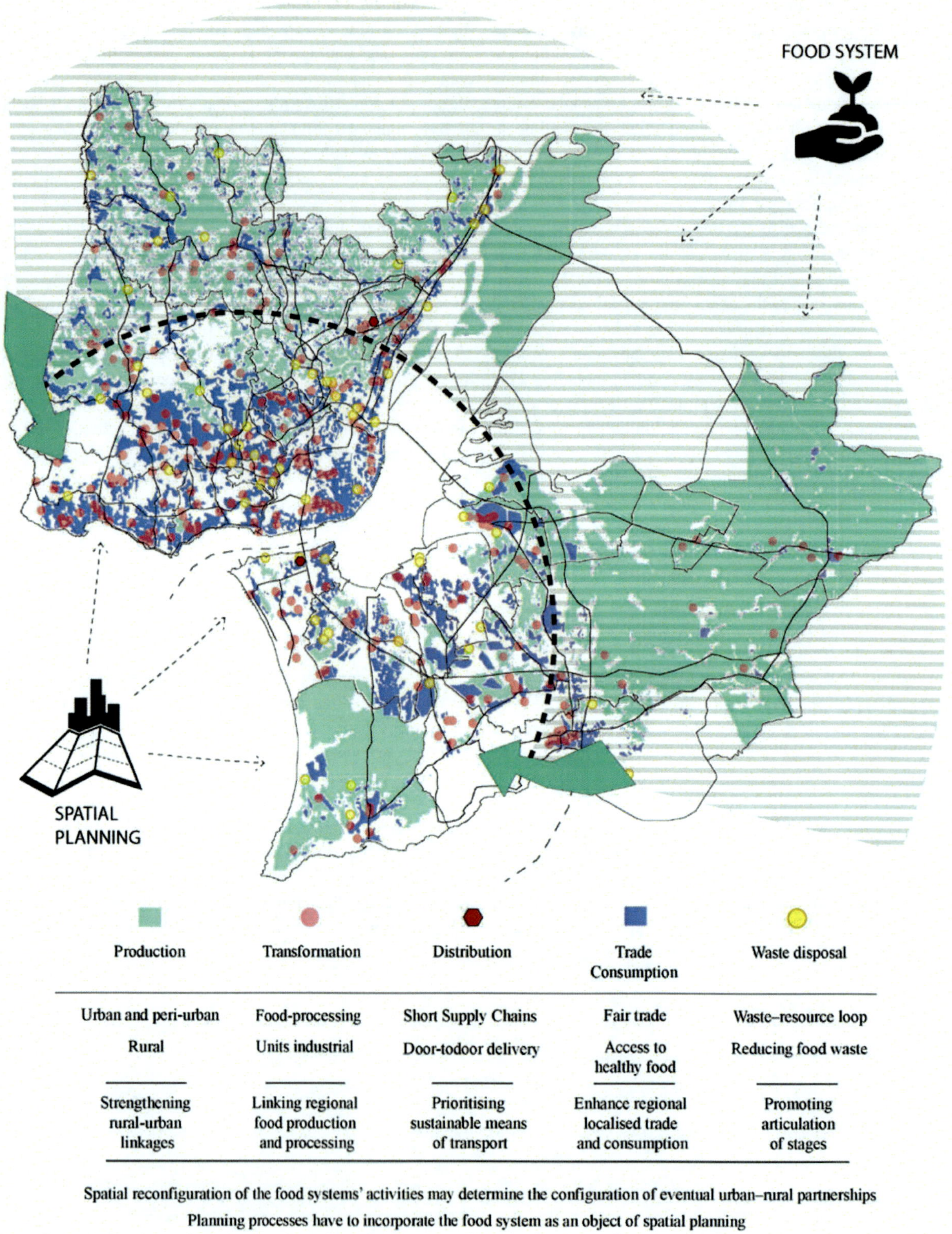

Fig. 10.5 Linking the food system through spatial planning. Source: Authors

References

AAVV (1961) [2004]) Arquitectura Popular em Portugal. Ordem dos Arquitectos, Lisbon

Banham R, Barker P, Hall P, Price C (1969) Non Plan: an experiment in freedom. New Soc 338:435–443

Banzo M (2007) Structure Verte Et Mise En Cohérence De La Périphérie Métropolitaine De Lisbonne. Revue Géographique Des Pyrénées Et Du Sud-Ouest. Sud-Ouest Européen 24:109–124

Baptista A, Henriques J (1985) Os custos sociais de urbanização e o financiamento dos municípios da Área Metropolitana de Lisboa. Sociedade e Território 3:31–37

Barreto A (2017) Anatomia de uma revolução – a Reforma Agrária em Portugal 1974–1976. D. Quixote, Lisbon

Battersby J (2017) Food system transformation in the absence of food system planning: the case of supermarket and shopping mall retail expansion in Cape Town, South Africa. Built Environ 43(3):417–430

Benjamin W (1940[2003]). Paralipomena to "On the Concept of History". In: Eiland H, Jennings MW (eds) Selected writings volume 4:1938–1940. Belknak Press, Cambridge, Massachusetts, pp 401–408

Borges JC, Marat-Mendes T (2019) When Lisbon met the team 10 cluster city. Joelho 10:86–99. https://doi.org/10.14195/1647-8681_10_5

Borges JC, Marat-Mendes T, Lopes SS (2020) Chelas Zone J revisited: urban morphology and change in a recovering neighbourhood. In: Strappa G, Carolitti P, Leva M (eds) Urban Substrata and City regeneration. Proceedings of the 5th ISUFitaly International Conference, Rome, 19–22 Februray 2020, p 751–760

Borges JC, Lopes SS, Fernandes RDP, Marat-Mendes T (2022) Planning at the edge: urbanism and socio-political transition in Chelas Lisbon. Planning Perspectives 1–33. https://doi.org/10.1080/02665433.2021.2001364

Brinkley C (2013) Avenues into food planning: a review of scholarly food system research. Int Plan Stud 18(2):243–266. https://doi.org/10.1080/13563475.2013.774150

Broadbent G (1990) Introduction. In: Taveira T (ed) Tomas Taveira: architectural works and designs. Academy Editions, London

Cabannes Y, Raposo I (2013) Peri-urban agriculture, social inclusion of migrant population and Right to the City—practices in Lisbon and London. City 17(29):235–250. https://doi.org/10.1080/13604813.2013.765652

Cabrita MA (2008) Território e identidade. MSc Dissertation. Iscte Instituto Universitário de Lisboa, Lisbon

Cachinho H (2007) L'espace public dans la banlieue de Lisbonne: de la rue au centre commercial. Sud-Ouest Européen 24:51–62

Campos V, Ferrão J (2015) O Ordenamento do território: uma perspetiva genealógica. ICS working papers 1, 2015. Instituto de Ciências Sociais da Universidade de Lisboa, Lisboa. http://www.ics.ul.pt/flipping/wp2015_1/index.html

Castro JD (1946 [1984]) Geografia da fome - O dilema brasileiro: pão ou aço, 10th edition, Antares, Rio de Janeiro

Cuchí A, Marat-Mendes T, Pérez MJ, Teira R, Albareda E (2008) Informe previo a la actuación urbanística en las Brañas de Sar en Santiago de Compostela. Universidade de Barcelona-Consórcio de Santiago, Santiago de Compostela

Cuchí A, Marat-Mendes T, Mourão J (2010) Urban material analysis and sustainability: a new methodological approach towards urban planning. In: Pinho P, Oliveira V (eds) Planning in times of uncertainty. FEUP-CITTA, Porto, pp 109–122

D'Alisa G, Demaria F, Kallis G (2015) Degrowth: a vocabulary for a new era. Routledge, London

d'Almeida PB, Marat-Mendes T (2021) A atualidade da investigação científica em arquitetura e urbanismo desenvolvida no LNEC (1961–1979) face ao desafio da sustentabilidade. Cidades Comunidades & Territórios 15–40

d'Almeida P, Marat-Mendes T, Toussaint M (2020) Portugal's rising research in architecture and urbanism: the influence of international research centers and authors. J Urban Hist 1–28. https://doi.org/10.1177/0096144220968078

d'Almeida PB (2015) Bairro(s) do Restelo. Panorama Urbanístico e Arquitectónico. Caleidoscópio, Casal de Cambra

Delgado C (2017) Mapping urban agriculture in Portugal: Lessons from practice and their relevance for European post-crisis contexts. Moravian Geographical Reports 25(3) 139–153. https://doi.org/10.1515/mgr-2017-0013

Dias AM, Marat-Mendes T (2020) The morphological impact of municipal planning instruments on urban agriculture. Cid Comunidades Territ 41. http://journals.openedition.org/cidades/2991

Dias AM (2018) The shape of food: an analysis of urban agricultural shapes in Lisbon's Greater Area. MSc dissertation, Iscte - Instituto Universitário de Lisboa, Lisbon. http://hdl.handle.net/10071/19300

Dominges A, Travasso N (eds) (2015) Território, casa comum: morfologias e dinâmicas do território. Circo de Ideias, Porto

European Union (2017) Food 2030. https://ec.europa.eu/info/research-and-innovation/research-area/environment/bioeconomy/food-systems/food-2030_en#documents

European Union (2018) Cohesion Policy 2021–2027. https://ec.europa.eu/regional_policy/en/2021_2027/

FAO (Food and agriculture organization) (2019) FAO framework for the Urban Food Agenda. FAO, Rome. https://doi.org/10.4060/ca3151en

Faraud C (2017) Urban metabolism in practice. DPU WORKING PAPER NO. 186. University College London, London. https://www.ucl.ac.uk/bartlett/development/sites/bartlett/files/wp186.pdf. Accessed 22 Jan 2019

Ferrão J (1999) Elementos para um Programa Nacional de Política de Ordenamento do Território – uma visão de síntese. In: MEPAT/SEALOT. Seminário Internacional Território para o século XXI – Ordenamento, Competitividade e Coesão. Programa Nacional da Política de Ordenamento do Território (PNPOT): 209–212

Ferreira JP, Isidoro C, Moura Sá F, Baptista Da Mota JC (2020) The economic value for cycling—a methodological assessment for Starter Cities. Hábitat y Sociedad 13(13):29–45

Ferreira, BM (2010) (In)Formar a cidade contemporânea. MSc dissertation, Iscte Instituto Universitário de Lisboa, Lisbon

Figueira J (2016) Arquitectanic – os dias da troika. Note, Lisbon

Filipe M (2014) Aldeias Abandonadas, Património Imaterial e Desenvolvimento Local: estudos de caso na AML. Novas Edições Académicas, Chinisao

Fischer-Kowalski M (1998) Society's metabolism. The intellectual history of material flow analysis, Part I, 1860–1970, Industrial Ecology 1:61–72

Frayne B, Battersby-Lennard J, Fincham R, Haysom G (2009) Urban food security in South Africa: case study of Cape Town, Msunduzi and Johannesburg. Development planning division working paper series no.15, DBSA, Midrand

Gandy M (2004) Rethinking urban metabolism: water, space and the modern city. City 8:363–379

Gandy M (2014) The fabric of space water, modernity, and the urban imagination. The MIT Press, Cambridge, Massachusetts, London, England

Hardman M, Larkham PJ (2016) Informal Urban Agriculture. The Secret Lives of Guerrilla Gardeners. Springer, London

Henriques CN, Pinto TC, Costa P, Marat-Mendes T (2021) Tracing Lisbon metropolitan area's foodscape. Cid, Comunidades Territ 42. http://journals.openedition.org/cidades/4065

Ibañez D, Katsikis N (eds) (2014) Grounding metabolism. New geographies 6. Harvard University, Graduate School of Design, Harvard, Cambridge, Massachusetts

INE - Instituto Nacional de Estatística (2018) Estatísticas Agrícolas 2017. INE, Lisboa. https://www.ine.pt/xurl/pub/320461359. ISSN: 0079-4139. ISBN 978-989-25-0445-2

INE - Instituto Nacional de Estatística (2021) Recenseamento Agrícola 2019. Análise dos principais resultados. INE, Lisboa. https://www.ine.pt/xurl/pub/437178558. ISBN 978-989-25-0562-6

Kennedy C, Pinctel S, Bunje P (2011) The study of urban metabolism and its applications to urban planning and design. Environ Polution 159:1965–1973

Lages J (2017) À conquista da justiça espacial - ntervenções liminares em territórios autoproduzidos da área metropolitana de Lisboa. PhD Thesis, Universidade de Lisboa, Faculdade de Arquitetura, Lisbon

Lôbo MS (1995) Planos de urbanização – a época de Duarte Pacheco. FAUP Edições, Porto

Lopes S, Borges JC, Fernandes R; Marat-Mendes T (2021) Rediscovering modern 'civil architecture': the case of Bela Vista neighbourhood in Setúbal. Paper presentes at the international conference optimistic Suburbia II, Iscte Instituto Universitário de Lisboa, 16–18 June 2021

Manning J (ed) (2016) Food and the city. Urban Design 140:14–39

Marat-Mendes T (2002) The sustainable urban form. A comparative study in Lisbon, Edinburgh and Barcelona. PhD thesis, University of Nottingham, Nottingham UK

Marat-Mendes T (2004) Dimensioning the sustainable urban fabric. In: Babalis D (ed) Ecological design for an effective urban regeneration. Firenze University Press, Firenze, pp 27–40

Marat-Mendes T (2008) Environment and urban form: the real scale of its morphological anatomy. In: Proceedings of the 3rd annual congress of the council for European Urbanism. CEU 2008, Climate change and urban form. University of Oslo, Norway, pp 28–41

Marat-Mendes, T (2009) O PUCS e os vazios planeados. Novas oportunidades para o ordenamento sustentado da Costa do Sol. In: Pereira M (ed) O Plano de Urbanização da Costa do Sol - Uma visão inovadora para o território. Câmara Municipal de Oeiras, Oeiras, pp 92–122

Marat-Mendes T (2015a) Adaptabilidade, continuidade, flexibilidade e resiliência. Algumas considerações sobre as propriedades das formas urbanas. Revista de Morfologia Urbana 3(2):133–134

Marat-Mendes T (2015b) O ensino da Morfologia Urbana no ISCTE-IUL. Revista De Morfologia Urbana 3(1):77–80

Marat-Mendes T (2016) Physical, social and cultural dimensions of Urban Morphology: redressing the balance? Urban Morphol 20 (2):167–168

Marat-Mendes T (2018a) Lx design with food. In: Bonacho RMC, De Sousa AP, Viegas C, Martins JP, Pires MJ, Estêvão SV (eds) Experiencing food: designing dialogues: Proceedings of the 1st international conference on food design and food studies (EFOOD 2017). CRC Press, London, pp 61–84

Marat-Mendes T (2020) Designing for sustainability: retrieving a systemic role for urban form. Urban Morphol 24(2):235–238

Marat-Mendes T (2021) Urban form and urban metabolism. Recent research and academic trends conducted at the Lisbon Metropolitan Area. In: Llop C, Cervera M, Peremiquel F (eds) Forma urbis Y territorios metropolitanos. Metrópolis en recomposición. Prospectivas proyectuales en el siglo XXI. Universitat Politècnica di Catalunya, Barcelona, pp 148–155

Marat-Mendes T, Borges JC (2019a) The urban planning transition history of the Lisbon Metropolitan Area. In: Proceedings of ASEOP 2019 Congress, Venice, 9–13 July 2019

Marat-Mendes T, Oliveira V (2013) Urban planners in Portugal in the middle of the 20th century: Étienne de Gröer and Antão Almeida Garrett. Plan Perspect 28(1):91–111

Marat-Mendes T, Cuchí A (2007) The role of resources management on shaping the landscape patterns: the water in the Royal Estates of Lisbon region. In: Proocedings of the international traditional Mediterranean architecture conference. Present and future. Collegi d'Aparelledors I Arquitectes Tècnic de Barcelona for the Consortium RehabiMed, Barcelona, pp 36–38

Marat-Mendes T, Cuchí A (2008) The Royal Estates and the landscape patterns of Lisbon region. In: Amoêda R, Lira S, Pinheiro C, Pinheiro F, Pinheiro J (eds) World heritage and sustainable development, volume 2. Green Lines Institute for Sustainable Development, Barcelos, pp 441–446

Marat-Mendes T, Oliveira MJ (2012) A Morfologia Urbana no Projecto Urbano. Contributos de uma prática metodológica e pedagógica no ISCTE-IUL. In: André P, Marat-Mendes T, Sampayo M (eds) Proceedings of PNUM 2012 - Morfologia Urbana nos Países Lusófonos. ISCTE-IUL, Lisbon, pp 1782–1810

Marat-Mendes T (cord), Mourão J, d'Almeida PB, Niza S (2015) Atlas da Água e da Agricultura. Região de Lisboa 1900-1940. Water and Agriculture Atlas: Lisbon Region in 1900-1940. Iscte Instituto Universitário de Lisboa/ DINAMIA'CET-IUL, Lisboa

Marat-Mendes T, Mourão J, d'Almeida PB (2016) Access to water in the Lisbon region in 1900. Water History 8(2):159–189. https://doi.org/10.1007/s12685-016-0160-3

Marat-Mendes, T (2017) Lx Design with food. Food Design and Food Studies, Experiencing Food: Designing Dialogues. In: Bonacho R, de Sousa AP, Viegas C, Martins, JP, Pires MJ, Estêvão SV (eds) Experiencing Food: Designing Dialogues: Proceedings of the 1st International Conference on Food Design and Food Studies (EFOOD 2017), CRC Press, London, p 61–84

Marat-Mendes T, d'Almeida PB, Borges JC (2021a) Concepts and definitions for a sustainable planning transition: lessons from moments of change. Eur Plan Stud. https://doi.org/10.1080/09654313.2021.1894095

Marat-Mendes T, d'Almeida PB, Borges JC (2021b) Food system photographic surveys: a necessary urban design agenda. In: Bonacho R, Pires MJ, Lamy ECCS (eds) eFood—experiencing food: designing sustainable and social practices. Routledge, London

Marat-Mendes T, Borges J, Dias A, Lopes R (2021c) Planning for a sustainable food system. The potential role of urban agriculture in Lisbon Metropolitan Area. J Urban Int Res Placemaking Urban Sustain. ID: 1880960. https://doi.org/10.1080/17549175.2021.1880960

Marat-Mendes T, Isidoro I, Catela J, Pereira J, Borges J, Lopes SS, Henriques CN (2021d) Drivers of change: how the food system of the Lisbon Metropolitan Area is being shaped by activities, initiatives and citizens needs towards a sustainable transition. Cidades Comunidades &Territórios, Special Issue: 41–62

Marat-Mendes T, Isidoro I, Borges JC, Lopes SS (2021e) Adapting morphological models to approach spatial typologies of food related human imagination. Paper presented at the 28th international seminar on urban form - ISUF, Glasgow, 29th June–2nd July 2021

Meadows DH, Meadows D, Randers J, Behrens WW III (1972) The limits to growth. Universe Books, New York

Moorcroft C (ed) (1972) Designing for survival. Arch Des 413–433

Modesto ML (1982[1990]) Cozinha tradicional Portuguesa. Verbo, Lisbon

Mónica MF (2020) O olhar do outro – estrangeiros em Portugal do século XVIII ao século XX. Relógio d'Água, Lisboa

Morgan K (2013) The rise of urban food planning. Int Plan Stud 18 (1):1–4. https://doi.org/10.1080/13563475.2012.752189

Mourão J, Marat-Mendes T (2016) Urban planning and territorial management in Portugal: Antecedents and impacts of the 2008 financial and economic crisis. In Othengrafen F, Knieling J (eds) Cities in crisis. Socio-spatial impacts of the economic crisis in Southern European cities. Routledge, London, pp 157–171

Napawan NC (2016) Complexity in urban agriculture: the role of landscape typologies in promoting urban agriculture's growth. J Urban Int Res Placemaking Urban Sustain 9(1):19–38. https://doi.org/10.1080/17549175.2014.950317

Niza S (2017) O sistema alimentar no contexto do metabolismo urbano da Área Metropolitana de Lisboa. In: Oliveira R, Amâncio S, Fadigas L (eds) Alfaces na avenida. Estratégias para (bem) alimentar a cidade. Universidade de Lisboa, Colégio Food, Farming and Forest, Lisboa, pp 42–47

Nunes JPS (2007) O programa Habitações de Renda Económica e a constituição da metrópole de Lisboa (1959-1969). Análise Social 206 (48):83–100

Nuwer R (2014). The world's most endangered food. BBC Future, April 1st. https://www.bbc.com/future/article/20140401-the-worlds-most-endangered-food. Accessed 11 June 2020

Oliveira V (2006) The morphological dimension of municipal plans. Urban Morphology 10(2):101–113

Ordem dos Arquitectos, Afonso J, Roseta H, Tostões A, Wilfried W, Portas N, Carapinha A, Rodeia JB (2006) IAP XX: Inquérito à Arquitectura do Sécula XX em Portugal, OA – CDN, Lisboa

Oswald F, Baccini P (2003) Netzstadt. Designing the urban. Birkhauser, Berlin

Parham S (2015) Food and urbanism—the convivial city and a sustainable future. Bloomsbury, London

Pinto TC, Guerra I (2019) Housing policies, market and home ownership in Portugal: Beyond the cultural model. Cid Comunidades Territ 39:101–114

Portas N, Domingues A, Cabral J (2007) Políticas urbanas – tendências, estratégias e oportunidades. Fundação Calouste Gulbenkian, Lisboa

Purcell M, Tyman S (2015) Cultivating food as a right to the city. Local Environ 20(10):1132–1147. https://doi.org/10.1080/13549839.2014.903236

Purdy J (2015) After Nature: A Politics for the Anthropocene. Harvard University Press, Cambridge, Massachusetts

Reisch L, Eberle U, Lorek S (2013) Sustainable food consumption: an overview of contemporary issues and policies. Sustain Sci Pract Policy 9(2):7–25

Ribeiro O (1987) A formação de Portugal. Ministério da Educação, Lisbon

Rogers R (1998) Cities for a small planet. Basic Books, London

Salvador MS (2019) Shaping the city through food: the historic foodscape of Lisbon as case study. Urban Des Int 24:80–93

Sonnino R (2016) The new geography of food security: exploring the potential of urban food strategies. Geogr J 182(2):190–200

Steel C (2020) Sitopia. Chatto & Windus, London

Steel C (2008[2013]) Hungry city—how food shapes our lives. Vintage, London

UN (United Nations) (2015) Transforming our world: the 2030 agenda for sustainable development. United Nations, New York

Viljoen A, Bohn K, Howe J (2005) CPUL—continuous productive urban landscapes. Routledge, London

Viljoen A, Bohn K (2012) Planning and designing food systems, moving to the physical. In: Viljoen A, Wiskerke JCS (eds) Sustainable food planning: evolving theory and practice. Wageningen Academics, Gelderland

Viljoen A, Bohn K (2014) Second Nature Urban Agriculture: Designing Productive Cities. Routledge, London.

Wekerle GR, Classens M (2015) Food production in the city: (Re) negotiating land, Food Prop. Local Environ 20(10):1175–1193

Index

T. Marat-Mendes et al., *Atlas of the Food System*,
https://doi.org/10.1007/978-3-030-94833-7